D1154676

CRC Handbook on Nondestructive Testing of Concrete

Editors

V. M. Malhotra
Head, Concrete Technology Section
Canada Centre for Minerals and Energy Technology (CANMET)
Energy, Mines, and Resources
Ottawa, Ontario, Canada

N. J. Carino
Research Civil Engineer
Structures Division
National Institute of Standards and Technology
Gaithersburg, Maryland

CRC Press
Boca Raton Ann Arbor Boston

Library of Congress Cataloging-in-Publication Data

CRC handbook on nondestructive testing of concrete / editors, V. M. Malhotra, N. J. Carino.
p. cm.
Includes bibliographical references and index.
ISBN 0-8493-2984-1
1. Concrete—Testing—Handbooks, manuals, etc. 2. Nondestructive testing—Handbooks, manuals, etc. I. Malhotra, V. M. II. Carino, Nicholas J. III. Title: Handbook on nondestructive testing of concrete.
TA440.C72 1991
620.1'367—dc20

90-2698
CIP

This book represents information obtained from authentic and highly regarded sources. Reprinted material is quoted with permission, and sources are indicated. A wide variety of references are listed. Every reasonable effort has been made to give reliable data and information, but the author and the publisher cannot assume responsibility for the validity of all materials or for the consequences of their use.

Direct all inquiries to CRC Press, Inc., 2000 Corporate Blvd., N. W., Boca Raton, Florida, 33431.

International Standard Book Number 0-8493-2984-1

Library of Congress Card Number 90-2698
Printed in the United States 2 3 4 5 6 7 8 9 0

Printed on acid-free paper

PREFACE

In the inspection of metals and metal-based materials, nondestructive testing is an accepted practice. For example, radiographic and ultrasonic techniques are routinely used to identify anomalies in steel pipelines, and there are recognized national and international standards on their use. However, in the inspection of concrete the use of nondestructive testing is relatively new. The slow development of nondestructive testing techniques for concrete is because, unlike steel, concrete is a highly non-homogeneous composite material. Apart from precast concrete units which, like steel products, are fabricated at a plant, most concrete is produced in ready-mixed plants and delivered to the construction site. The placing, consolidation and curing of concrete takes place in the field using labor which is relatively unskilled. The resulting product is, by its very nature and construction method, highly variable, and does not lend itself to testing by nondestructive methods as easily as steel products.

Despite the above drawbacks, there has been progress in the development of nondestructive methods for testing concrete, and in recent years several methods have been standardized by the American Society for Testing and Materials (ASTM), the International Standards Organization (ISO), and the British Standards Institute (BSI). The direct determination of mechanical properties requires that concrete specimens must be tested destructively; therefore, nondestructive methods cannot yield absolute values of these properties. Methods have been developed to measure other properties of concrete from which estimates of mechanical properties are obtained.

Broadly speaking, there are two types of nondestructive test methods for concrete. The first type consists of those methods which are used to estimate strength. The surface hardness, penetration, pullout, breakoff, and maturity techniques belong to this category. Some of these methods are not truly nondestructive because they cause some surface damage, which is, however, minor compared with that produced by drilling a core. The second type includes those methods which measure other properties of concrete such as moisture content, density, thickness, pulse velocity, and dynamic elastic modulus. The latter two properties are used to determine changes in concrete when exposed to an aggressive environment such as cycles of freezing and thawing or aggressive chemicals. Also included in the second category are the stress wave, radar, and infrared thermography techniques which are used to locate delaminations, voids, and cracks in concrete.

The opening chapter of the handbook deals with the surface hardness test methods, followed by chapters on the penetration, pullout, breakoff, maturity, vibration, and pulse velocity techniques. These chapters are followed by a chapter on the combined methods, in which more than one technique is used to estimate strength of concrete. The remaining chapters of the book deal with magnetic, electrical, radioactive, nuclear, radar, stress wave, infrared thermography, and acoustic emission techniques.

This handbook is written primarily for practicing engineers engaged in quality control or investigations of hardened concrete. The chapters are written by individuals who are recognized specialists in the subject area, and are active participants in technical committees on nondestructive testing of concrete. Each chapter discusses the basic principles of the methods and provides practical information for their use. In-depth mathematical treatment and derivations have been kept to a minimum. Those interested in more detailed information about the development of these methods are referred to the original papers cited at the end of each chapter.

Some of the test methods described in this handbook are fairly simple and are easy to carry out, whereas others are complex and need sophisticated equipment and trained personnel to perform the tests. Regardless of which test is used, it is strongly recommended that interpretation of the test results be performed by persons who are thoroughly familiar with the principles and limitations of the method. Interpretation should not be delegated to unqualified field technicians.

It is hoped that this handbook will meet the growing needs of practicing engineers and technologists in the area of nondestructive testing of concrete. Graduate students in concrete technology should also find this handbook useful as a comprehensive state-of-the-art document, and as a source of reference material on the subject.

V. Mohan Malhotra
Nicholas J. Carino
Editors
December 1989

THE EDITORS

V. Mohan Malhotra, P. Eng., DDL (Hon), internationally known researcher, author and speaker, is Head of the Concrete Technology Section, Canada Centre for Mineral and Energy Technology (CANMET), Energy, Mines and Resources Canada, Ottawa, Canada.

Dr. Malhotra received his B.Sc. degree in 1951 from Delhi University, India and B.E. (Civil) in 1957 from the University of Western Australia, Perth. In 1984 he was awarded the honorary degree of Doctor of Laws (DDL) by the University of Dundee, Dundee, Scotland.

Dr. Malhotra has been actively engaged in research in all aspects of concrete technology, including nondestructive testing, for the past 25 years. From 1975 to 1990 he was Chairman of ASTM subcommittee C09.02.05 on Nondestructive Testing of Concrete.

Dr. Malhotra is Honorary Member and Charter Fellow of the American Concrete Institute, Fellow of the American Society for Testing and Materials, Fellow of the Canadian Society for Civil Engineering, Fellow of the Engineering Institute of Canada, Honorary Member of the Concrete Society, U.K., and Honorary Fellow of the Institute of Concrete Technology, U.K. He has received numerous awards and honors from the American Concrete Institute and American Society for Testing and Materials.

Dr. Malhotra is on the Editorial Board of several international journals on concrete technology. He has published more than 100 technical papers on concrete technology including nondestructive testing. He is the author/co-author of several books, including the one on Condensed Silica Fume published by CRC Press in 1988. He has edited numerous special publications for the American Concrete Institute and CANMET.

Dr. Malhotra has organized and chaired numerous international conferences on concrete technology in North America and Europe. In 1984, he organized and chaired the CANMET/ACI International Conference on In-situ/Non-destructive Testing of Concrete in Ottawa, Canada.

Nicholas J. Carino, Ph.D., is a research civil engineer in the Structures Division of the National Institute of Standards and Technology (formerly National Bureau of Standards) in Gaithersburg, Maryland.

Dr. Carino received his undergraduate and graduate education from Cornell University (B.S. 1969, M.S. 1971, and Ph.D. 1974). Upon receiving the Ph.D. degree, Dr. Carino accepted a teaching position at The University of Texas at Austin, where he received several awards for teaching excellence.

In 1979, Dr. Carino accepted a research position at the National Bureau of Standards. His research has dealt with methods for in-place testing of concrete for strength and with nondestructive methods for flaw detection in construction materials. His work on the maturity method gained national and international recognition for which he received a U.S. Dept. of Commerce Bronze Medal and the Center for Building Technology Communicator Award in 1983. In 1986, he was awarded the Wason Medal for Materials Research by the American Concrete Institute (ACI) for a paper describing the impact-echo method for flaw detection which he developed, along with co-worker Mary Sansalone, at NBS.

In addition to research, Dr. Carino has been involved in structural investigations. In 1987, he was the project leader in the structural assessment of the new office building of the U.S. Embassy in Moscow. He also participated in the investigation of the 1981 condominium collapse in Cocoa Beach, Florida; the 1982 highway ramp failure in East Chicago, Indiana; and the 1989 Loma Prieta earthquake.

Dr. Carino is a member of the American Society of Civil Engineers, the American Concrete Institute, and the American Society for Testing and Materials (ASTM). He served as Chairman of ACI Committee 228 on Nondestructive Testing for Concrete and of ACI

Commitee 306 on Cold Weather Concreting. In 1988 he was elected Fellow of ACI for contributions to the advancement of concrete technology. He is an active participant in the work of ASTM and has provided leadership in the development of several standards.

CONTRIBUTORS

Georges G. Carette
Senior Materials Engineer
Concrete Technology Section
CANMET
Department of Energy, Mines & Resources
Ottawa, Ontario, Canada

Gerardo G. Clemeña
Senior Research Scientist
Virginia Transportation Research Council
Charlottesville, Virginia

Nicholas J. Carino
Research Civil Engineer
Structures Division
National Institute of Standards and Technology
Gaithersburg, Maryland

Kenneth R. Lauer
Professor, Civil Engineering
University of Notre Dame
Notre Dame, Indiana

V. Mohan Malhotra
Head
Concrete Technology Section
CANMET
Department of Energy, Mines & Resources
Ottawa, Ontario, Canada

Sidney Mindess
Professor, Civil Engineering
University of British Columbia
Vancouver, B.C., Canada

Terry M. Mitchell
Research Materials Engineer
Federal Highway Administration
U.S. Department of Transportation
McLean, Virginia

Tarun R. Naik
Director, Center for By-Products Utilization
Associate Professor, Civil Engineering
College of Engineering and Applied Science
The University of Wisconsin-Milwaukee
Milwaukee, Wisconsin

Aleksander Samarin
Director of Research
BORAL
Sydney, Australia

Mary Sansalone
Assistant Professor
Structural Engineering
Cornell University
Ithaca, New York

Vasanthy Sivasundaram
Research Engineer
Concrete Technology Section
CANMET
Department of Energy, Mines & Research
Ottawa, Ontario, Canada

Gary J. Weil
President
EnTech Engineering, Inc.
St. Louis, Missouri

TABLE OF CONTENTS

Chapter 1

SURFACE HARDNESS METHODS*

V. M. Malhotra

ABSTRACT

The chapter deals with surface hardness methods for nondestructive testing of concrete. These methods consist of the indentation type and those based on the rebound principle. The rebound method is described in detail, and a procedure is given for the preparation of correlation curves between compressive strength and rebound number. The advantages and limitations of the surface hardness methods are discussed. It is concluded that these methods must not be regarded as substitutes for standard compression tests, but as a means for determining the uniformity of concrete in a structure and comparing one concrete against another.

INTRODUCTION

The increase in the hardness of concrete with age and strength has led to the development of test methods to measure this property. These methods consist of the indentation type and those based on the rebound principle. The indentation methods consist principally of impacting the surface of concrete by means of a given mass having a given kinetic energy and measuring the width and or depth of the resulting indentation. The methods based on the rebound principle consist of measuring the rebound of a spring-driven hammer mass after its impact with concrete.

INDENTATION METHODS

According to Jones,[1] the indentation methods originated in Germany in 1934 and were incorporated in the German standards in 1935.[2,3] The use of these methods has also been reported in the U.K.[4] and the USSR.[5] There is little apparent theoretical relationship between the strength of concrete and its surface hardness. However, several researchers have published empirical correlations between the strength properties of concrete and its surface hardness as measured by the indentation methods.

The three known historical methods based on the indentation principle are

- Testing Pistol by Williams
- Spring Hammer by Frank
- Pendulum Hammer by Einbeck

TESTING PISTOL BY WILLIAMS

In 1936 Williams[4] reported the development of a testing pistol that uses a ball as an indenter. The diameter of the impression made by the ball is measured by a graduated magnifying lens or other means. The impression is usually quite sharp and well defined, particularly with concrete of medium and high strength. The depth of indentation is only about 1.5 mm for concrete with compressive strengths as low as 7 MPa.

The utility of the method according to Williams[4] depends on the approximate relationship

SURFACE HARDNESS METHODS

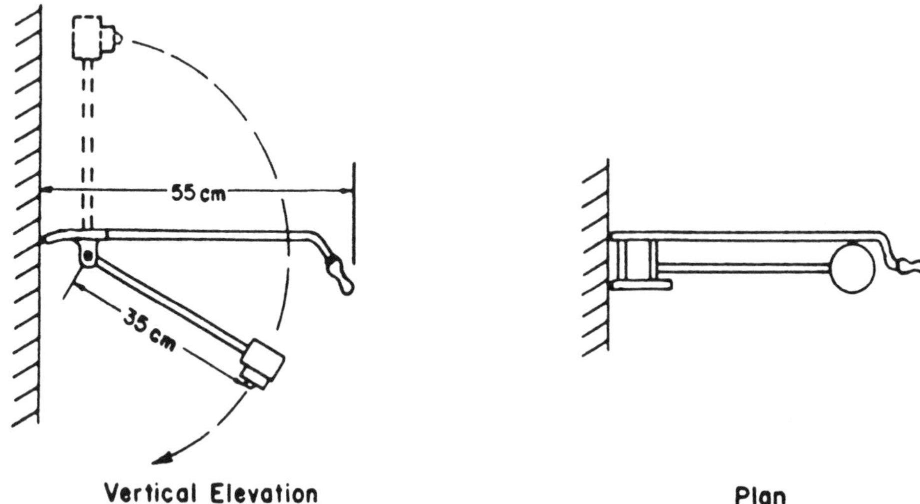

FIGURE 1. Vertical elevation and plan of Einbeck pendulum hammer. (Adapted from Reference 6.)

found to exist between the compressive strength of concrete and the resistance of its surface to indentation during impact.

Skramtaev and Leshchinszy[5] have also reported the use of a pistol in the testing of concrete in the USSR.

SPRING HAMMER BY FRANK

The device developed by Frank consists of a spring-controlled mechanism housed in a tubular frame. The tip of the hammer can be fitted with balls having different diameters, and impact is achieved by placing the hammer against the surface under test and manipulating the spring mechanism. Generally about 20 impact readings are taken at short distances from one another and the mean of the results is considered as one test value. The diameter and/or depth of the indentation is measured, and this, in turn, is correlated with the compressive strength of concrete.[6] The spring mechanism can be adjusted to provide an energy of 50 kg/cm or of 12.5 kg/cm so that the indentation on the concrete surface is within 0.3 to 0.7 times the diameter of the steel ball.

PENDULUM HAMMER BY EINBECK

A line diagram of the pendulum hammer developed by Einbeck is given in Figure 1.[6] The hammer consists of horizontal leg, at the end of which is pivoted an arm with a pendulum head with a mass of about 2.26 kg. The indentation is made by holding the horizontal leg against the concrete surface under test and allowing the pendulum head to strike the concrete. The height of fall of the pendulum head can be varied from full impact (180°) to half impact (90°). The diameter and depth of indentation are measured, and these are correlated with the compressive strength of concrete.

The biggest drawback to this hammer is that it can be used only on vertical surfaces and is, therefore, less versatile than the spring hammer by Frank.

REBOUND METHOD

REBOUND HAMMER BY SCHMIDT

In 1948 a Swiss engineer, Ernst Schmidt,[7-9] developed a test hammer for measuring the

FIGURE 2. Schmidt rebound hammer.

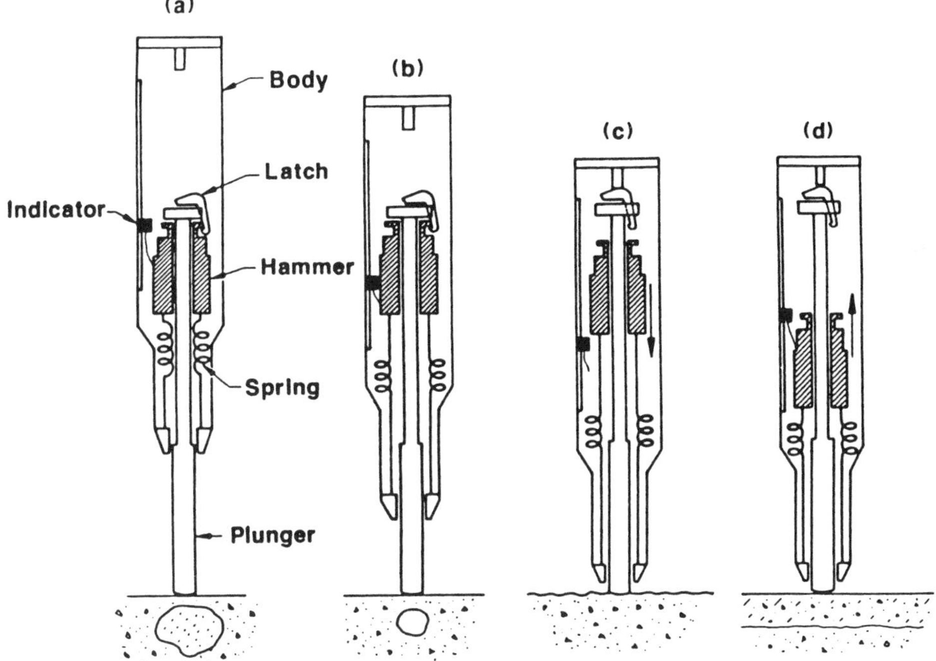

FIGURE 3. A cutaway schematic view of the Schmidt rebound hammer.

hardness of concrete by the rebound principle. Results of his work were presented to the Swiss Federal Materials Testing and Experimental Institute of Zurich, where the hammer was constructed and extensively tested. About 50,000 Schmidt rebound hammers had been sold by 1986 on a worldwide basis.

Principle—The Schmidt rebound hammer is principally a surface hardness tester with little apparent theoretical relationship between the strength of concrete and the rebound number of the hammer. However, within limits, empirical correlations have been established between strength properties and the rebound number. Further, Kolek[10] has attempted to establish a correlation between the hammer rebound number and the hardness as measured by the Brinell method.

Description—The Schmidt rebound hammer is shown in Figure 2. The hammer weighs about 1.8 kg and is suitable for use both in a laboratory and in the field. A schematic cutaway view of the rebound hammer is shown in Figure 3. The main components include the outer body, the plunger, the hammer mass, and the main spring. Other features include a latching mechanism that locks the hammer mass to the plunger rod and a sliding rider to measure the rebound of the hammer mass. The rebound distance is measured on an arbitrary scale marked from 10 to 100. The rebound distance is recorded as a ''rebound number'' corresponding to the position of the rider on the scale.

Method of Testing—To prepare the instrument for a test, release the plunger from its locked position by pushing the plunger against the concrete and slowly moving the body away from the concrete. This causes the plunger to extend from the body and the latch engages the hammer mass to the plunger rod (Figure 3a). Hold the plunger perpendicular to the concrete surface and slowly push the body toward the test object. As the body is pushed, the main spring connecting the hammer mass to the body is stretched (Figure 3b). When the body is pushed to the limit, the latch is automatically released, and the energy stored in the spring propels the hammer mass toward the plunger tip (Figure 3c). The mass impacts the shoulder of the plunger rod and rebounds. During rebound, the slide indicator travels with the hammer mass and records the rebound distance (Figure 3d). A button on the side of the body is pushed to lock the plunger in the retracted position, and the rebound number is read from the scale.

The test can be conducted horizontally, vertically upward or downward, or at any intermediate angle. Due to different effects of gravity on the rebound as the test angle is changed, the rebound number will be different for the same concrete and will require separate calibration or correction charts.

Correlation Procedure—Each hammer is furnished with correlation curves developed by the manufacturer using standard cube specimens. However, the use of these curves is not recommended because material and testing conditions may not be similar to those in effect when the calibration of the instrument was performed. A typical correlation procedure is given below.

1. Prepare a number of 150 × 300-mm cylinders* covering the strength range to be encountered on the job site. Use the same cement and aggregates as are to be used on the job. Cure the cylinders under standard moist-curing room conditions,** keeping the curing period the same as the specified control age in the field.
2. After capping, place the cylinders in a compression-testing machine under an initial load of approximately 15% of the ultimate load to restrain the specimen. Ensure that cylinders are in a saturated surface-dry condition.
3. Make 15 hammer rebound readings, 5 on each of 3 vertical lines 120° apart, against the side surface in the middle two thirds of each cylinder. Avoid testing the same spot twice. For cubes, take 5 readings on each of the 4 molded faces without testing the same spot twice.
4. Average the readings and call this the rebound number for the cylinder under test.***
5. Repeat this procedure for all the cylinders.
6. Test the cylinders to failure in compression and plot the rebound numbers against the compressive strengths on a graph.
7. Fit a curve or a line by the method of least squares.

A typical curve established by Zoldners[13] for limestone aggregate concrete is shown in Figure 4. This curve was based on tests performed at 28 days using different concrete mixtures.

Figure 5 shows four calibration curves obtained by research workers in four different countries.[10] It is important to note that some of the curves deviate considerably from the curve supplied with the hammer.

* In countries where a cube is the standard specimen, use 150-mm cube specimens.

** Temperature 73.4 ± 3°F (23 ± 1.7°C) and 100% relative humidity.

***Some erratic rebound readings may occur when a test is performed directly over an aggregate particle or an air void. Accordingly, the outliers should be discarded and ASTM C 805 has a procedure for discarding these test results.

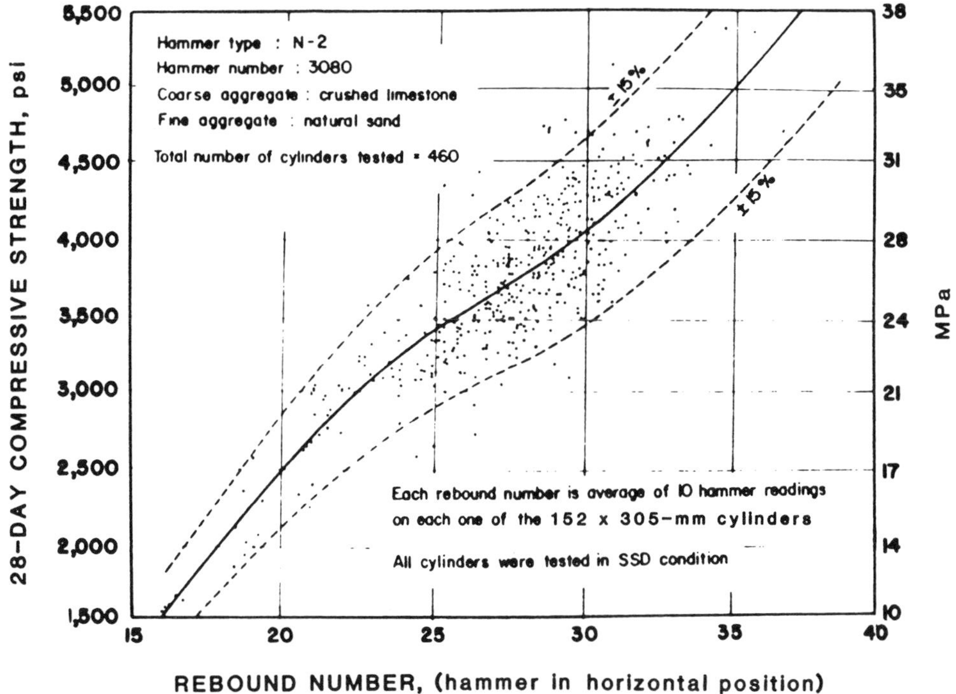

FIGURE 4. Relationship between 28-day compressive strength and rebound number for limestone aggregate concrete obtained with Type N-2 hammer. (Adapted from Reference 13.)

To gain a basic understanding of the complex phenomena involved in the rebound test, Akashi and Amasaki[14] have studied the stress waves in the plunger of a rebound hammer at the time of impact. Using a specially designed plunger instrumented with strain gauges, the authors showed that the impact of the hammer mass produces a large compressive wave σi and a large reflected stress wave σr at the center of the plunger. The ratio $\sigma r/\sigma i$ of the amplitudes of these waves and the time T between their appearance was found to depend upon the surface hardness of hardened concrete. The rebound number was found to be approximately proportional to the ratio of the two stresses, and was not significantly affected by the moisture condition of the concrete. A schematic diagram of the equipment used for observing stress waves is shown in Figure 6, and Figure 7 is an oscilloscope trace of the impact stresses in the plunger showing the initial and reflected waves. From their research, the authors concluded that to correctly measure the rebound number of hardened concrete, the Schmidt hammer should be calibrated by testing a material with a constant hardness and measuring the resulting impact stress waves. Thus, by measuring the impact stress waves in the plunger, the surface hardness of concrete can be measured with a higher accuracy. A typical relationship between the rebound number R and stress ratio $\sigma r/\sigma i$ is shown in Figure 8.

LIMITATIONS

Although the rebound hammer provides a quick, inexpensive means of checking the uniformity of concrete, it has serious limitations and these must be recognized. The results of the Schmidt rebound hammer are affected by:

1. Smoothness of test surface
2. Size, shape, and rigidity of the specimens
3. Age of test specimen

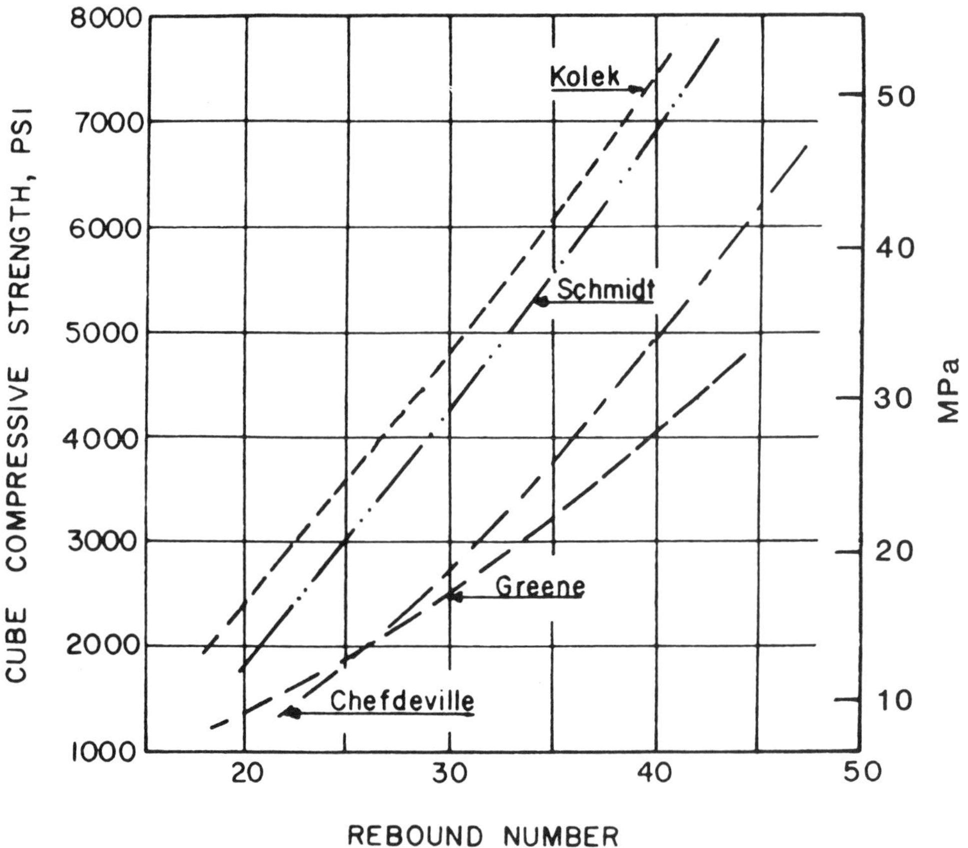

FIGURE 5. Correlation curves obtained by different investigators with a Schmidt rebound hammer Type N-2. Curve by Greene was obtained with Type N. (Adapted from Reference 5.)

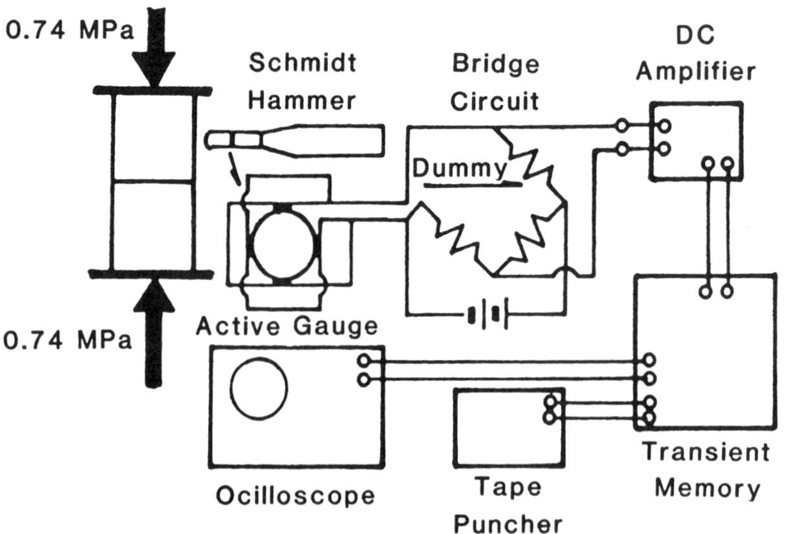

FIGURE 6. Schematic diagram of the equipment used for observing stress waves. (Adapted from Reference 14.)

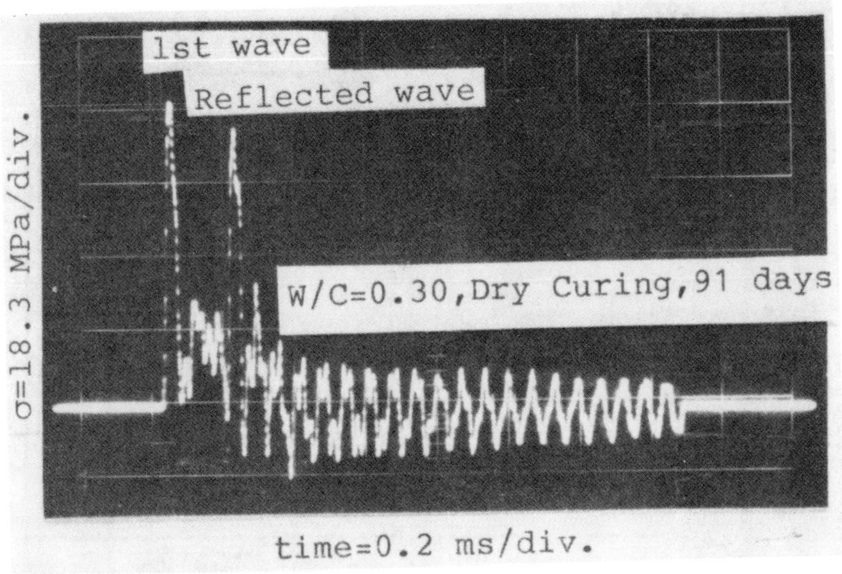

FIGURE 7. Oscilloscope trace of stress waves in the test plunger when testing concrete. (Adapted from Reference 14.)

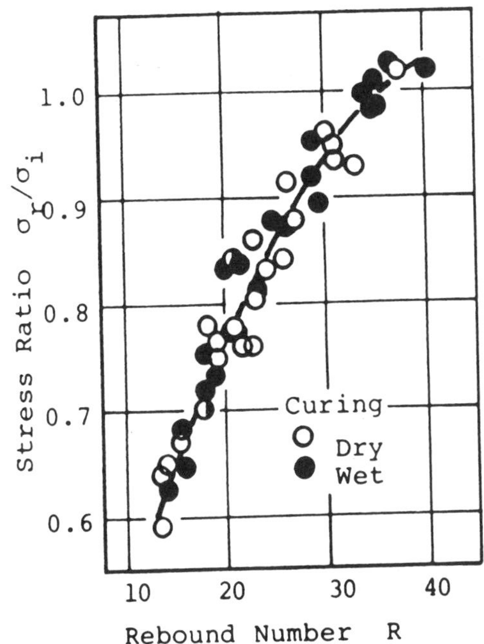

FIGURE 8. Stress ratio vs. rebound number. (Adapted from Reference 14.)

4. Surface and internal moisture conditions of the concrete
5. Type of coarse aggregate
6. Type of cement
7. Type of mold
8. Carbonation of the concrete surface

These limitations are discussed in the foregoing order.

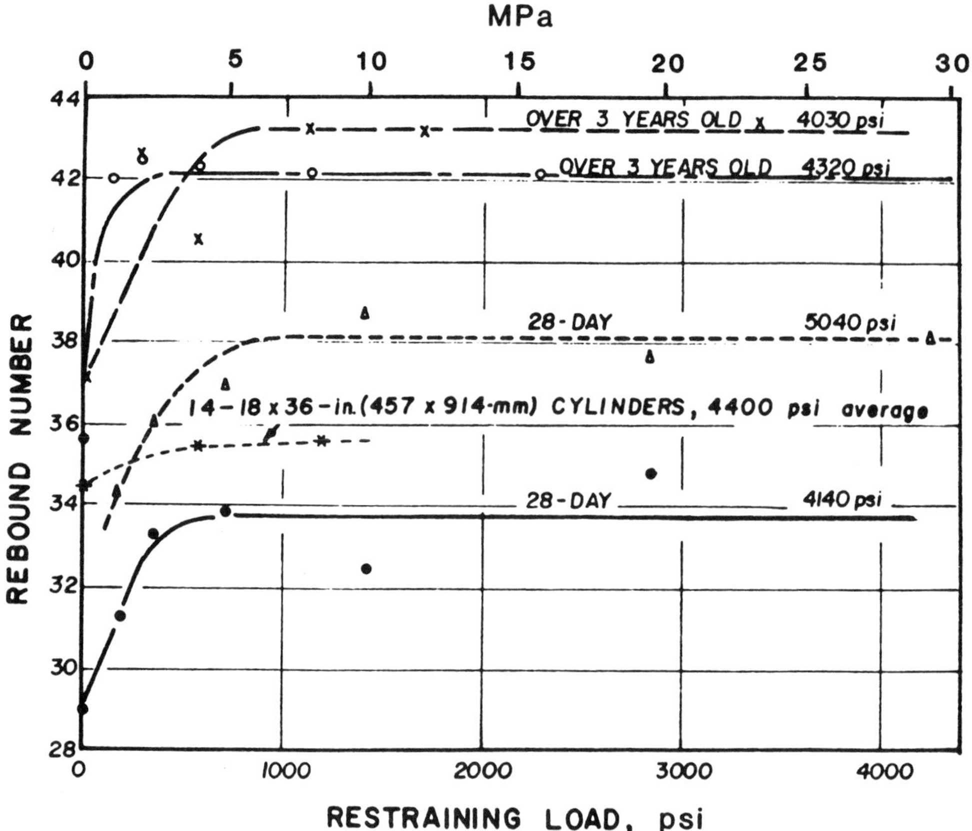

FIGURE 9. Restraining load vs. rebound readings for 6 × 12 in. (152 × 305 mm) cylinders. (Adapted from Reference 16.)

SMOOTHNESS OF SURFACE UNDER TEST

Surface texture has an important effect on the accuracy of the test results. When a test is performed on a rough textured surface, the plunger tip causes excessive crushing and a reduced rebound number is measured. More accurate results can be obtained by grinding a rough surface to uniform smoothness with a carborundum stone. It has been shown by Kolek[10] and Greene[15] that trowelled surfaces or surfaces made against metal forms yield rebound numbers 5 to 25% higher than surfaces made against wooden forms. This implies that if such surfaces are to be used, a special correlation curve or correction chart must be developed. Further, trowelled surfaces will give a higher scatter of individual results and, therefore, a lower confidence in estimated strength.

SIZE, SHAPE, AND RIGIDITY OF TEST SPECIMENS

If the concrete section or test specimen is small, such as a thin beam, wall, 152-mm cube, or 150 × 300-mm cylinder, any movement under the impact will lower the rebound readings. In such cases the member has to be rigidly held or backed up by a heavy mass.

It has been shown by Mitchell and Hoagland[16] that the restraining load for test specimens at which the rebound number remains constant appears to vary with the individual specimen. However, the effective restraining load for consistent results appears to be about 15% of the ultimate strength of 152 × 305-mm cylinders (Figure 9). Zoldners,[13] Greene,[15] and Grieb[17] have indicated effective stress of 1, 1.7, and 2.0 MPa, respectively, and these are considerably lower than the 15% value obtained by Mitchell and Hoagland.

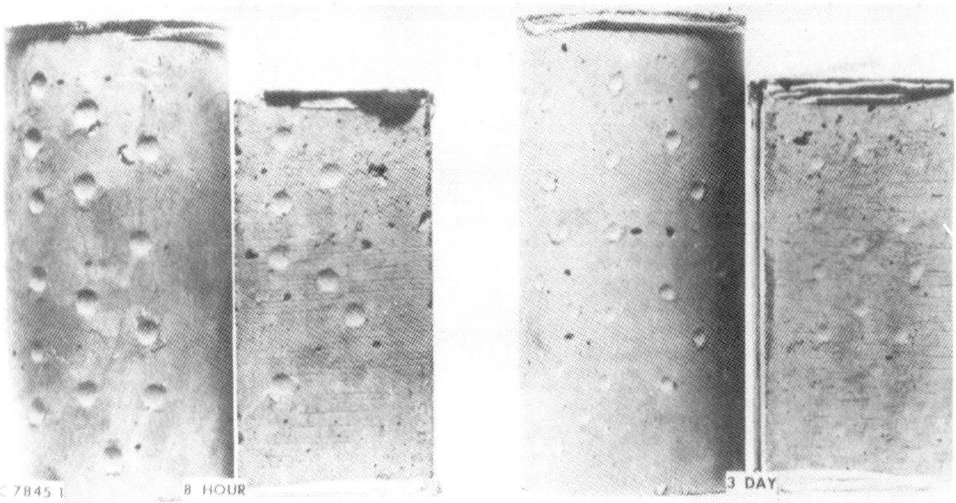

FIGURE 10. Eight-hour old (left) and the three-day old (right) specimens showing surface blemishes after Schmidt hammer impact. (Adapted from Reference 16.)

AGE OF TEST SPECIMEN

Kolek[10] has indicated that the rate of gain of surface hardness of concrete is rapid up to the age of 7 days, and following which there is little or no gain in the surface hardness; however, for a properly cured concrete, there is significant strength gain beyond 7 days. It has been confirmed by Zoldners[13] and Victor[18] that for equal strength, higher rebound values are obtained on 7-day-old concrete than on 28-day-old concrete. It is emphasized that when old concrete is to be tested, direct correlations are necessary between the rebound numbers taken on the structure and the compressive strength of cores taken from the structure.

The use of the Schmidt hammer for testing low-strength concrete at early ages, or where concrete strength is less than 7 MPa, is not recommended because rebound numbers are too low for accurate reading and the test hammer badly damages the concrete surface.[16] Figure 10 shows blemishes caused by rebound tests on surfaces of 8-h-old and 3-day-old concrete cylinders.

SURFACE AND INTERNAL MOISTURE CONDITION OF THE CONCRETE

The degree of saturation of the concrete and the presence of surface moisture have a decisive effect on the evaluation of test hammer results.[13,18,19] Zoldners[13] has demonstrated that well-cured, air-dried specimens, when soaked in water and tested in the saturated surface-dried condition, show rebound readings 5 points lower than when tested dry. When the same specimens were left in a room at 70°F (21.1°C) and air dried, they recovered 3 points in 3 days and 5 points in 7 days. Klieger[20] has shown that for a 3-year-old concrete differences up to 10 to 12 points in rebound numbers existed between specimens stored in a wet condition and laboratory-dry samples. This difference in rebound numbers represents approximately 14 MPa difference in compressive strength.

It is suggested that, whenever the actual moisture condition of the field concrete or specimens is not known, it would be desirable to presaturate the surface several hours prior to testing and use the correlation for tests on saturated surface-dried specimens.

TYPE OF COARSE AGGREGATE

It is generally agreed that the rebound number is affected by the type of aggregate used. According to Klieger,[20] for equal compressive strengths, concrete made with crushed limestone coarse aggregate show rebound numbers approximately 7 points lower than those for

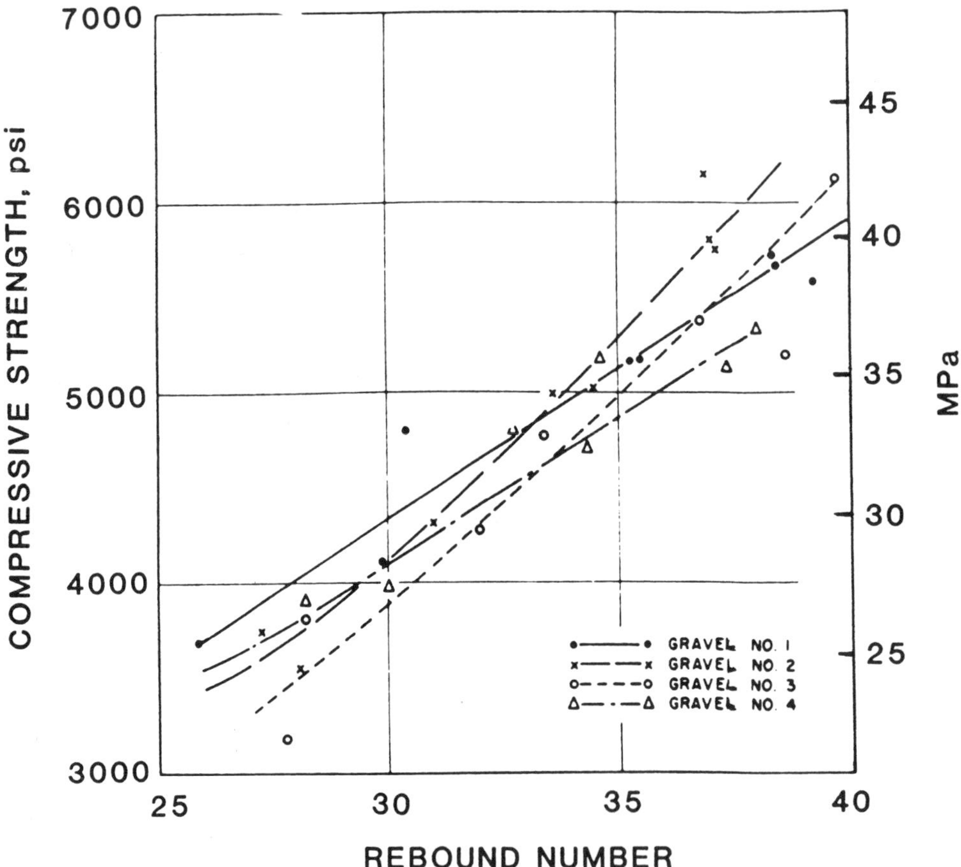

FIGURE 11. Effect of gravel from different sources on rebound numbers of concrete cylinders. (Adapted from Reference 17.)

concretes made with gravel coarse aggregate, representing approximately 7 MPa difference in compressive strength.

Grieb[17] has shown that, even though the type of coarse aggregate used is the same, if it is obtained from different sources different correlation curves would be needed. Figure 11 shows results of one such study where four different gravels were used in making the concrete cylinders tested. At equal rebound numbers, the spread in compressive strength among the correlation curves varied from 1.7 to 3.9 MPa.

Greene[15] found that the use of the test hammer on specimens and structures made of lightweight concrete showed widely differing results. For example, lightweight concrete made with expanded shale aggregate yielded, at equal compressive strengths, different rebound numbers from concrete made with pumice aggregate. But for any given type of lightweight aggregate concrete, the rebound numbers proved to be proportional to the compressive strength.

TYPE OF CEMENT

According to Kolek,[21] the type of cement significantly affects the rebound number readings. High-alumina cement concrete can have actual strengths 100% higher than those obtained using a correlation curve based on concrete made with ordinary portland cement. Also, supersulfated cement concrete can have 50% lower strength than obtained from the ordinary portland cement concrete correlation curves.

TYPE OF MOLD

Mitchell and Hoagland[16] have carried out studies to determine the effect of the type of concrete mold on the rebound number. When cylinders cast in steel, tin can, and paper carton molds were tested, there was no significant difference in the rebound readings between those cast in steel molds and tin can molds, but the paper carton-molded specimens gave higher rebound numbers. This is probably due to the fact that paper molds withdraw moisture from the fresh concrete, thus lowering the water-cement ratio at the surface and resulting in a higher strength. As the hammer is a surface hardness tester, it is possible in such cases for the hammer to indicate an unrealistically high strength. It is therefore suggested that if paper carton molds are being used in the field, the hammer should be correlated against the strength results obtained from test cylinders cast in similar molds.

CARBONATION OF CONCRETE SURFACE

Surface carbonation of concrete significantly affects the Schmidt rebound hammer test results. The carbonation effects are more severe in older concretes when the carbonated layer can be several millimeters thick and in extreme cases up to 20-mm thick.[21] In such cases, the rebound numbers can be up to 50% higher than those obtained on an uncarbonated concrete surface. Suitable correction factors should be established in such cases, otherwise overestimation of concrete strength will result.

REBOUND NUMBER AND ESTIMATION OF COMPRESSIVE STRENGTH

According to Kolek[10] and Malhotra[11,12] there is a general correlation between compressive strength of concrete and the hammer rebound number. However, there is a wide degree of disagreement among various researchers concerning the accuracy of the estimation of strength from the rebound readings and the correlation relationship. Coefficients of variation for compressive strength for a wide variety of specimens averaged 18.8% and exceeded 30% for some groups of specimens.[7] The large deviations in strength can be narrowed down considerably by developing a proper correlation curve for the hammer, which allows for various variables discussed earlier. By consensus, the accuracy of estimation of compressive strength of test specimens cast, cured, and tested under laboratory conditions by a properly calibrated hammer lies between ±15 and ±20%. However, the probable accuracy of estimation of concrete strength in a structure is ±25%.

Boundy and Hondros[22] have suggested the use of the rebound hammer in conjunction with some method of accelerated curing to provide a rapid and convenient method for estimating the expected strength and quality of concrete test specimens. For *in situ* applications, Facaoaru[23] has suggested combined methods based on rebound number and pulse velocity measurements (see Chapter 5).

Carette and Malhotra[24] have investigated the within-test variability of the rebound hammer test at test ages of 1 to 3 days and to the ability of the test to determine early-age strength development of concrete for formwork removal purposes. The rebound tests were performed at 1, 2, and 3 days on plain concrete slabs, 300 × 1270 × 1220 mm in size. Also, companion cylinders and cores taken from the slabs were tested in compression. The mixture proportioning data, and within-test variation for the rebound hammer test and compression strength tests are shown in Tables 1 to 5.

From the analyses of the test data, the authors concluded that because of the large within-test variation, the rebound hammer test was not a satisfactory method for predicting strength development of concrete at early ages.

TABLE 1
Properties of Fresh and Hardened Concrete[a]

		Properties of fresh concrete			Properties of hardened concrete			
					Compressive strength,[b] MPa			
Mix no.	Nominal cement content, (kg/m³)	Unit mass (kg/m³)	Slump (mm)	Air content (%)	1d	2d	3d	28d
1	250	2345	75	5.5	10.4	17.2	19.2	26.5
2	300	2390	100	4.4	15.3	21.6	24.6	33.3
3	350	2410	145	3.3	20.1	24.8	26.1	34.5
4	350	2410	75	3.8	20.1	25.0	26.3	35.1

[a] Determined on 150 × 300-mm cylinders
[b] From Reference 24.

TABLE 2
Summary of Compressive Strength[a] and Rebound Hammer Test Results at Ages 1, 2, and 3 Days (Mix No. 1)

		Average			Standard deviation			Coefficient of variation (%)		
Test	n[b]	1d	2d	3d	1d	2d	3d	1d	2d	3d
Compressive strength of cylinders										
150 × 300-mm cylinders, MPa	3	10.4	17.2	19.2	0.39	0.25	0.14	3.80	1.48	0.73
100 × 200-mm cylinders, MPa	5	9.5	16.0	18.4	0.53	0.83	0.40	5.58	5.19	2.17
Compressive strength of cores										
100 × 200-mm cores, MPa	3	11.1	16.9	18.3	0.36	0.57	0.17	3.21	3.38	0.92
Rebound hammer										
Rebound number	20	15.1	20.5	22.2	2.49	1.93	2.60	16.49	9.43	11.74

[a] Determined on 150 × 300-mm cylinders.
[b] From Reference 24.

TABLE 3
Summary of Compressive Strength[a] and Rebound Hammer Test Results at Ages 1, 2, and 3 Days (Mix No. 2)

Test	n[b]	Average			Standard deviation			Coefficient of variation (%)		
		1d	2d	3d	1d	2d	3d	1d	2d	3d
Compressive strength of cylinders										
150 × 300-mm cylinders, MPa	3	15.3	21.6	24.6	0.57	0.21	0.25	3.71	0.98	1.03
100 × 200-mm cylinders, MPa	5	14.2	20.8	24.1	0.46	0.71	0.61	3.24	3.41	2.54
Compressive strength of cores										
100 × 200-mm cores, MPa	3	15.9	18.1	18.7	0	0	0.42	0	0	2.24
Rebound hammer										
Rebound Number	20	18.9	21.9	21.5	1.89	3.18	2.24	9.99	14.51	10.40

[a] From Reference 24.
[b] Number of test determinations.

TABLE 4
Summary of Compressive Strength[a] and Rebound Hammer Test Results at Ages 1, 2, and 3 Days (Mix No. 3)

Test	n[b]	Average			Standard deviation			Coefficient of variation (%)		
		1d	2d	3d	1d	2d	3d	1d	2d	3d
Compressive strength of cylinders										
150 × 300-mm cylinders, MPa	3	20.1	24.8	26.1	0.81	0.31	0.19	4.04	1.24	0.71
100 × 200-mm cylinders, MPa	5	18.8	25.8	28.1	0.50	1.03	0.60	2.65	3.98	2.14
Compressive strength of cores										
100 × 200-mm cores, MPa	3	19.1	24.7	24.5	0.13	0.27	0.32	0.66	1.11	1.29
Rebound hammer										
Rebound Number	20	19.8	21.2	22.0	2.61	2.17	3.46	13.17	10.23	15.75

[a] From Reference 24.
[b] Number of test determinations.

TABLE 5
Summary of Compressive Strength[a] and Rebound Hammer Test Results at Ages 1, 2, and 3 Days (Mix No. 4)

		Average			Standard deviation			Coefficient of variation (%)		
Test	n[b]	1d	2d	3d	1d	2d	3d	1d	2d	3d
Compressive strength of cylinders										
150 × 300-mm cylinders, MPa	3	20.1	25.0	26.3	0.53	0.58	0.31	2.64	2.31	1.17
100 × 200-mm cylinders, MPa	5	19.8	25.8	27.5	0.94	1.24	1.52	4.74	4.80	2.94
Compressive strength of cores										
100 × 200-mm cores, MPa	3	19.7	20.6	21.6	0.16	0.82	0.63	0.81	4.00	2.94
Rebound hammer										
Rebound Number	20	19.9	21.3	20.2	2.46	1.79	1.87	12.37	8.36	9.29

[a] From Reference 24.
[b] Number of test determinations.

REBOUND NUMBER AND FLEXURAL STRENGTH

Greene[15] and Klieger[20] have established correlation relationships between the flexural strength of concrete and the hammer rebound number. They have found that the relationships are similar to those obtained for compressive strength, except that the scatter of the results is greater. Further, they found that the rebound numbers for tests conducted on the top of finished surface of a beam were 5 to 15% lower than those conducted on the sides of the same beam.

The effects of moisture condition and aggregate type on the flexural strength are similar to those found for the compressive strength.

REBOUND NUMBER AND MODULUS OF ELASTICITY

Mitchell and Hoagland[16] have attempted to correlate hammer rebound number with the modulus of elasticity of the concrete specimens. They concluded that no generally valid correlation could be made between the rebound number and the static modulus of elasticity; however, a satisfactory relationship between the two might be possible if the hammer were to be calibrated for each individual type of concrete.

Petersen and Stoll[25] and Klieger[26] have established an empirical relationship between dynamic* modulus of elasticity and rebound number. They have shown that the relationships are affected by both moisture condition and aggregate type in the same manner as for compressive and flexural strengths.

NORTH AMERICAN SURVEY ON THE USE OF THE REBOUND HAMMER

In the early 1980s, a survey of concrete testing laboratories in Canada and the U.S. revealed that despite its limitations, the rebound hammer was the most often used nondestructive test method by those surveyed.[27]

STANDARDIZATION OF SURFACE HARDNESS METHODS

The indentation methods though once included in the German Standards are no longer being used in the industry, and no major international standardization organization has issued any standards on the subject. On the contrary, the rebound method has won considerable acceptance, and standards have been issued both by the ASTM and ISO and by several other countries for determining the rebound number of concrete.

ASTM Standard C 805 "Rebound Number of Hardened Concrete" was issued in 1985; the significance and use statement of the test method as given in the Standard is as follows:

"4.1 The rebound number determined by this method may be used to assess the uniformity of concrete in situ, to delineate zones or regions (areas) of poor quality or deteriorated concrete in structures, and to indicate changes with time in characteristics of concrete such as those caused by the hydration of cement so that it provides useful information in determining when forms and shoring may be removed.

4.2 This test method is not intended as an alternative for strength determination of concrete.

4.3 Optimally, rebound numbers should be correlated with core testing information. Due to the difficulty of acquiring the appropriate correlation data in a given instance, the rebound hammer is most useful for rapidly surveying large areas of similar concretes in the construction under consideration".

* Modulus of elasticity obtained by flexural vibration of a cylindrical or prismatic specimen.

LIMITATIONS AND USEFULNESS

The rebound hammer developed by Schmidt provides an inexpensive and quick method for nondestructive testing of concrete in the laboratory and in the field.

The limitations of the Schmidt hammer should be recognized and taken into account when using the hammer. It cannot be overstressed that the hammer must not be regarded as a substitute for standard compression tests but as a method for determining the uniformity of concrete in the structures, and comparing one concrete against another. Estimation of strength of concrete by the rebound hammer within an accuracy of ± 15 to $\pm 20\%$ may be possible only for specimens cast, cured, and tested under similar conditions as those from which the correlation curves are established.

REFERENCES

1. **Jones, R.,** A review of the non-destructive testing of concrete, *Proc. Symp. Non-destructive Testing of Concrete and Timber,* Institution of Civil Engineers, London, June 1969, 1.
2. Ball Impact Test for Normal Concrete (Kugelschlagprufung von Beton mit dichtem Gefuge), Standard Code DIN 4240, No. 6, German Committee for Reinforced Concrete, 1966, 311.
3. Ball Test for Cellular Concrete (Kugelschlagprufung von Gas-und Schaumbeton), Draft Code of Practice (June 1955), *Beton Stahlbetonbau (Berlin),* 50(8), 224, 1955.
4. **Williams, J. F.,** A method for the estimation of compressive strength of concrete in the field, *Struct. Eng. (London),* 14(7), 321, 1936.
5. **Skramtaev, B. G. and Leshchinszy, M. Yu,** Complex methods of non-destructive tests of concrete in construction and structural works, *RILEM Bull. (Paris),* New Series No. 30, 99, 1966.
6. **Gaede, K.,** Ball Impact Testing for Concrete (Die Kugelschlagprufung von Beton), Bull, No. 107, Deutscher Ausschuss fur Stahlbeton, Berlin, 1952, 15.
7. **Schmidt, E.,** The Concrete Test Hammer (Der Beton-Prufhammer), *Schweiz. Bauz. (Zurich),* 68(28), 378, 1950.
8. **Schmidt, E.,** Investigations with the New Concrete Test Hammer for Estimating the Quality of Concrete (Versuche mit den neuen Beton-Prufhammer zur Qualitatsbestimmung des Beton), *Schweiz. Archiv. angerwandte Wissenschaft Technik (Solothurn),* 17(5), 139, 1951.
9. **Schmidt, E.,** The concrete sclerometer, *Proc. Int. Symp. Non-destructive Testing on Materials and Structures,* Vol. 2, RILEM, Paris, 1954, 310.
10. **Kolek, J.,** An appreciation of the Schmidt rebound Hammer, *Mag. Concr. Res. (London),* 10(28, 27, 1958.
11. **Malhotra, V. M.,** Non-destructive Methods for Testing Concrete, Mines Branch Monogr. No. 875, Department of Energy Mines and Resources, Ottawa, 1968.
12. **Malhotra, V. M.,** Testing of hardened concrete: non-destructive methods, ACI Monogr. No. 9, 1976, 188.
13. **Zoldners, N. G.,** Calibration and use of impact test hammer, *ACI J. Proc.,* 54(2), 161, 1957.
14. **Akashi, T. and Amasaki, S.,** Study of the Stress Waves in the plunger of a Rebound Hammer at the Time of Impact, Malhotra, V. M., Ed., American Concrete Institute Special Publication SP-82, 1984, 17.
15. **Greene, G. W.,** Test hammer provides new method of evaluating hardened concrete, *ACI J. Proc.,* 51(3), 249, 1954.
16. **Mitchell, L. J. and Hoagland, G. G.,** Investigation of the Impact Tube Concrete Test Hammer, Bull. No. 305, Highway Research Board, 1961, 14.
17. **Grieb, W. E.,** Use of Swiss hammer for establishing the compressive strength of hardened concrete, *Public Roads,* 30(2), 45, 1958.
18. **Victor, D. J.,** Evaluation of Hardened Field Concrete with Rebound Hammer, *Indian Conc. J. (Bombay),* 37(11), 407, 1963.
19. **Willetts, C. H.,** Investigation of the Schmidt Concrete Test Hammer, Miscellaneous Paper No. 6-627, U.S. Army Engineer Waterways Experiment Station, Vicksburg, MS. June 1958.
20. **Klieger, P., Anderson, A. R., Bloem, D. L., Howard, E. L., and Schlintz, H.,** Discussion of "Test Hammer Provides New Method of Evaluating Hardened Concrete" by Gordon W. Greene, *ACI J. Proc.,* 51(3), 256-1, 1954.

21. **Kolek, J.,** Non-destructive testing of concrete by hardness methods, *Proc. Symp. on Non-destructive Testing of Concrete and Timber,* Institution of Civil Engineers, London, June 1969, 15.
22. **Boundy, C. A. P. and Hondros, G.,** Rapid field assessment of strength of concrete by accelerated curing and Schmidt rebound hammer, *ACI J.,* Proc. 61(9) 1185, 1964.
23. **Facaoaru, I.,** Report on RILEM Technical Committee on Non-destructive Testing of Concrete, Materials and Structure/Research and Testing (Paris), 2(10), 251, 1969.
24. **Carette, G. G. and Malhotra, V. M.,** In-situ tests: variability and strength prediction of concrete at early ages, Malhotra, V. M., Ed., American Concrete Institute, Special Publication SP-82, 111.
25. **Petersen, H. and Stoll, U. W.,** Relation of rebound hammer test results to sonic modulus and compressive strength data, *Proc. Highway Res. Board,* 34, 387, 1955.
26. **Klieger, P.,** Discussion of "Relation of Rebound Hammer Test Results to Sonic Modulus and Compressive Strength Data", by Perry H. Petersen and Ulrich W. Stoll, *Proc. Highw. Res. Board,* 34, 392, 1955.
27. ACI 228 Committee Report, In-place methods for determination of strength of concrete, *ACI J. Mater.,* September/October 1988.

Chapter 2

PENETRATION RESISTANCE METHODS*

V. M. Malhotra and G. G. Carette

ABSTRACT

This report reviews the development of penetration resistance methods for testing concrete nondestructively. These are being increasingly used for quality control and strength estimation of *in situ* concrete. Among the penetration techniques presently available, the most well known and widely used is the Windsor probe test. The principle of this method, the test equipment and procedures, and the preparation of calibration charts are described in detail. Factors affecting the variability of the test are discussed. Correlations that have been developed between the Windsor probe test results and the compressive strength of concrete are presented. A new pin penetration test recently developed in Canada for the purpose of determining safe form removal times is described. The advantages, limitations, and applications of the penetration methods are outlined. The report is concluded by providing a list of pertinent references.

INTRODUCTION

Penetration resistance methods are based on the determination of the depth of penetration of probes (steel rods or pins) into concrete. This provides a measure of the hardness or penetration resistance of the material that can be related to its strength.

The measurement of concrete hardness by probing techniques was reported by Voellmy[1] in 1954. Two techniques were used. In one case, a hammer known as Simbi was used to perforate concrete, and the depth of the borehole was correlated to the compressive strength of concrete cubes. In the other technique, the probing of concrete was achieved by Spit pins, and the depth of penetration of the pins was correlated with the compressive strength of concrete.

Apart from the data reported by Voellmy, there is little other published work available on these tests, and they appear to have received little acceptance in Europe or elsewhere. Perhaps the introduction of the rebound method around 1950 was one of the reasons for the failure of the above tests to achieve general acceptance.

In the 1960s, the Windsor probe test system was introduced in the U.S. and this was followed by a pin penetration test in Canada in the 1980s.

PROBE PENETRATION TEST SYSTEM

Between 1964 and 1966, a device known as the Windsor probe was advanced for penetration testing of concrete in the laboratory as well as *in situ*. The device was meant to estimate the quality and compressive strength of *in situ* concrete by measuring the depth of penetration of probes driven into the concrete by means of a powder-actuated driver. The development of this technique was the joint undertaking of the Port of New York Authority, New York, and the Windsor Machinery Co., Connecticut. This development was closely related to studies reported by Kopf.[2] Results of the investigations carried out by the Port of New York Authority were presented by Cantor[3] in 1970. Meanwhile, a number of other organizations had initiated exploratory studies of this technique,[4-8] and a few years later,

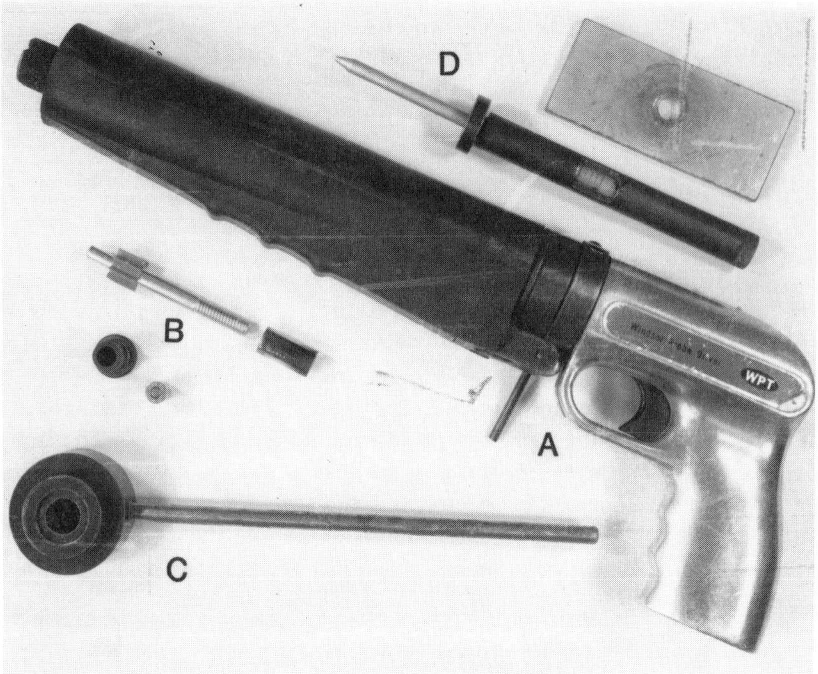

FIGURE 1. A view of the Windsor probe equipment. (A) Driver unit. (B) Probe for normal weight concrete. (C) Single probe template. (D) Calibrated depth gage. (Adapted from Reference 11.)

Arni[9,10] reported the results of a detailed investigation on the evaluation of the Windsor probe, while Malhotra[11-13] reported the results of his investigations on both 150 × 300-mm cylinders and 610 × 610 × 200-mm concrete slabs.

In 1972, Klotz[14] stated that extensive application of the Windsor probe test system had been made in investigations of in-place compressive strength of concrete and in determinations of concrete quality. The Windsor probe had been used to test reinforced concrete pipes, highway bridge piers, abutments, pavements, and concrete damaged by fire. In the 1970s, several U.S. federal agencies and state highway departments reported investigations on the assessment of the Windsor probe for *in situ* testing of hardened concrete.[15-19] In 1984 Swamy and Al-Hamed[20] in the U.K. published results of a study on the use of the Windsor probe system to estimate the *in situ* strength of both lightweight and normal weight concretes.

PRINCIPLE

The Windsor probe, like the rebound hammer, is a hardness tester, and its inventors claim that the penetration of the probe reflects the precise compressive strength in a localized area is not strictly true.[21] However, the probe penetration relates to some property of the concrete below the surface, and, within limits, it has been possible to develop empirical correlations between strength properties and the penetration of the probe.

DESCRIPTION

The Windsor probe consists of a powder-actuated gun or driver (Figure 1), hardened alloy-steel probes, loaded cartridges, a depth gauge for measuring the penetration of probes, and other related equipment. The probes have a tip diameter of 6.3 mm, a length of 79.5 mm, and a conical point (Figure 2). Probes of 7.9 mm diameter are also available for the testing of concretes made with lightweight aggregates. The rear of the probe is threaded and screws into a probe-driving head, which is 12.7 mm in diameter and fits snugly into the

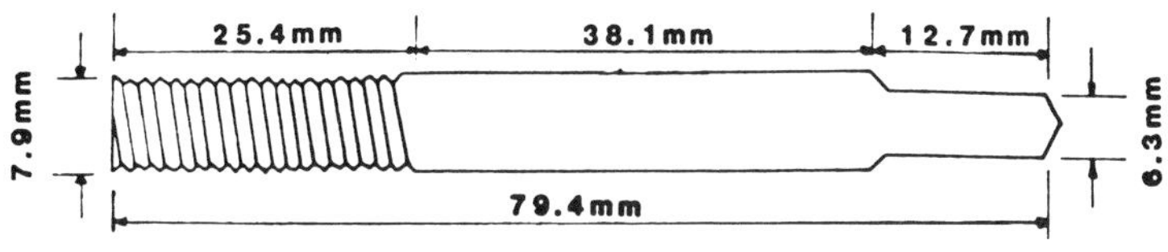

PROBE BEFORE ASSEMBLY

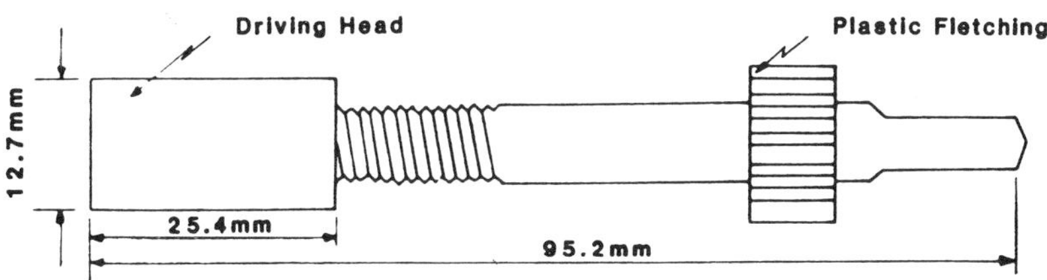

ASSEMBLED DRIVING HEAD AND PROBE

FIGURE 2. A view of probe for normal-weight concrete before and after assembly. (Adapted from Reference 11.)

bore of the driver. The probe is driven into the concrete by the firing of a precision powder charge that develops an energy of 79.5 m-kg. For the testing of relatively low-strength concrete, the power level can be reduced by pushing the driver head further into the barrel.

METHOD OF TESTING

The method of testing is relatively simple and is given in the manual supplied by the manufacturer. The area to be tested must have a brush finish or a smooth surface. To test structures with coarse finishes, the surface first must be ground smooth in the area of the test. Briefly, the powder-actuated driver is used to drive a probe into concrete. If flat surfaces are to be tested, a suitable locating template to provide 178-mm equilateral triangular pattern is used, and three probes are driven into the concrete, one at each corner. The exposed lengths of the individual probes are measured by a depth gauge. The manufacturer also supplies a mechanical averaging device for measuring the average exposed length of the three probes fired in a triangular pattern. The mechanical averaging device consists of two triangular plates. The reference plate with three legs slips over the three probes and rests on the surface of the concrete. The other triangular plate rests against the tops of the three probes. The distance between the two plates, giving the mechanical average of exposed lengths of the three probes, is measured by a depth gauge inserted through a hole in the center of the top plate. For testing structures with curved surfaces, three probes are driven individually using the single probe locating template. In either case, the measured average value of exposed probe length may then be used to estimate the compressive strength of concrete by means of appropriate correlation data.

CORRELATION PROCEDURE

The manufacturer of the Windsor probe test system has published tables relating exposed length of the probe with compressive strength of concrete. For each exposed length value, different values for compressive strength are given, depending on the hardness of the aggregate as measured by the Mohs' scale of hardness. The tables provided by the manufacturer are based on empirical relationships established in his laboratory. However, investigations

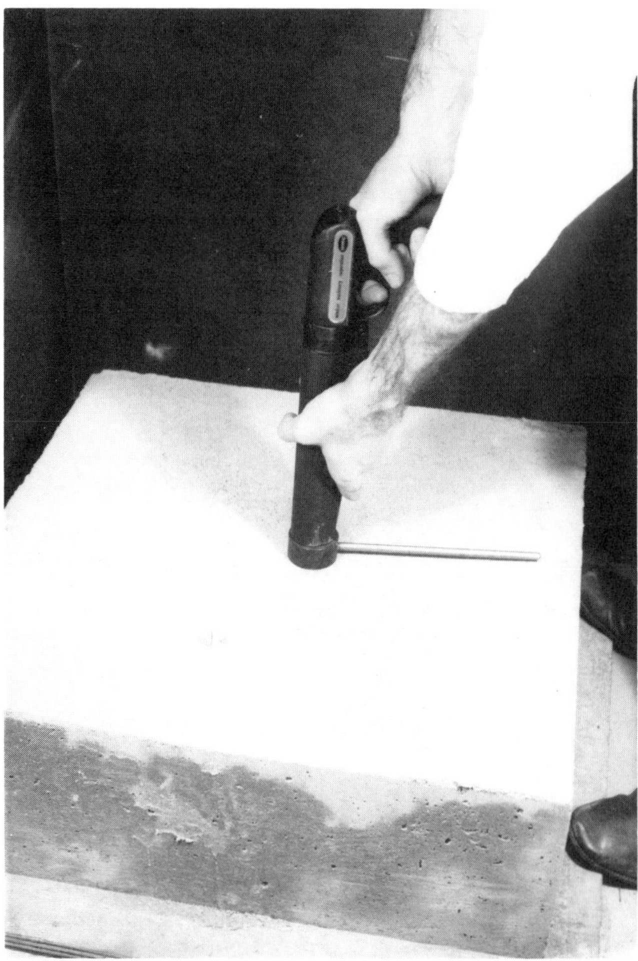

FIGURE 3. A view of the Windsor probe in operation: a 600 × 600 × 200-mm slab under test for correlation purposes. (Adapted from Reference 21.)

carried out by Gaynor,[7] Arni,[9] Malhotra,[11-13] and several others[8,16,22-24] indicate that the manufacturer's tables do not always give satisfactory results. Sometimes they considerably overestimate the actual strength[11,20,22] and in other instances they underestimate the strength. It is, therefore, imperative for each user of the probe to correlate probe test results with the type of concrete being used. Although the penetration resistance technique has been standardized, the standard does not provide a procedure for developing a correlation. A practical procedure for developing such a relationship is outlined below.

1. Prepare a number of 150 × 300-mm cylinders, or 150-mm cubes, and companion 600 × 600 × 200-mm concrete slabs covering a strength range that is to be encountered on a job site. Use the same cement and the same type and size of aggregates as those to be used on the job. Cure the specimens under standard moist-curing conditions, keeping the curing period the same as the specified control age in the field.
2. Test three specimens in compression at the age specified, using standard testing procedure. Then fire three probes into the top surface of the slab at least 150 mm apart and at least 150 mm from the edges (Figure 3). If any of the three probes fails to properly penetrate the slab, remove it and fire another. Make sure that at least three

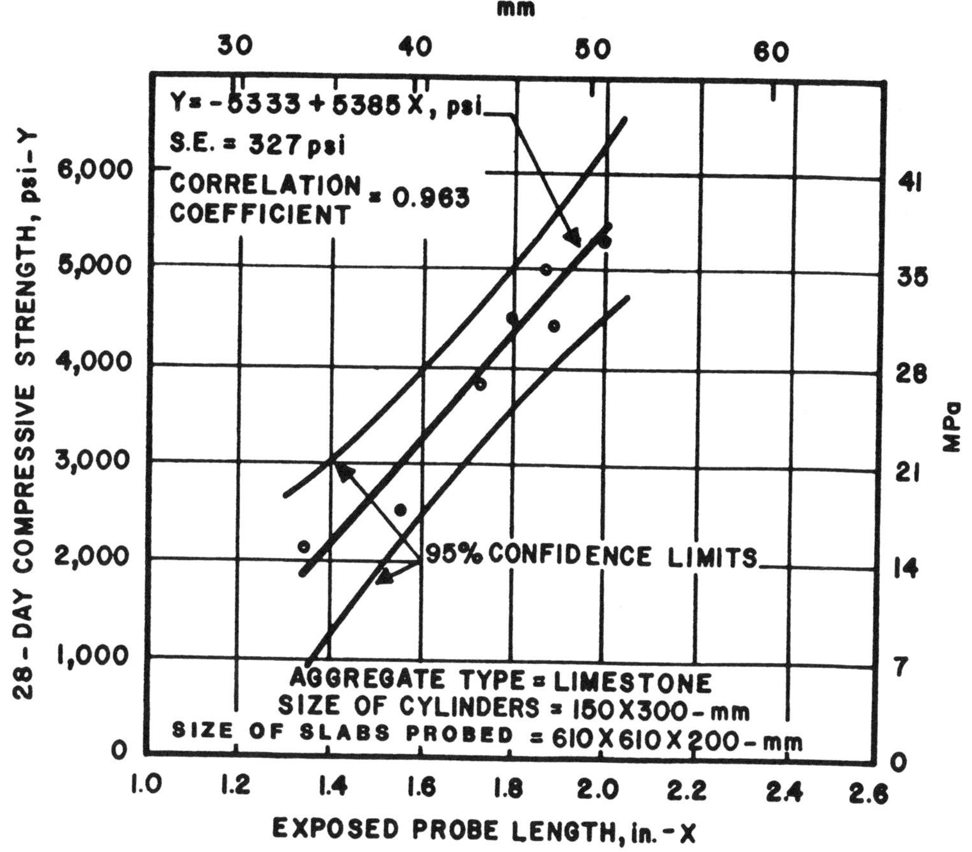

FIGURE 4. Relationship between exposed probe length and 28-day compressive strength of concrete. (Adapted from Reference 12)

valid probe results are available. measure the exposed probe lengths and average the three results.

3. Repeat the above procedure for all test specimens.
4. Plot the exposed probe length against the compressive strength, and fit a curve or line by the method of least squares. The 95% confidence limits for individual results may also be drawn on the graph. These limits will describe the interval within which the probability of a test result falling is 95%.

A typical correlation curve is shown in Figure 4, together with the 95% confidence limits for individual values. The correlation published by several investigators for concretes made with limestone gravel, chert, and traprock aggregates are shown in Figure 5. Note that different relationships have been obtained for concretes with aggregates having similar Mohs' hardness numbers.

EVALUATION OF THE PROBE PENETRATION TEST

MECHANISM OF CONCRETE FAILURE

There is no rigorous theoretical analysis of the probe penetration test available. Such analysis may, in fact, not be easy to achieve in view of the complex combinations of dynamic stresses developed during penetration of the probe, and the heterogeneous nature of concrete. The test involves a given initial amount of kinetic energy of the probe which is absorbed

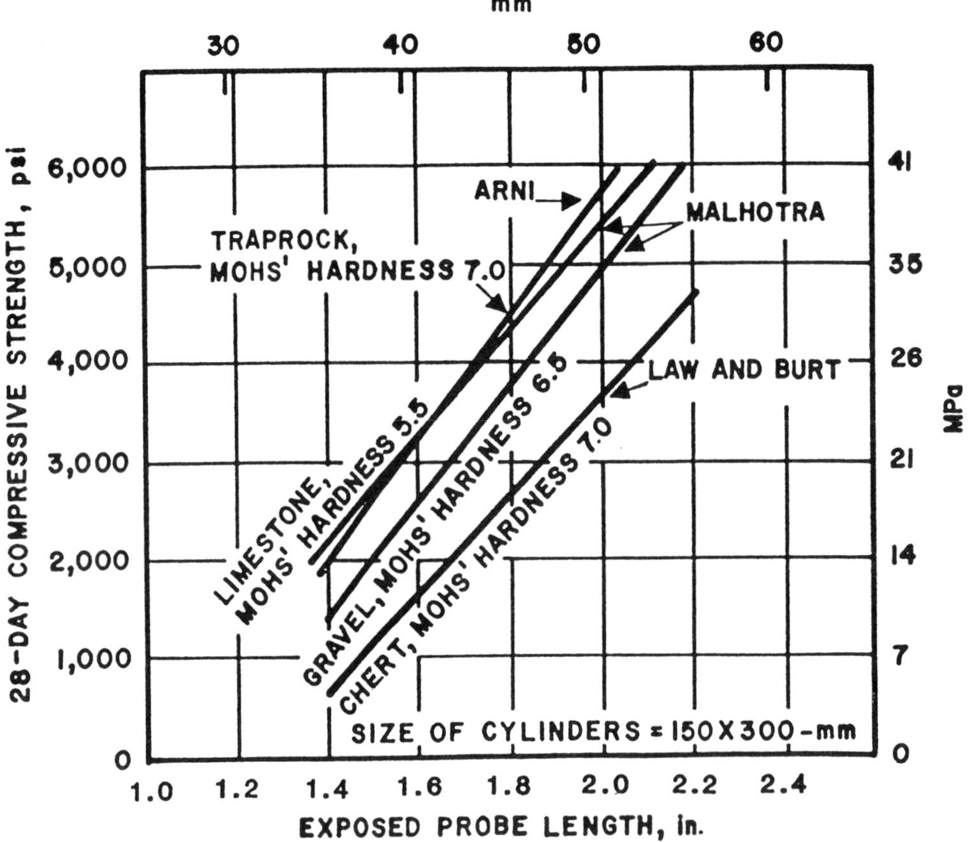

FIGURE 5. Relationship between exposed probe length and 28-day compressive strength of concrete as obtained by different investigators. (Adapted from References 8, 9, and 12.)

during penetration, in large part through crushing and fracturing of the concrete, and in lesser part through friction between the probe and the concrete.

Penetration of the probe causes the concrete to fracture within a cone-shaped zone below the surface with cracks propagating up to the surface (Figure 6). Further penetration below this zone is, in large part, resisted by the compression of the adjacent material, and it has been claimed[25] that the Windsor probe test measures the compressibility of a localized area of concrete by creating a subsurface compaction bulb. Further, it has been claimed that the energy required to break pieces of aggregate is a low percentage of the total energy of the driven probe, and the depth of penetration is not significantly affected. However, these claims have never been proven.

Notwithstanding the extent to which the above claims may be true, it nevertheless appears clear that the probe penetrations do relate to some strength parameter of the concrete below the surface, which makes it possible to establish useful empirical relationships between the depth of penetration and compressive strength.

CORRELATIONS BETWEEN PROBE TEST RESULTS AND COMPRESSIVE STRENGTH

The usefulness of the probe penetration test lies primarily upon its being able to establish sufficiently accurate relationships between probe penetration and compressive strength. A factor long recognized which affects this relationship is the hardness of the coarse aggregate, and this is taken into account in the correlation tables provided by the equipment manufac-

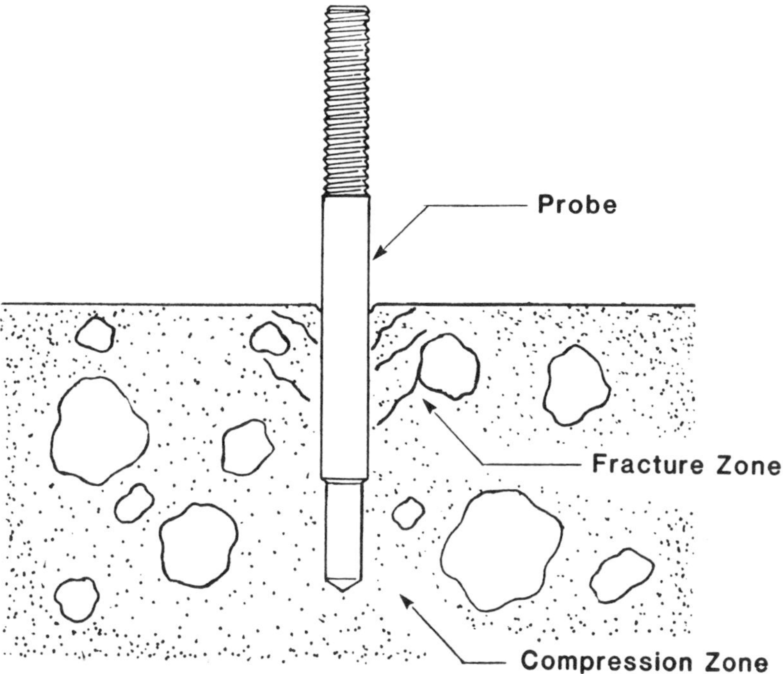

FIGURE 6. Typical failure of mature concrete during probe penetration.

turer. However, as previously mentioned, the use of the manufacturer's tables has been found by several investigators not to be satisfactory. This is probably because factors other than aggregate hardness which also affect probe penetration have not been considered.

There appears to have been no systematic attempts to determine the relative influences of these factors that could affect the probe penetration test results. However, it is generally agreed that the largest influence comes from the coarse aggregate. Apart from its hardness, the type and size of coarse aggregate used have been reported to have a significant effect on probe penetration.[9,20,23] The considerable differences shown in Figure 5 between correlations obtained by various investigators tend to support the important influence of the aggregate type. However, other parameters such as mixture proportions, moisture content, curing regime, and surface conditions are likely to have affected these correlations to some extent, and could explain some of the observed differences.

Other significant parameters that may also affect the accuracy of the probe penetration-strength relationships include the degree of carbonation and the age of concrete.[4,20,21] Carbonation may change the physical and chemical characteristics of concrete through a substantial depth below the surface, and can clearly have an important influence on the depth of penetration of the probe. On the other hand, the age of concrete has been found[20] to considerably affect the accuracy of strength prediction in some instances. In particular, for very old concrete, the probe test may indicate higher strength than actually exists in the structure. It has been suggested that the higher values may relate to microcracking between the aggregates and the paste which affect compressive strength but do not affect probe penetration.[26] Similarly, the stress history of concrete has been identified as a potential factor influencing the relationship between probe penetration and *in situ* strength.[27] This is due to the cracking from service loading again affecting the cylinder strength test, but not affecting the probe penetration test.

In view of the above, when the probe penetration test is to be used for strength estimation, it is advisable to prepare a correlation curve based on the particular type of concrete to be investigated. Further, any other factors which may limit the accuracy of the correlation, such as related to particular testing conditions, would need to be recognized, and taken into account in the use of these correlation curves.

VARIABILITY OF THE PROBE PENETRATION TEST

The within-batch variability in the probe test results as obtained by various investigators is shown in Table 1. Variability is reported in terms of standard deviations and coefficients of variation with values in the latter case being calculated from the exposed probe readings, although more correctly they should be based on the embedded lengths of the probe. These data show that for concrete with a maximum size of aggregate of 19 mm, a typical value for the within-test coefficient of variation (based on depth of penetration) is about 5%.[28] Statistically, for such concrete, the minimum number of individual penetration tests required to insure that the average penetration is known with the same degree of confidence as the average standard cylinder strength (assuming a coefficient of variation of 4% based on two cylinders) would be three. This number, however, would not insure that the *in situ* strength is known with the same degree of confidence, since, obviously, the above within-test coefficient of variation for the probe penetration test does not take into account the uncertainty of the correlation relationship, which also affects the reliability of the estimated strength.

In general, the within-batch variability of probe penetration tests can be attributed partly to operator and equipment errors, and partly to the heterogeneous nature of concrete. In the former case, the operator error has been reported to be generally minimal, and the variations come rather from the test equipment. For example, the variations may be due to the degree of precision achieved in measuring the length of the exposed portion of the probe. In this regard, it has been suggested that the development of more accurate devices for measuring exposed probe length could possibly reduce the within-test variations.[29] On the other hand, the heterogeneous nature of concrete, which on a macro-scale can be regarded as the result of phases such as coarse aggregate and air-voids distributed in the mortar matrix, is likely the major contributor to variations in the probe test results. In particular, the random distribution of larger size aggregates tends to magnify the effect of heterogeneity over the relatively small area affected by the test, and it has indeed been shown that the variability of the probe penetration test increases with an increase in the maximum size of aggregates.[9] As already mentioned, the effect of the aggregate hardness on the probe penetration is generally well recognized. However, its effect on the variability of the results has not been made clear. Some investigations have indicated a possible increase of variability with harder aggregates.[22] In their investigations, Swamy and Al-Hamed[20] have observed a slightly higher variability of the exposed probe length readings for normal weight aggregate concrete than for lightweight aggregate concrete.

VARIATIONS IN THE ESTIMATED STRENGTH VALUES

Although the variability of the probe penetration test provides a useful index of precision for the test, it is however the magnitude of variation in the strength values estimated from the probe readings which provides a basis for the interpretation of the probe test results. The uncertainty in the estimated strength is a function of both the variability of the penetration measurements, and the degree of sensitivity of the penetration test in detecting small changes in strength. There is relatively little information concerning the variability of the estimated strength, though there has been mention of its being relatively large.[3,30] It can nevertheless be readily demonstrated from typical correlations developed by various investigators (Figure 5) that variations in probe readings of the order shown in Table 1 would correspond to variabilities in strength significantly larger than those normally observed with the standard cylinder test. For the proper interpretation of the probe test results, it is, therefore, necessary to make use of statistical procedures that do take into account the variability of the penetration readings, and the uncertainty of the correlation relationship.[28]

TABLE 1
Within-Batch Standard Deviation and Coefficient of Variation of Probe Penetration Measurements

Investigation reported by	Type of aggregate used	Maximum aggregate size (mm)	Type of specimens tested	Total number of probes	Number of probes per test	Age of test, (days)	Average standard deviation (mm)	Average coefficient of variation,[a] %
Arni (Ref. 9)	Gravel, limestone, trap rock	50	410 × 510 × 200-mm slabs	136	9	3, 7, 14, and 28	3.62	7.1
		25	410 × 510 × 200-mm slabs	198	9	3, 7, 14, and 28	2.66	5.4
Malhotra (Ref. 11)	Limestone	19	152 × 305-mm cylinders	20	2	7 and 28	3.14	7.7
		19	610 × 610 × 200-mm slabs	48	3	7 and 28	1.37	3.4
	Gravel	19	150 × 150 × 1690-mm prisms	28	2	35	1.57	3.4
		19	610 × 610 × 200-mm slabs	48	3	7 and 28	2.21	5.5
Gaynor (Ref. 7)	Quartz	25	150 × 580 × 1210-mm walls	384	16	3 and 91	4.05	—
	Semi-lightweight expanded shale as coarse aggregate	25	150 × 580 × 1210-mm walls	256	9	3 and 91	4.30	—
Carette, Malhotra (Ref. 29)	Limestone	19	300 × 1220 × 1220-mm slabs	72	6	1, 2, and 3	2.52	8.3 (5.4)
Keiller (Ref. 22)	Limestone, gravel	19	250 × 300 × 1500-mm prisms	45	3	7 and 28	1.91	3.5

[a] Based on exposed length of probe, except for value in bracket which is based on depth of penetration.

NONDESTRUCTIVE NATURE OF THE PROBE PENETRATION TEST

The probe penetration test is generally considered as nondestructive in nature; however, this is not exactly true. The probe leaves a minor disturbance on a very small area with an 8-mm hole in the concrete for the depth of penetration of the probe. In the case of mature concrete, there is also around the probe a cone-shaped region where the concrete may be heavily fractured, and which may extend to the depth of probe penetration.

This damage would be of little consequence if testing were being carried out on the side of a wall that is to be backfilled or on a foundation slab that is to be covered. However, on an exposed face the damage would be unsightly. The probes would have to be removed and the surface patched at added cost. The test may be considered nondestructive to the extent that concrete can be tested *in situ*, and the strength of structural members is not affected significantly by the test.

USE OF THE PROBE PENETRATION TEST FOR EARLY FORM REMOVAL

An increasingly important area of the application of nondestructive techniques is in the estimation of early-age strength of concrete for the determination of safe form removal times.

Relatively little information has been published in regard to the performance of the probe penetration test at early ages. However, by the late 1970s, it had been reported that the probe penetration test was probably the most widely used nondestructive method for the determination of safe stripping times.[18] One main advantage cited was the great simplicity of the test: "One simply fires the probes into the concrete and compares penetration to previously established criteria. If the probes penetrate too far, the contractor knows the concrete is not yet strong enough".

Carette and Malhotra[29] have investigated in the laboratory the within-test variability at the ages of 1 to 3 days of the probe penetration test, and the ability of the test to indicate the early-age strength development of concrete for formwork removal purposes. The penetration tests were performed at 1, 2, and 3 days on plain concrete slabs, 300 × 1220 × 1220 mm, along with compression tests on standard cylinders and cores taken from the slabs. The mixture proportioning data along with the probe penetration and strength results are shown in Tables 2 to 6. Excellent correlations between compressive strength and probe penetration were observed at these ages for each concrete. From the analysis of the test data, the authors concluded that unlike the rebound method, the probe penetration test can estimate the early-age strength development of concrete within a reasonable degree of accuracy, and thus can be applied to determine safe stripping times for the removal of formwork in concrete constructions.

PROBE PENETRATION TEST VS. CORE TESTING

The determination of the strength of concrete in a structure may become necessary when standard cylinder strength test results fail to comply with specified values, or the quality of the concrete is being questioned because of inadequate placing or curing procedures. It may also be required in the case of older structures where changes in the quality of the concrete are being investigated. In these instances, the most direct and common method of determining the strength of concrete is through drilled core testing; however, some nondestructive techniques such as the probe penetration test have been gaining acceptance as a means to estimate the *in situ* strength of concrete.[18,31]

It has been claimed that the probe penetration test is superior to core testing and should be considered as an alternative to the latter for estimating the compressive strength of concrete.[25] It is true that the probe test can be carried out in a matter of minutes, whereas cores, if from exposed areas and if they have to be tested in accordance with ASTM C 42-87, must be soaked for 40 h;[32] also, the cores may have to be transported to a testing laboratory, causing further delay in getting the results. However, the advantages of the probe

TABLE 2
Properties of Fresh and Hardened Concrete[a]

Mix no.	Nominal cement content, kg/m^3	Properties of fresh concrete: Unit mass, kg/m^3	Slump, mm	Air content, %	Properties of hardened concrete: Compressive strength,[a] MPa: 1 day	2 days	3 days	28 days
1	250	2345	75	5.5	10.4	17.2	19.2	26.5
2	300	2390	100	4.4	15.3	21.6	24.6	33.3
3	350	2410	145	3.3	20.1	24.8	26.1	34.5
4	350	2410	75	3.8	20.1	25.0	26.3	35.1

[a] Determined on 150 × 300-mm cylinders

Source: Reference 29.

TABLE 3
Summary of Compressive Strength and Windsor Probe Test Results at Ages 1, 2, and 3 Days (Mix No. 1)

Test	n[a]	Average			Standard deviation			Coefficient of variation (%)		
		1 day	2 days	3 days	1 day	2 days	3 days	1 day	2 days	3 days
Compressive strength of cylinders										
150 × 300-mm cylinders, MPa	3	10.4	17.2	19.2	0.39	0.25	0.14	3.80	1.48	0.73
100 × 200-mm cylinders, MPa	5	9.5	16.0	18.4	0.53	0.83	0.40	5.58	5.19	2.17
Compressive strength of cores										
100 × 200-mm cores, MPa	3	11.1	16.9	18.3	0.36	0.57	0.17	3.21	3.38	0.92
Penetration resistance (Windsor probe)										
Exposed length of probe, mm	6	17.9	29.0	33.1	2.12	3.69	2.20	11.84	12.72	6.65
Embedded length of probe, mm		61.5	50.4	46.3	2.12	3.69	2.20	3.45	7.32	4.76

[a] Determined on 150 × 300-mm cylinders.

Source: Reference 29.

TABLE 4
Summary of Compressive Strength and Windsor Probe Test Results at Ages 1, 2, and 3 Days (Mix No. 2)

Test	n[a]	Average			Standard deviation			Coefficient of variation (%)		
		1 day	2 days	3 days	1 day	2 days	3 days	1 day	2 days	3 days
Compressive strength of cylinders										
150 × 300-mm cylinders, MPa	3	15.3	21.6	24.6	0.57	0.21	0.25	3.71	0.98	1.03
100 × 200-mm cylinders, MPa	5	14.2	20.8	24.1	0.46	0.71	0.61	3.24	3.41	2.54
Compressive strength of cores										
100 × 200-mm cores, MPa	3	15.9	18.1	18.7	0	0	0.42	0	0	2.24
Penetration resistance (Windsor probe)										
Exposed length of probe, mm	6	29.1	33.2	38.0	2.21	2.05	2.78	7.59	6.17	7.32
Embedded length of probe, mm		50.3	46.2	41.4	2.21	2.05	2.78	4.39	4.44	6.71

[a] Number of test determinations.

Source: Reference 29.

TABLE 5
Summary of Compressive Strength and Windsor Probe Test Results at Ages 1, 2, and 3 Days (Mix No. 3)

Test	n[a]	Average			Standard deviation			Coefficient of variation (%)		
		1 day	2 days	3 days	1 day	2 days	3 days	1 day	2 days	3 days
Compressive strength of cylinders										
150 × 300-mm cylinders, MPa	3	20.1	24.8	26.1	0.81	0.31	0.19	4.04	1.24	0.71
100 × 200-mm cylinders, MPa	5	18.8	25.8	28.1	0.50	1.03	0.60	2.65	3.98	2.14
Compressive strength of cores										
100 × 200-mm cores, MPa	3	19.1	24.7	24.5	0.13	0.27	0.32	0.66	1.11	1.29
Penetration resistance (Windsor probe)										
Exposed length of probe, mm	6	23.2	34.1	36.3	2.64	1.78	2.48	11.38	5.22	6.83
Embedded length of probe, mm		56.2	45.3	43.1	2.64	1.78	2.48	4.70	3.93	5.76

[a] Number of test determinations.

Source: Reference 29.

TABLE 6
Summary of Compressive Strength and Windsor Probe Test Results at Ages 1, 2, and 3 Days (Mix No. 4)

Test	n[a]	Average			Standard deviation			Coefficient of variation (%)		
		1 day	2 days	3 days	1 day	2 days	3 days	1 day	2 days	3 days
Compressive strength of cylinders										
150 × 300-mm cylinders, MPa	3	20.1	25.0	26.3	0.53	0.58	0.31	2.64	2.31	1.17
100 × 200-mm cylinders, MPa	5	19.8	25.8	27.5	0.94	1.24	1.52	4.74	4.80	2.94
Compressive strength of cores										
100 × 200-mm cores, MPa	3	19.7	20.6	21.6	0.16	0.82	0.63	0.81	4.00	2.94
Penetration resistance (Windsor probe)										
Exposed length of probe, mm	6	31.8	36.5	38.3	3.43	3.28	1.64	10.79	8.99	4.28
Embedded length of probe, mm		47.6	42.9	41.1	3.43	3.28	1.64	7.21	7.65	3.99

[a] Number of test determinations.

Source: Reference 29.

penetration test should be judged against the precision of its test results, and the following statement by Gaynor[7] should be of interest in this regard:

> Based on these tests, the probe system does not supply the accuracy required if it is to replace conventional core tests. However, it will be useful in much the same manner that the rebound hammer is useful. In these tests, neither the probe system nor the rebound hammer provides precise quantitative estimates of compressive strength of marginal concretes. Both should be used to locate areas of relatively low- or relatively high-strength concretes in structures.

On the other hand, it has been shown by Malhotra and Painter[12] that the standard error of estimate of 28 day compressive strength of concrete cylinders is of the same order for both the probe and the core tests. More recently, Swamy and Al-Hamed[20] have compared the results of the probe penetration test and core strength tests, and examined how these related to the results of the standard wet-cube strength test. Their work was carried out on slabs, 1800 × 890 × 125 mm, and on 50-mm cores and 100-mm cubes, and covered both normal-weight and lightweight concretes. One of their conclusions was that the probe system, as a general method of nondestructive testing, estimated the wet-cube strength better than the small diameter cores at ages up to 28 days while the cores estimated the strength of older concrete better, particularly in the lower range of strength.

Carette and Malhotra[29] have also observed that, at early ages, probe penetration test results showed better correlation with standard cylinder strengths than with core strengths. This was attributed by the investigators to the variations in the temperature history of the large test slabs used.

It must be stressed that in cases where standard cylinder or cube strength is strictly the parameter of interest because of the specifications being expressed primarily in these terms, the core test which provides a direct measure of compressive strength clearly remains the most reliable means of estimating *in situ* strength. In many situations, however, it has been found possible to establish, within certain limits of material composition and testing conditions, relationships between probe penetration and strength that are accurate enough so that the probe test can be used as a satisfactory substitute for the core test.[31]

PROBE PENETRATION TEST VS. REBOUND HAMMER TEST

Both the probe penetration test and the rebound hammer test provide means of estimating the relative quality of concrete under investigation. Correlations between the rebound numbers and the exposed probe lengths as obtained by Malhotra and Painter[12] are shown in Figure 7. Because the probe can penetrate up to about 50 mm in concrete, the probe penetration results are more meaningful than the results of the rebound hammer, which is a surface hardness tester only. Because of the greater penetration in concrete, the probe test results are influenced to a lesser degree by surface moisture, texture, and carbonation effects.[4] However, size and distribution of coarse aggregate in the concrete affect the probe test results to a much greater degree than those obtained by the rebound hammer. Where cost is a critical factor, the above advantages of the probe penetration test may be offset by a higher initial cost of the equipment compared with the rebound hammer, and the recurring expenses for the probes. Both tests damage the concrete surface to varying degrees. The rebound hammer leaves surface blemishes on young concrete, whereas the probe leaves a hole 8 mm in diameter for the depth of the probe and may cause minor cracking.

NORTH AMERICAN SURVEY ON THE USE OF THE PROBE PENETRATION TEST

In the early 1980s, a survey of concrete testing laboratories in Canada and the U.S. indicated that the Windsor probe penetration technique was the second most often used method for *in situ* strength testing of concrete.[28] The survey included methods such as rebound hammer, probe penetration, pullout, pulse velocity, maturity, and cast-in-place

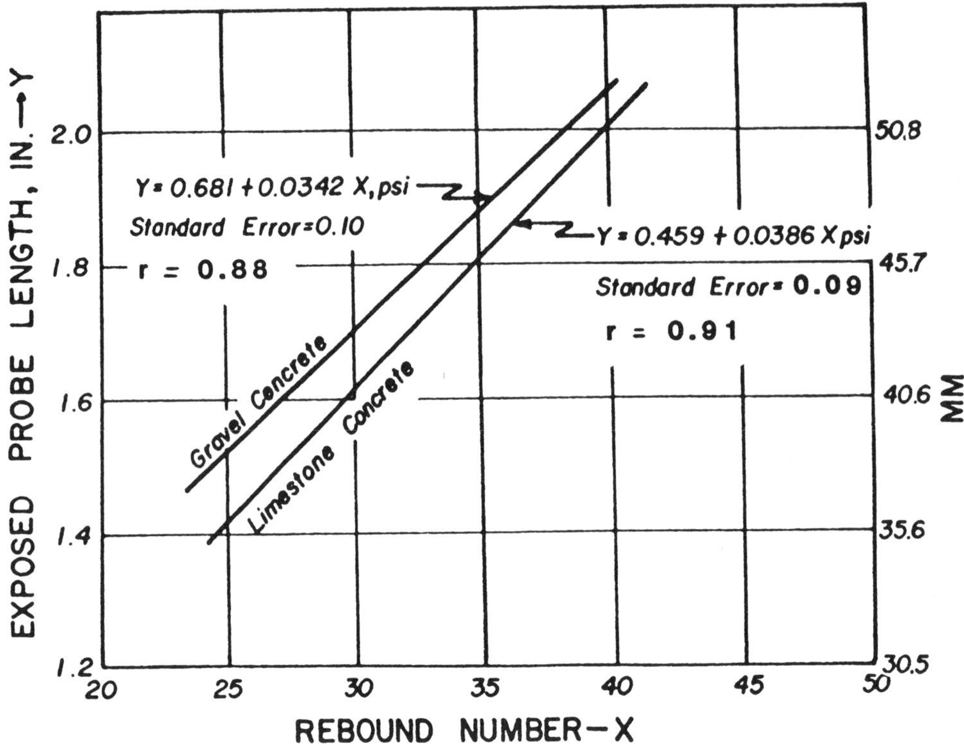

FIGURE 7. Relationship between rebound number and exposed length of probe. (Adapted from Reference 12.)

cylinder. In terms of reliability, simplicity, accuracy, and economy, the probe test was given one of the best combined ratings.

ADVANTAGES AND DISADVANTAGES OF THE PROBE PENETRATION TEST

The probe penetration test system is simple to operate, rugged and needs little maintenance except for occasional cleaning of the gun barrel. The system has a number of built-in safety features that prevent accidental discharge of the probe from the gun. However, wearing of safety glasses is required. In the field, the probe penetration test offers the main advantages of speed and simplicity, and that of requiring only one surface for the test. Its correlation with concrete strength is affected by a relatively small number of variables which is an advantage over some other methods for *in situ* strength testing.

However, the probe test has limitations which must be recognized. These include minimum size requirements for the concrete member to be tested. The minimum acceptable distance from a test location to any edges of the concrete member or between two given test locations is of the order of 150 to 200 mm, while the minimum thickness of the members is about three times the expected depth of penetration. Distance from reinforcement can also have an effect on depth of probe penetration especially when the distance is less than about 100 mm.[27]

As previously indicated, the uncertainty of the estimated strength value, in general, is relatively large and the test results may lack the degree of accuracy required for certain applications. The test is limited to a certain range of strength (<40 MPa), and the use of two different power levels to accommodate a larger range of concrete strength within a given investigation complicates the correlation procedures. Finally, as noted above, the test causes some minor damage to the surface which generally needs to be repaired.

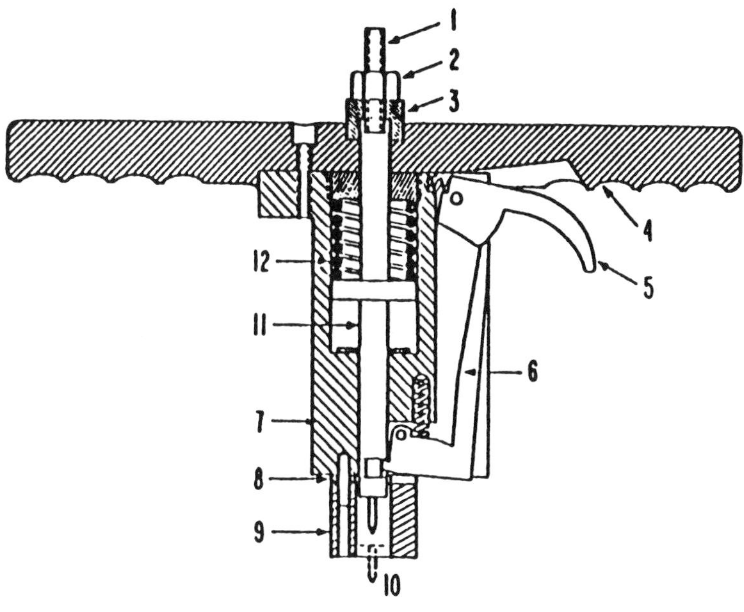

FIGURE 8. Diagram of pin penetration testing apparatus. (Adapted from Reference 33.)

PIN PENETRATION TEST

In the late 1980s, Nasser and Al-Manaseer[33,34] reported the development of a simple pin penetration test for the determination of early-age strength of concrete for removal of concrete formwork. Briefly, this apparatus consists of a device which grips a pin having a length of 30.5 mm, a diameter of 3.56 mm and a tip machined at an angle of 22.5 degrees (Figure 8). The pin is held within a shaft which is encased within the main body of the tester. The pin is driven into the concrete by a spring which is mechanically compressed when the device is prepared for a test. The spring is reported to have a stiffness of 49.7 N/mm and stores about 10.3 N.m. of energy when compressed.

When ready for testing, the apparatus is held against the surface of concrete to be tested, and a triggering device is used to release the spring forcing the pin into the concrete. Following this, the apparatus is removed, and the small hole created in the concrete is cleared by means of an air blower. A dial gauge is then inserted in the pin hole to measure the penetration depth.

A typical correlation between the pin penetration (depth of the hole) and the compressive strength of concrete is shown in Figure 9.

The test, though simple in concept, has limitations. The pin penetrates only a small depth into the concrete, and therefore the results can be seriously affected by the conditions

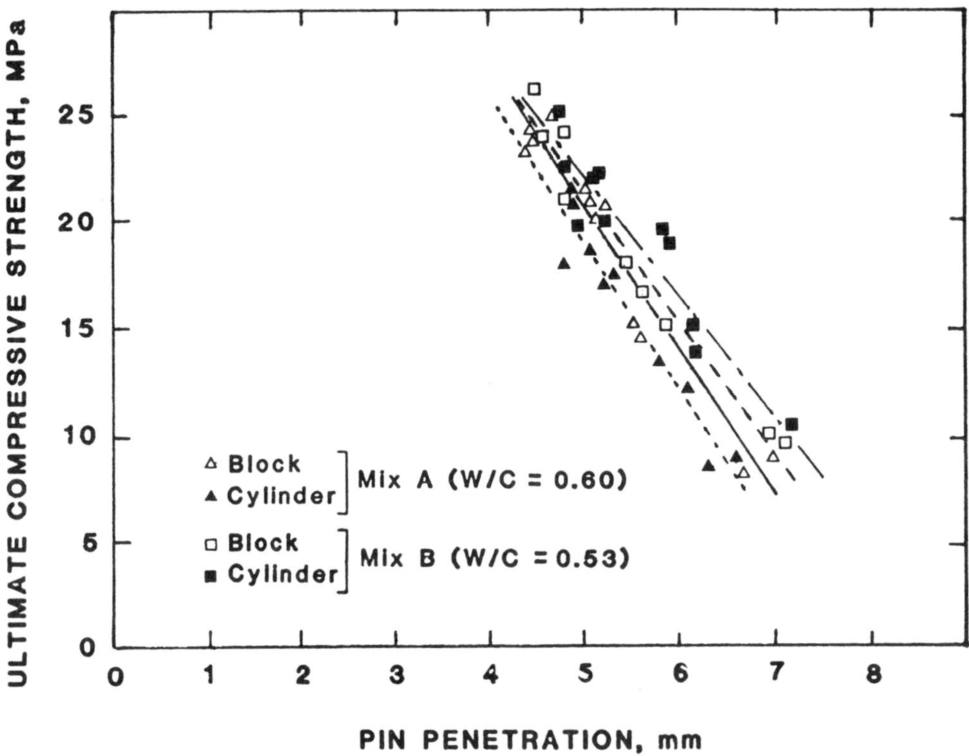

FIGURE 9. Relationship between pin penetration and compressive strength of concrete. (Adapted from Reference 33.)

of the material at the surface. For a similar reason, the variability of the test for which no documentation is yet available, is expected to be relatively large. The tests results are invalid when an aggregate particle is struck. The simplicity of the test makes it possible to obtain as many readings as necessary at little extra cost, and this could somewhat overcome some of the variations in the test results. The equipment is rather heavy for field use, and because of the nature of the spring mechanism, cannot be used for concrete with strength greater than about 30 MPa. Calibration of the equipment is important and its frequent verification may be found necessary.

STANDARDIZATION OF PENETRATION RESISTANCE TECHNIQUES

ASTM Committee C-9 initiated the development of a standard for these techniques in 1972 and a tentative test method covering their use was first issued in 1975. A standard test method designated ASTM C 803-82 "Penetration Resistance of Hardened Concrete" was later issued in 1982.[35] The significance and use statement of the test method as given in the present standard is as follows:

1. This method may be used to assess the uniformity of concrete, to delineate zones of poor quality or deteriorated concrete in structures, and to indicate strength development of concrete. This method is intended to supplement and not supplant strength determination.
2. For a given concrete and a given test apparatus, a relationship between penetration resistance and strength may be established and used to assess in-place concrete strength.

Such a relationship may change with curing and exposure conditions, type and size of aggregate, and level of strength developed in the concrete.

Note 1—Penetration resistance test results have been correlated with results of strength of drilled cores taken from the same structure. The statistical correlation studies and additional penetration tests have been used as the basis for estimating the core strength of similar concrete in other locations in the structure.

At the present time, the only equipment known to be available and to meet the requirements of ASTM C 803 is the Windsor probe test system. However, the standard is currently under revision and consideration is being given for modifications to cover the use of a pin penetration technique such as that recently developed in Canada.[33]

LIMITATIONS AND USEFULNESS OF PENETRATION RESISTANCE METHODS

Penetration resistance methods are basically hardness methods, and like other hardness methods, should not be expected to yield absolute values of strength of concrete in a structure. However, like surface hardness tests, penetration tests provide an excellent means of determining the relative strength of concrete in the same structure, or relative strengths in different structures without extensive correlation with specific concretes. The probe penetration test can also be used to estimate strength of concrete *in situ*, however this requires accurate correlations. The correlation curves provided by the test equipment manufacturer do not appear to be satisfactory. It is, therefore, recommended for each user of the probe equipment to prepare his own correlation curves for the type of concrete under investigation. The test is particularly sensitive to certain characteristics of the aggregate and with a change in source of aggregates, the establishment of new correlations become mandatory.

REFERENCES

1. **Voellmy, A.,** Examination of Concrete by Measurements of superficial Hardness, *Proc. Int. Symp. on Nondestructive Testing of Materials and Structures, RILEM Paris,* 2, 323, 1954.
2. **Kopf, R. J.,** Powder Actuated Fastening Tools for Use in the Concrete Industry, Mechanical Fasteners for Concrete, SP-22, American Concrete Institute, Detroit, 1969, 55.
3. **Cantor, T. R.,** Status Report on the Windsor Probe Test System, presented to Highway Research Board Committee A2-03, Mechanical Properties of Concrete, at 1970 Annual Meeting, Washington, D.C., January 1970.
4. **Freedman, S.,** Field Testing of Concrete Strength, *Modern Concr.,* 14(2), 31, 1969.
5. Arizona Aggregate Association, Report on Windsor Probe, Available from Portland Cement Association, Phoenix, AZ, 1969.
6. Probe quickly shows concrete strength, *Eng. News-Rec.,* 183, 43, 1969.
7. **Gaynor, R. D.,** In-Place Strength of Concrete—A Comparison of Two Test Systems, presented at 39th Annual Convention of the National Ready Mixed Concrete Association (New York, January 28, 1969). Published with NRMCA Tech. Information Letter No. 272, November 4, 1969.
8. **Law, S. M. and Burt, W. T., III,** Concrete Probe Strength Study, Research Report No. 44, Research Project No. 68-2C(B), Louisiana HPR (7), Louisiana Department of Highways, December 1969.
9. **Arni, H. T.,** Impact and penetration tests of portland cement concrete, *Highw. Res. Rec.,* 378, 55, 1972.
10. **Arni, H. T.,** Impact and Penetration Tests of Portland Cement Concrete, Federal Highway Administration Rep. No. FHWA-RD-73-5, 1973.
11. **Malhotra, V. M.,** Preliminary Evaluation of Windsor Probe Equipment for Estimating the Compressive Strength of Concrete, Mines Branch Investigation Rep. IR 71-1, Department of Energy, Mines and Resources, Ottawa, December 1970.
12. **Malhotra, V. M. and Painter, K. P.,** Evaluation of the Windsor Probe Test for Estimating Compressive strength of Concrete, Mines Branch Investigation Rep. IR 71-50, Ottawa, Canada, 1971.

13. **Malhotra, V. M.,** Evaluation of the Windsor Probe Test for Estimating Compressive Strength of Concrete; RILEM Materials and Structures, Paris; 7:37:3-15; 1974.
14. **Klotz, R. C.,** Field investigation of concrete quality using the Windsor probe test system, *Highw. Res. Rec.*, 378 50, 1972.
15. **Keeton, J. R. and Hernandez, V.,** Calibration of Windsor Probe Test System for Evaluation of Concrete in Naval Structures, Technical Note N-1233, Naval Civil Engineering Laboratory, Port Hueneme, CA, 1972.
16. **Clifton, J. R.,** Non-Destructive Tests to Determine Concrete Strength—A Status Report, NBSIR 75-729, Natl. Bur. of Stand., Washington, D.C.
17. **Bowers, D. G. G.,** Assessment of Various Methods of Test for Concrete Strength'', Connecticut Department of Transportation/Federal Highway Administration, December 1978 (available though National Technical Information Service, NTIS No. PB 296317, Springfield, VA).
18. **Bartos, M. J.,** Testing concrete in Place, Civil Engineering, Am. Soc. Civ. Engrs., 1979, 66.
19. **Strong, H.** In-Place Testing of Hardened Concrete with the Use of the Windsor Probe, New Idaho Test Method T-128-79, Division of Highways, State of Idaho, 1979, 1.
20. **Swamy, R. N. and Al-Hamed, A. H. M. S.,** Evaluation of the Windsor Probe Test to Assess In-Situ Concrete Strength, Proc. Inst. Civ. Eng., Part 2, June 1984, 167.
21. **Malhotra, V. M.,** Testing Hardened Concrete: Non-destructive Methods, American Concrete Institute Monogr. No. 9, Iowa State University Press/American Concrete Institute, Detroit, MI, 1976, 188.
22. **Keiller, A. P.,** A Preliminary Investigation of Test Methods for the Assessment of Strength of In-Situ Concrete, Tech. Rep. No. 551, Cement and Concrete Association, Wexham Springs, September 1982.
23. **Bungey, J. H.,** *The Testing of Concrete in Structures,* Chapman and Hall, New York, 1982.
24. **Keiller, A. P.,** Assessing the Strength of In-Situ Concrete, *Concr. Int.*, ACI, February 1985, 15.
25. Compressive Strength Testing of Concrete, Windsor Probe Test System, Inc., Elmwood, CT, 1970.
26. ACI 207-79, Practices for Evaluation of Concrete in Existing Massive Structures for Service Conditions, ACI Manual of Concrete Practice, American Concrete Institute, Detroit, MI, 1983.
27. **Lee, S. L., Tam, C. T., Paramasivam, P., Ong., K. C. G., Swaddiwudhipong, S., and Tan, K. H.** Structural Assessment in In-Situ Testing and Interpretation of Concrete Strength, Department of Civil Engineering, National University of Singapore, July 1988.
28. ACI Committee 228 on: In-place methods for determination of strength of concrete, *ACI Mater. J.*, 85(5), 446, 1988.
29. **Carette, G. G. and Malhotra, V. M.,** In-Situ Tests: Variability and Strength Prediction of Concrete at Early Ages, Malhotra, V. M., Ed., American Concrete Institute, Spec. Publ. SP-82, 1984, 111.
30. **Jenkins, R. S.,** Non-destructive testing—an evaluation tool, *Concr. Int.*, ACI, 1985, 22.
31. **Kopf R. J., Cooper, C. G., and Williams, F. W.,** In-Situ Strength Evaluation of Concrete Case Histories and Laboratory Investigations, *Concr. Int. Design and Construction,* ACI, March 1981, 66.
32. ASTM C 42-87, Standard Test Method for Obtaining and Testing Drilled Cores and Sawed Beams of Concrete, Annual Book of ASTM Standards, 1988, American Society for Testing and Materials, Philadelphia.
33. **Nasser, K. W. and Al-Manaseer, A.,** New Non-Destructive Test for Removal of Concrete Forms, *Concr. Int. ACI,* 9(1), 41, 1987.
34. **Nasser, K. W. and Al-Manaseer, A.,** Comparison of non-destructive testers of hardened concrete, *ACI J.*, 84(5), 374, 1987.
35. ASTM C 803-82, Standard Test Method for Penetration Resistance of Hardened Concrete, Annual Book of ASTM Standards, 1988, American Society for Testing and Materials, Philadelphia.

Chapter 3

PULLOUT TEST*

Nicholas J. Carino

The pullout test measures the force needed to extract an embedded insert from a concrete mass. Using a correlation relationship, the measured ultimate pullout load is used to estimate the in-place compressive strength of the concrete. This chapter reviews the history of the development of this test method, including the various analytical studies conducted to understand the underlying failure mechanism for the test. Statistical characteristics of the method, such as within-test variability and the nature of the correlation relationship, are discussed. It is shown that the characteristics of the coarse aggregate play an important role in the statistical properties of the test. Some of the requirements of the ASTM standard on the pullout test are discussed, and recommendations for developing correlation relationships and interpreting tests results presented. The chapter concludes with a review of test methods that can be performed on existing construction.

INTRODUCTION

The pullout test measures the force required to pull an embedded metal insert with an enlarged head from a concrete specimen or a structure. Figure 1 illustrates the configuration of a pullout test. The insert is pulled by a loading ram seated on a bearing ring which is concentric with the insert shaft. The bearing ring transmits the reaction force to the concrete. As the insert is pulled out, a conical-shaped fragment of concrete is extracted from the concrete mass. The idealized shape of the extracted conic frustum is shown in Figure 1. Frustum geometry is controlled by the inner diameter of the bearing ring (D), the diameter of the insert head (d), and the embedment depth (h). The apex angle (2α) of the idealized frustum is given by:

$$2\alpha = 2 \tan^{-1}\left(\frac{D - d}{2h}\right) \tag{1}$$

The pullout test is widely used during construction to estimate the in-place strength of concrete to help decide whether critical activities such as form removal, application of post-tensioning, or termination of cold weather protection can proceed. Since the compressive strength is usually required to evaluate structural safety, the ultimate pullout load measured during the in-place test is converted to an equivalent compressive strength by means of a previously established correlation relationship.

Unlike some other tests used to estimate the in-place strength of concrete, the pullout test subjects the concrete to a slowly applied load and measures an actual strength property of the concrete. However, the concrete is subjected to a complex three-dimensional state of stress, and the pullout strength is not likely to be related simply to uniaxial strength properties. Nevertheless, by use of correlation curves the pullout test can be used to make reliable estimates of in-place strength.

The purpose of this chapter is to provide basic and practical information about the pullout test. The chapter begins with a brief historical review of the developments leading to the current test method. This is followed by discussions of the failure mechanism and the

* Contribution of the National Institute of Standards and Technology and not subject to copyright in the U.S.

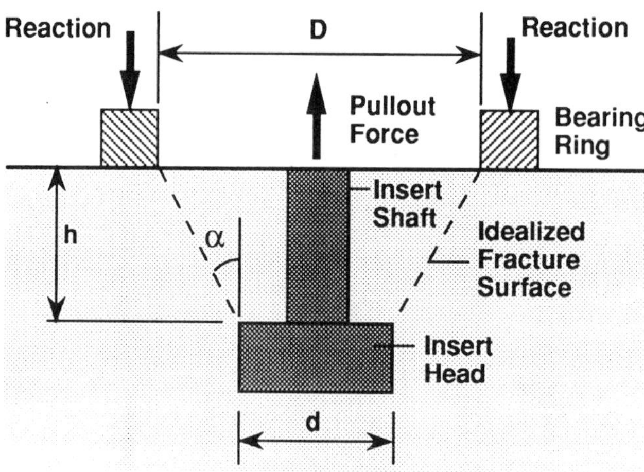

FIGURE 1. Schematic of the pullout test.

statistical characteristics of the test. The chapter concludes with a discussion of practical matters in the application of the test.

HISTORICAL BACKGROUND

DEVELOPMENT IN THE SOVIET UNION

The earliest known description of the pullout test method was reported by Skramtajew[1] of the Central Institute for Industrial Building Research in the Soviet Union. His article reviewed a variety of proposed methods to measure the in-place strength of concrete. One of these methods involved embedding a rod with a spherical end into the fresh concrete and measuring the force required to extract the rod from the hardened concrete. The method was reportedly developed simultaneously by two engineers, I. V. Volf and O. A. Gershberg. Volf's insert was described as follows:[1]

> The rod is of steel, its overall length is 48 mm, of which 38 mm are embedded in concrete, its throat diameter is 8 mm, and the diameter of the spherical end — 12 mm.

Figure 2 shows the geometry of the rod based upon this description and the sketches provided in Reference 1. The insert was attached to the formwork by a screw. At the time of test, the holding screw was removed, an eye-bolt was screwed into the insert, a mechanical jack was attached, and the rod was extracted.

Skramtajew noted that when the rod was pulled, the concrete was subjected to tensile and shear stresses and failure occurred by pulling out a cone of concrete with an apex angle of approximately 90°. In this test the bearing plate of the jack was sufficiently large so that it did not interfere with concrete fracture during pullout of the cone. The large diameter of the extracted cone was reported to be between 100 and 120 mm. The exact geometries of the extracted cones were not described, but it is assumed that they were similar to that shown in Figure 2.

It was reported[1] that, for concrete strengths less than about 10 MPa (1500 psi), there was nearly a constant ratio between ultimate pullout load and the compressive strength of companion cubes. The ratios of ultimate pullout load to cube strength varied between ±9% of the average value. Skramtajew concluded that his pullout test had the advantages of simplicity, availability, and precision. The disadvantages included the need to attach the inserts to formwork at planned locations and the need to repair the holes created by the test.

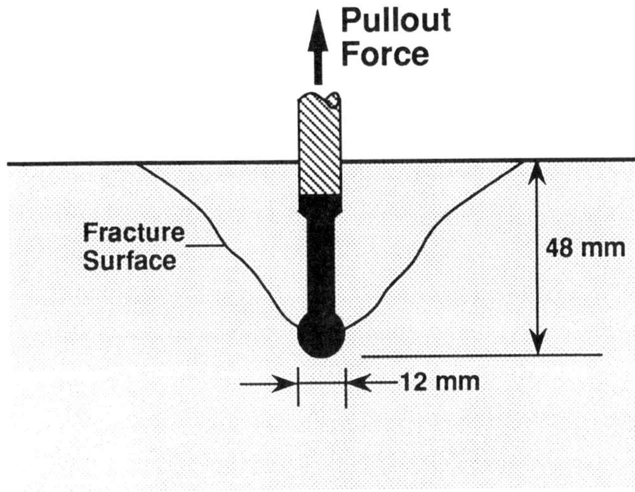

FIGURE 2. Configuration of first pullout test developed by Volf.[1]

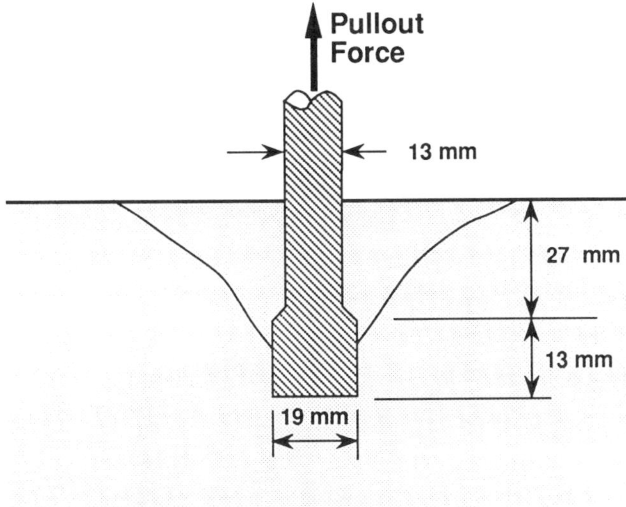

FIGURE 3. Configuration of pullout test developed by Tremper.[2]

However, he concluded that the advantages were so significant that the method should be widely used.

DEVELOPMENT IN THE UNITED STATES

Six years after the publication of Skramtajew's paper, Tremper became the first American to report research results of pullout tests.[2] To simplify fabrication of the inserts, a cylindrical head was used rather than a spherical head as in the Soviet design. Figure 3 shows the approximate dimensions of Tremper's insert. The shoulder of the enlarged head was machined at a 45-degree angle. The bearing ring had a diameter of 152 mm (6 in.). According to Tremper:[2]

> Failure occurred by the formation of a cone 4 to 6 in. in diameter extending to a point about midway in the length of the enlarged end.

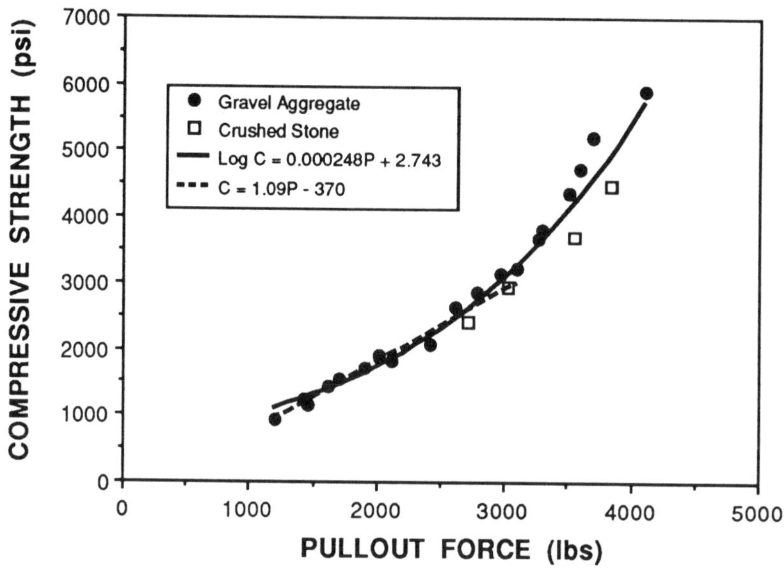

FIGURE 4. Compressive strength and pullout force results reported by Tremper.[2]

The approximate shape of the failure cone, based on this description, is shown in Figure 2. Because of the large diameter, the bearing ring did not interfere with the failure cone. Hence, Tremper's test was similar to the earlier Soviet test.

Tremper did pullout tests on prisms and companion compression tests on cylinders. Six concrete mixtures were used, and the compressive strength varied between about 6 to 40 MPa (1000 to 6000 psi). Five mixtures were made with rounded gravel, and the sixth mixture was made with 50% crushed gravel and 50% rounded gravel. The nominal maximum aggregate size was 28 mm (1 $^1/_4$) in.).

At each test age, Tremper performed five replicate pullout tests and five replicate compression tests. Figure 4 shows the average test results. Over the complete range of concrete strength, the relationship between compressive strength and ultimate pullout load is not linear. The equation of the nonlinear relationship is shown in Figure 4. For concrete with a compressive strength below 20 MPa (3000 psi), Tremper used a straight line to approximate the relationship as shown in Figure 4. An important observation in Figure 4 is that the data for the five concrete mixtures made with gravel appear to obey the same relationship. However, the concrete with 50% crushed gravel appears to have higher pullout strength for the same level of compressive strength. Thus, as early as 1944, there was evidence that the ultimate pullout load was influenced by the characteristics of the coarse aggregate.

Tremper found that the within-test coefficient of variation of the pullout tests was 9.6% while that for the cylinder tests was 8.4%.* Thus he concluded that the pullout strength could be measured with as much precision as the cylinder strength. A seen in Figure 4, data scatter about the correlation curve increased for compressive strength above 25 MPa (3500 psi). As a final conclusion, Tremper stated:[2]

> Data from laboratory tests indicate that the pull-out test can be applied to concrete in structures with less error in estimating actual compressive strengths that are not above about 3500 psi than is often obtained through the use of test cylinders.

* By today's standards, a coefficient of variation of 8.4% for the strength of laboratory-prepared cylinders would be considered unusually high.

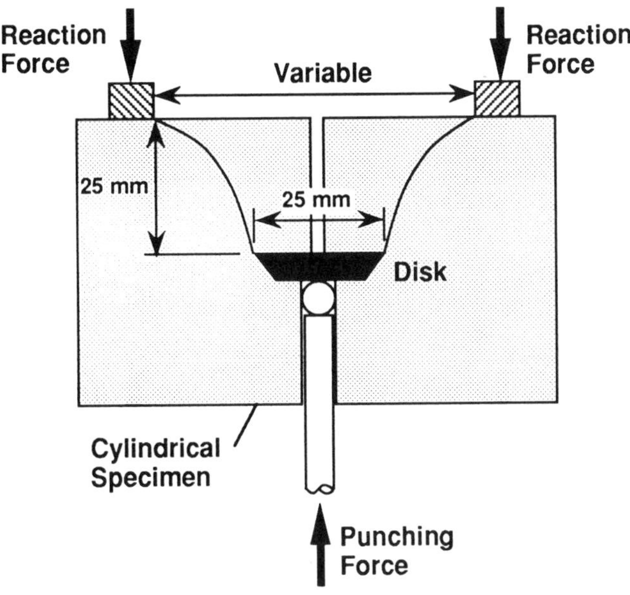

FIGURE 5. Testing configuration for pilot tests by Kierkegaard-Hansen.[3]

Despite Tremper's encouraging results, the use of the pullout test did not gain acceptance. Only after the work of Kierkegaard-Hansen in Denmark and that of Richards in the U.S. did the pullout test become recognized as a useful field technique.

DANISH TEST (LOK-STRENGTH)

In 1962, Kierkegaard-Hansen[3] initiated a research program to determine the optimum geometry for the pullout test so that it could be performed in the field with simple equipment and so that there would be a high correlation between ultimate pullout load and compressive strength. The results of his work led to the widely used test system known as LOK-TEST. Kierkegaard-Hansen's work is reviewed in detail because it is useful to understand the reasoning used to establish the values of the embedment depth, insert head diameter, and bearing ring diameter for this test system.

According to Kierkegaard-Hansen, the embedment depth should be sufficient to assure that more than the outermost "skin" of the concrete is tested and that some coarse aggregates is included within the failure cone. This would favor deep embedment. However, with increasing embedment, the force required to pullout the insert would increase, leading to bulky testing equipment and increasing the damage to the structure. Based on these factors, an embedment depth of 25 mm (1 in.) was chosen arbitrarily.

Kierkegaard-Hansen performed a series of pilot tests to establish the optimum diameters for the insert head and bearing ring. Figure 5 shows the test configuration used in these pilot tests. Since a suitable tension ram loading system did not exist, a laboratory compression testing machine was used to apply the load. The insert was extracted by applying a compressive load to the bottom of the embedded disk as shown in Figure 5. In this configuration, it is seen that the pullout test can also be considered as a punching-type test. The Danish word for punching is "lokning". Hence Kierkegaard-Hansen called the quantity measured by the test the "lok-strength" rather than the pullout strength.

In the first series of tests, the head diameter was varied from 20 to 40 mm (0.79 to 1.57 in.). For these tests, a large-diameter bearing ring was used so that the failure cone formed within the ring. He observed that the failure surface was not the idealized conic frustum shown in Figure 1. Instead, the extracted fragment was "trumpet-shaped", i.e., the inclination of the fracture surface, with respect to the load direction, increased with increasing distance from the insert. It was found that the pullout strength increased about 1% for each 1 mm increase in diameter. The insert head diameter was chosen arbitrarily to be 25 mm (1 in.).

The next series of pilot tests examined the relationship between compressive strength and ultimate pullout load. A bearing ring with a diameter of 130 mm (5.1 in.) was used. For this large diameter, failure occurred within the bearing ring, and the test was analogous to the earlier tests of Volf and Tremper. The compressive strength ranged from about 10 to 45 MPa (1500 to 6500 psi). In agreement with Tremper, Kierkegaard-Hansen found that the relationship between ultimate pullout load and compressive strength was nonlinear, and he stated:[3]

> It follows from this that the stress field in the rupture surface can not be equal to the stress field occurring during crushing of cylinders.

The relationship had a shape similar to that of tensile strength vs. compressive strength. It was concluded that, for a large bearing ring, the pullout strength is likely to be related to the concrete tensile strength. Because of the nonlinear relationship, test sensitivity decreased with increasing strength of concrete, i.e., large changes in compressive strength resulted in small changes in pullout strength (see Figure 4). Thus, Kierkegaard-Hansen decided to examine the effects of reducing the bearing ring diameter. Here is where the modern pullout test improved upon the earlier tests of Volf and Tremper.

As the ring diameter decreases, the fracture surface area decreases. It was reasoned that the ultimate pullout load would also decrease unless the presence of the ring alters the state of stress, in which case it could increase. Hence, in the next series of pilot tests, Kierkegaard-Hansen examined the relationship between ring diameter and ultimate pullout load. The ring diameter was varied between 130 and 50 mm (5.1 and 2.0 in.). He found that the ultimate pullout load increased gradually as the diameter decreased from 130 mm (5.1 in.) to about 80 mm (3.1 in.). As the diameter was reduced further, the ultimate pullout load increased rapidly. After these pilot tests were completed, a loading apparatus was developed for applying a tensile load to the insert as depicted in Figure 1.

The new loading apparatus was used to examine further the relationship between bearing ring diameter and pullout strength. For the next series of tests, the pullout strength was expressed as a stress by dividing the ultimate pullout load by the area of the idealized conic frustum defined by the embedment depth, insert head diameter, and bearing ring diameter. For reasons not clearly stated, Kierkegaard-Hansen concluded that the optimum bearing ring diameter should be 55 mm (2.2 in.).*

Having established the dimensions of the pullout test, Kierkegaard-Hansen studied the correlation relationship between ultimate pullout load and compressive strength. The results of these studies will be discussed in a subsequent section.

In 1970, Kierkegaard-Hansen obtained a U.S. patent** for *A Method for Testing the Strength of Cast Structures, Particularly Concrete Structures,* which described a pullout testing device composed of an embedded disk (called a "piston" in the patent), a pull rod, and a bearing ring. Specific dimensions of the test system were not given except that the apex angle of the conic frustum should be "at least about 60°".

* For a ring diameter of 55 mm, the correlation relationship between ultimate pullout load and compressive strength was such that the pullout load expressed in kilonewtons was about the same number as the compressive strength of the concrete expressed in MPa. Thus, if the ultimate pullout load was 20 kN, the compressive strength would be about 20 MPa. (Note that in Figure 4 there is similar numerical relationship between pullout load and compressive strength at the lower strength range.)

** U.S. Patent No. 3,541,845, November 24, 1970.

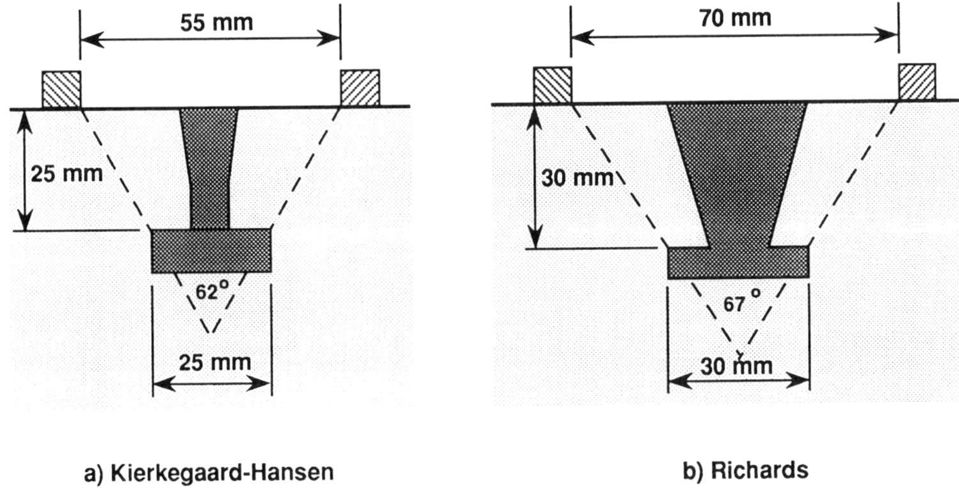

FIGURE 6. Comparison of pullout test configurations.

U.S. TEST BY RICHARDS

In the early 1970s, Owen Richards of the U.S. obtained a U.S. patent* on a pullout test system similar to Kierkegaard-Hansen's. One of the differences was that Richards' insert (described as a ''shank'' in the patent) consisted of an enlarged head that was integral with the insert shaft (Figure 1). The patent did not recommend an apex angle for the idealized conic frustum.

Rutenbeck[4] was the first to report the results of work based on Richards' ideas. In these early studies, the insert shaft was made from 19-mm (3/4 in.) threaded steel rod, and the insert head was formed by a steel washer brazed to a nut screwed onto the rod. The insert head diameter (d) was 57 mm (2.25 in.), the depth of embedment (h) was 53 mm (2.08 in.), and the bearing ring diameter (D) was 127 mm (5 in.). The apex angle for this configuration was 67°. A similar insert was used in evaluations by Malhotra.[5,6]

Richards' early test configuration produced a conic frustum with a surface area of about 18,320 mm^2 (28.4 in^2), which is about five times greater than the surface area of Kierkegaard-Hansen's system. The author believes that Richards chose this size so that the conic surface area would be approximately equal to the cross-sectional area of a 152 × 305-mm (6 × 12-in.) cylinder. Because of the large dimensions, the test equipment was bulky and not suited for field applications.

In 1977, Richards[7] reported on a smaller, machine-produced insert with a head diameter and embedment depth of 30 mm (1.18 in.). The bearing ring diameter was 70 mm (2.75 in.), thereby preserving a 67° apex angle. Richards' new configuration has a fracture surface that is about 50% greater than that of Kierkegaard-Hansen. These two pullout test configurations are compared in Figure 6. The shaft of Richards' insert is integral with the head. The pullout force is applied through a rod screwed into the shaft. On the other hand, Kierkegaard-Hansen's insert has a removable shaft, and a high strength pull-rod is screwed into the head for load application.

MODIFICATIONS BY KAINDL

In 1975, Franz Kaindl of Austria obtained a U.S. patent** (with O. Richards as assignee) describing an assortment of modifications to the basic pullout test system. Some of these included screens placed around the insert head to exclude coarse aggregates from

* U.S. Patent No. 3,595,072, July 27, 1971.
** U.S. Patent No. 3,861,201, January 21, 1975.

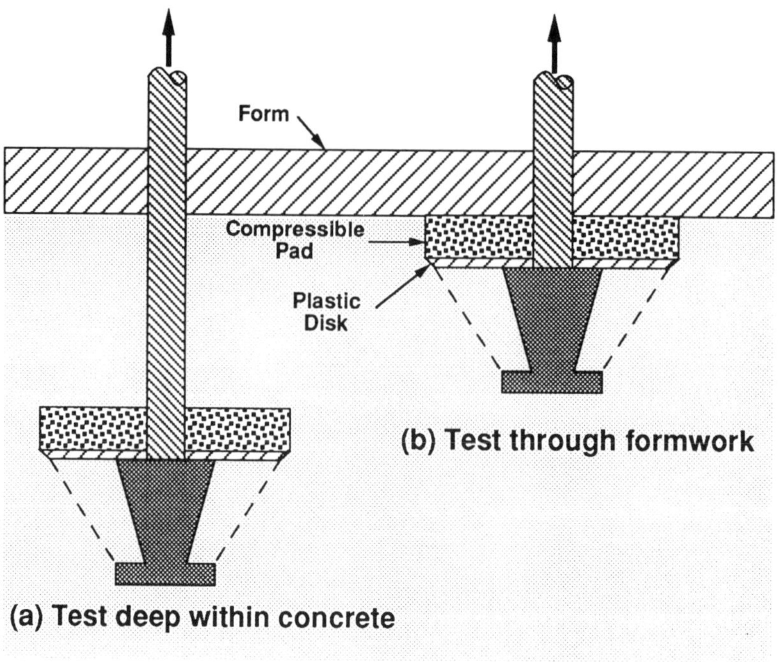

FIGURE 7. Techniques suggested by Richards for (a) performing a pullout test deep within a structure and (b) performing a surface pullout test without removing forms.

the failure surface. The objective was to reduce test variability by excluding coarse aggregate from the failure zone. These screens have not found widespread application.

Other modifications included the use of compressible pads to permit pullout testing without removing forms or to permit testing deep within the structure. Figure 7 shows some of these modifications as used by Richards.[7] The configuration in Figure 7a would be used to perform a pullout test deep within a structural member. The plastic disk is used to define the large diameter of the conic frustum, and the compressible pad allows pullout of the conical fragment. The configuration in Figure 7b would be used to perform a pullout test without having to remove formwork. There is no bearing ring in these configurations. There are few data on the performance of these modified pullout tests.

SUMMARY

The ideas for the modern pullout tests go back to a Soviet test described in 1938. This test, and the later test by Tremper, did not use a bearing ring, and the ultimate pullout load was believed to be controlled by the tensile strength of the concrete. As a result, the correlation between pullout strength and compressive strength was found to be nonlinear. In 1962, Kierkegaard-Hansen improved upon the original idea by introducing a bearing ring. This modification resulted in a failure cone with a well-defined geometry and he reported a linear correlation between pullout strength and compressive strength (over the strength range used). The apex angle in Kierkegaard-Hansen's test was 62°. Later, Richards introduced a similar pullout test to North America, but he suggested an apex angle of 67°.

FAILURE MECHANISM

The pullout test subjects the concrete to static load. Therefore, it should be possible to calculate the internal stresses in the concrete and to predict the onset of cracking and the ultimate pullout load. This is desirable so that the ultimate pullout load could be related to

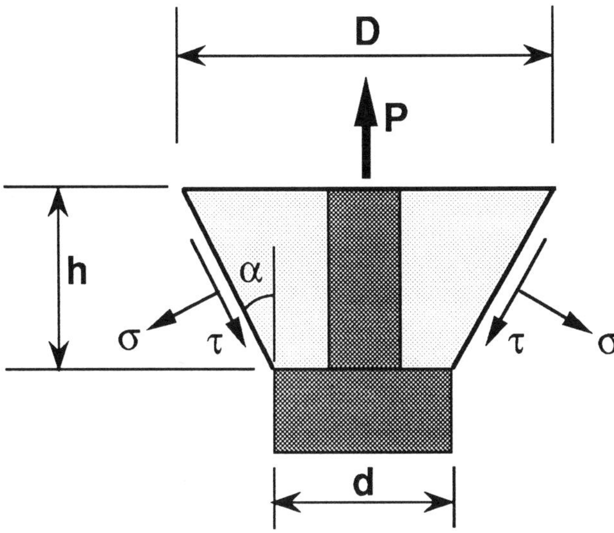

FIGURE 8. Freebody diagram showing normal and shearing stresses acting on failure surface of idealized frustum.

uniaxial strength properties of concrete. Unfortunately, the stress distribution is not easy to calculate, the state of stress is altered by the presence of coarse aggregate particles, and the fundamental failure criterion for concrete is not completely understood. This section reviews various theories about the failure mechanism for the pullout test. It will be shown that there is no consensus on this point.

QUALITATIVE EXPLANATIONS

In his paper, Skramtajew noted that:[1]

> In this case concrete is simultaneously in tension and shear, the generating lines of the cone running approximately at an angle of 45 degrees to the vertical.

Thus from the beginning it was recognized that the pullout test subjects the concrete to a complex state of stress. Figure 8 is a freebody diagram of the idealized conic frustum extracted during a pullout test. The pullout force (P) is resisted by normal (σ) and shearing (τ) stresses acting on the frustum surface. The normal stress acts perpendicular to the surface and is a tensile stress, while the shearing stress acts parallel to the surface in the direction shown. The vertical components of these stresses multiplied by the surface area (A) produces a vertical force to counteract the applied pullout force. Assuming that these stresses are uniformly distributed on the failure surface, one can show that

$$\sigma = \frac{P}{A} \sin \alpha \tag{2}$$

$$\tau = \frac{P}{A} \cos \alpha \tag{3}$$

The surface area of the frustum is calculated as follows:

$$A = \frac{\pi (D + d)}{4} \sqrt{4 h^2 + (D - d)^2} \tag{4}$$

In his patent disclosure, Kierkegaard-Hansen made the following statement about the failure cones;

> ...the fracture faces attain substantially the same shape as one half of the well known hour-glass-shaped fracture faces, which are produced in compressive strength tests of cylindrical specimens,...

This implies that the cones of the pullout test are related to the end cones typically observed during the testing of cylinders, and this was offered as an explanation for the correlation between pullout strength and compressive strength. This explanation is incorrect because the cones are formed due to entirely different factors. In the pullout test, the cone is extracted from the concrete mass under the action of the applied pullout force. In the compression test, the cones represent intact concrete that is prevented from failing due to triaxial compressive stresses introduced by friction between the cylinder and the solid loading platens.[8]

Malhotra and Carette[6] calculated a "pullout strength" by dividing the ultimate pullout load by the surface area of the idealized frustum. The pullout test geometry developed by Richards was used (apex angle = 67°). The ratio of this pullout strength to the compressive strength of companion cylinders or cores varied from 0.24 to 0.18, as the compressive strength varied from about 20 MPa (2900 psi) to 52 MPa (7500 psi). These ratios were similar to the reported ratios of shear strength to compressive strength obtained from triaxial tests.[9] Therefore, it was suggested that the pullout strength may be related to the direct shear strength of concrete. The criticism to this analysis is that the calculated "pullout strength" is not really a stress because the pullout force is inclined to the surface of the frustum. Dividing the pullout force by the surface area results in neither a normal stress nor a shearing stress. The author believes that the "pullout strength" was found to be approximately 20% of the compressive strength because of the particular value of the apex angle recommended by Richards, rather than because of an inherent relationship between shear strength and pullout strength. In addition, the so-called "direct shear" strengths in Reference 9 were obtained by assuming a straight line envelope to the Mohr's circles of failure stresses under triaxial loading.[10] These computed direct shear strengths are recognized as being larger than the true shear strength of concrete.[9]

In summary, these qualitative explanations do not provide insight into the actual failure mechanism during a pullout test.

ANALYTICAL STUDIES

Rigid-plastic analysis — Jensen and Braestrup[11] presented the first analytical study which attempted to provide a theoretical basis for the existence of a linear relationship between ultimate pullout load and compressive strength. They assumed that concrete obeys the modified Mohr-Coulomb failure theory (sliding or separation possible) and that the extracted cone has the shape of the idealized conic frustum. The analysis assumed "rigid-plastic" behavior and that the normal and shearing stresses were distributed uniformly on the failure surface. It was concluded that, if the friction angle of the concrete equals one half of the apex angle and if the tensile strength is a constant fraction of the compressive strength, there is a proportional relationship between the ultimate pullout load and compressive strength. The analysis has been criticized as not providing a true behavioral prediction because the conclusions are a direct result of the underlying assumptions rather than from a rigorous assessment of the true behavior during the test.[13,14]

Non-linear finite element analysis — In 1981, Ottosen was the first to use the finite-element method to analyze the state of stress and to attempt to determine the failure mechanism of the pullout test.[15] He used nonlinear material models, a three-dimensional failure criterion, and a smeared cracking approach to follow the progression of failure with increasing pullout load. The pullout test geometry developed by Kierkegaard-Hansen was used. The analysis considered the concrete as a homogeneous material, i.e., the presence of individual coarse aggregate particles was not modeled.

A significant finding of Ottosen's analysis was that, at about 65% of the ultimate load, a series of circumferential cracks had developed extending from the edge of the insert head to the bearing ring. Despite the circumferential cracks, additional load could be sustained by a highly stressed narrow band, or "strut", extending from the insert head to the bearing ring. Ultimate failure was attributed to "crushing", or compressive failure, of the concrete within this strut. Ottosen concluded that this was the reason for the good correlation between pullout strength and compressive strength.

Ottosen's analysis demonstrated that the pullout test subjects the concrete to a highly nonuniform, triaxial state of stress. Within the compression strut, Ottosen found that the state of stress was predominantly biaxial-compression "occasionally superposed by small tensile stresses."[15] Because of the tensile stresses, Ottosen concluded that the tensile strength of concrete had a secondary influence on the ultimate pullout load. He showed that, because the ratio of tensile strength to compressive strength decreases with increasing strength of concrete, the ratio of pullout strength to compressive strength would be expected to decrease for increasing concrete strength. This would explain the previous observations of Malhotra and Carette[6] and Richards.[7]

Yener and Vajarasathira[16] performed a plastic fracture analysis using the finite element method. Cracking was assumed to occur perpendicular to the direction of maximum tensile strain. "Crushing" failure was defined to occur if the maximum strain was compressive when an element cracked. It was noted that the high shearing stresses within the region between the insert head and bearing ring cause high tensile stresses, which result in circumferential cracking that defines the eventual failure surface. The analysis predicted that circumferential cracking began to form at the corner of the insert head at about 25% of the ultimate load. The circumferential crack propagated toward the bearing ring, but was arrested by high compressive stresses at about 50% of ultimate load. Another crack initiated at the corner of the insert head and propagated toward the bearing ring, so that at 70% of the ultimate load the trumpet-shaped frustum was completely formed. At this stage, the frustum was prevented from pulling out completely by frictional resistance due to high radial compressive stresses acting at the juncture of the frustum and the main body, just below the bearing ring perimeter. Additional load could be applied until crushing occurred around the perimeter of the frustum. Thus, while the crack patterns were similar to those in Ottosen's analysis, a different ultimate load carrying mechanism was hypothesized.

Linear-elastic finite element analysis — Stone and Carino[17] also performed finite element analyses for pullout tests with apex angles of 54° and 70°. However, because their analyses were linear-elastic, the results are applicable only until the formation of cracks. In agreement with Ottosen, Stone and Carino found that the pullout test subjects the concrete to a complex three-dimensional state of stress. High compressive stresses exist within the "strut" region between the bearing ring and insert head. Figure 9 shows the principal stress trajectories for the two test configurations that were analyzed (because of symmetry only one-half of the specimens are shown). For the 70° apex angle, there is close agreement between the compressive stress trajectories and the trumpet-shaped fracture surface observed in the companion test specimen.[12] Note that the principal tensile stresses act perpendicular to the compressive stress trajectories. For the 54° apex angle. the compressive stress trajectories and failure surface are less curved. The agreement between the directions of the stress trajectories and the shape of the failure surface led Stone and Carino to conclude that tensile strength is likely to play a greater role in the pullout test than was proposed by Ottosen.

The nonuniformity of the stresses along the idealized conic frustum is illustrated in Figure 10, which shows the variation of the principal stresses for the 70° apex angle at about 20% of the ultimate load.[17] There are very high tensile and compressive stresses at the edge

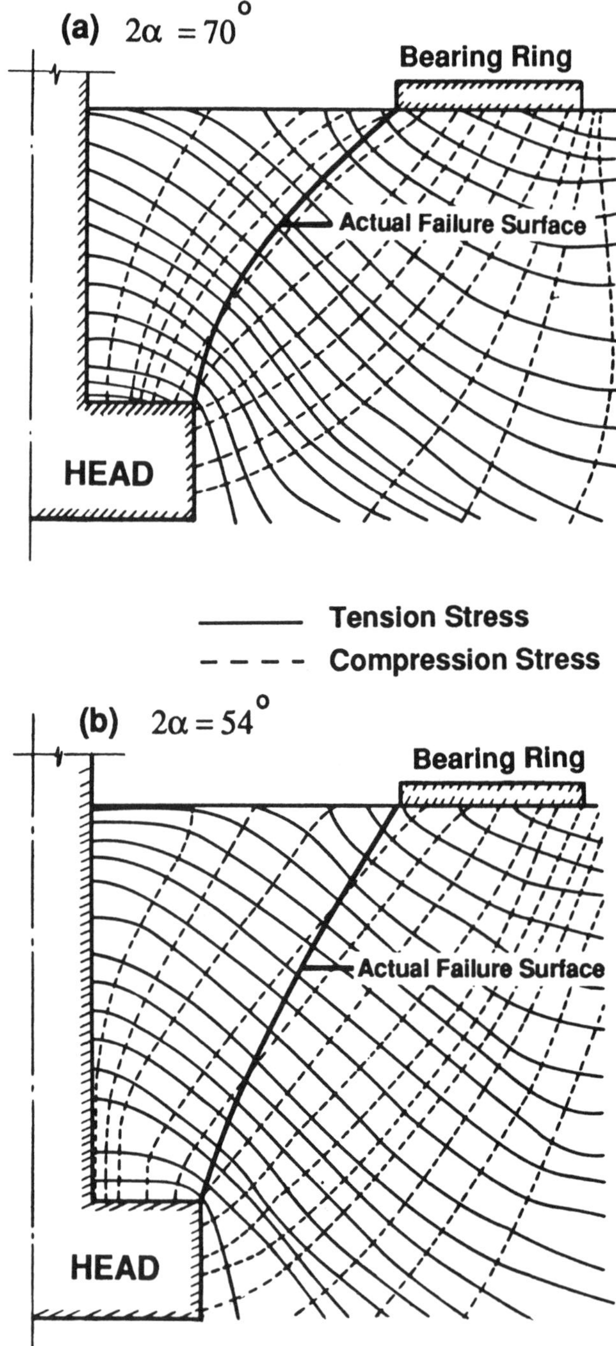

FIGURE 9. Tension and compression stress trajectories prior to formation of cracks.[17]

of the insert head due to the stress concentration effect at the sharp corner. The tensile stresses decrease with distance from the insert head and become compressive near the bearing ring. The compressive stresses are high near the insert head and near the bearing ring, and they are nearly uniform in the middle region of the idealized frustum. Thus, in contrast to Ottosen's conclusion, the frustum is subjected to significant tensile stresses.

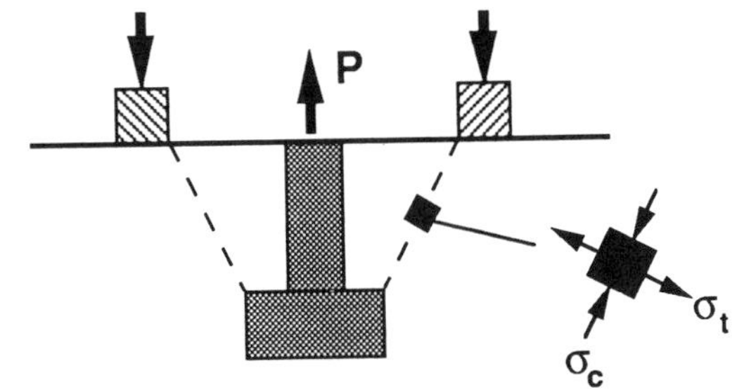

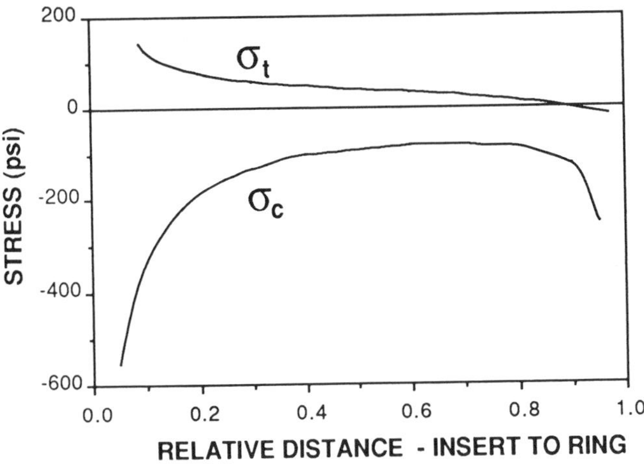

FIGURE 10. Variation of principal stresses along a line extending from insert head to bearing ring (70° apex angle).[17]

Fracture mechanics analysis — A more recent finite element analysis used the principles of nonlinear fracture mechanics to gain an understanding of the failure mechanism.[18] This analysis used a discrete cracking model with special elements to represent the behavior of cracked concrete. The pullout test configuration had the following geometry: D = 61 mm (2.4 in.); d = h = 25.4 mm (1 in.); and apex angle (2α) = 70°.

The nonlinear, fracture mechanics analysis revealed that two crack systems develop during the course of the pullout test. The first (primary) crack is a circumferential crack which initiates at the edge of the insert head and propagates into the concrete at angle of about 60° with the load axis. The primary crack begins at a low pullout load because of the large tensile stress concentration at the insert head (Figure 10), and it is arrested as it penetrates a region of low tensile stress below the bearing ring. Tensile and shearing stresses continue to be carried across the primary crack because of the small crack-opening displacement. According to the analysis, the formation of the primary crack results in high tensile stresses in the region between the insert head and the bearing ring. Thus a second (or secondary) crack initiates at a point between the insert head and bearing ring, and it propagates in two directions toward the ring and the insert head.

Figure 11 shows the deformed shape (highly exaggerated) of the finite element model after the secondary crack has propagated toward the ring and insert head. It is apparent that the secondary crack defines the shape of the conical fragment that is eventually extracted from the concrete. The ultimate load could not be determined in the analysis because the computer program would not permit the formation of a crack at the highly compressed node at the corner of the bearing ring. It was postulated that the failure surface would be formed

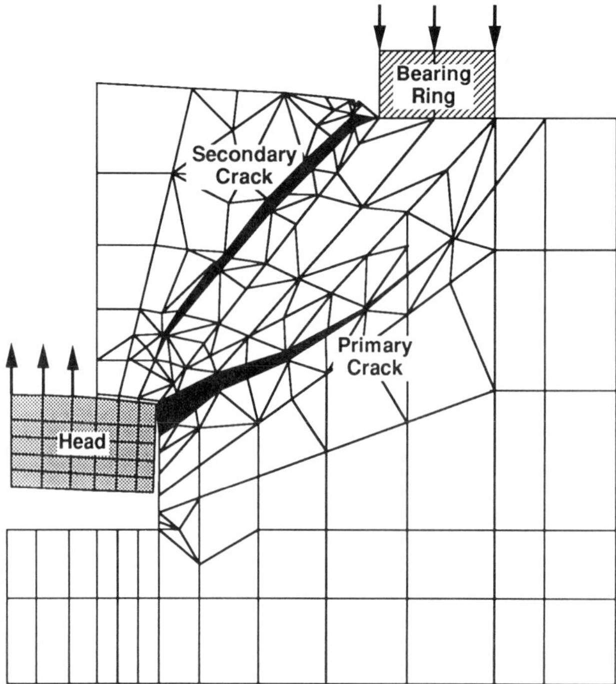

FIGURE 11. Deformed shape of finite element model showing the primary and secondary cracks.[18]

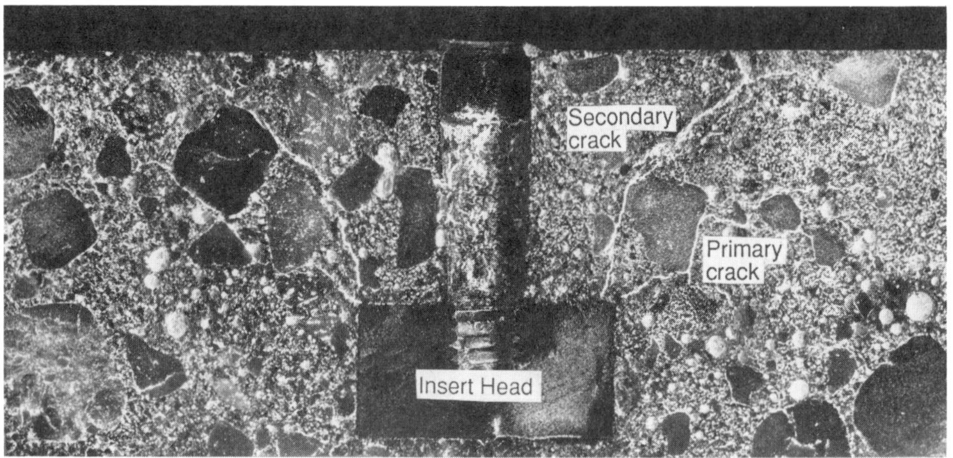

FIGURE 12. Internal cracking in pullout test loaded to 90% of ultimate load.[18]

completely by cracking of the final ligament between the insert head and the secondary crack tip. This final crack propagation would be primarily a shear failure along a surface inclined at small angle to the load axis.

The results of the analysis were compared with experimental crack patterns observed in sectioned pullout test specimens which had been loaded to various fractions of ultimate strength. For example, Figure 12 shows the crack pattern in a specimen loaded to approximately 90% of the expected ultimate pullout load. The primary and secondary crack systems are clearly visible, and it is seen that the crack trajectory at the insert head is nearly parallel to the load axis. This figure also shows that the trumpet-shaped failure surface is completely formed, yet additional load could have been applied to the specimen. The reason for this behavior is explained in the next section.

Ballarini et al.[19] reported on the results of a linear-elastic fracture analysis of a two-dimensional (as opposed to axi-symmetric) pullout test. A perfectly elastic, brittle material was assumed. Some of the conclusions of their analysis are as follows:

- Cracking begins at the edge of the insert as a tensile crack (as opposed to shearing).
- Initial cracking begins at an angle of about 75° with respect to the load axis, and the initial cracking is stable.
- With increasing load, the crack propagates toward the bearing ring at a decreasing angle with respect to the load axis.

Experiments using mortar specimens and different pullout configurations verified the analytical predictions. It was found that by expressing ultimate load in terms of fracture toughness, differences between the correlation relationships for various test configurations were greatly reduced. While not directly applicable to actual pullout tests, this study provided some insight into the crack propagation process.

EXPERIMENTAL STUDIES

Compared with extensive experimental work to establish the nature of the correlation relationship between pullout strength and compressive strength, there has been relatively little experimental work on the failure mechanism. Two of the notable studies were the large-scale tests performed at the National Bureau of Standards* (NBS) and the microcracking study at the Denmark Technical University.

Large-scale tests — Stone and Carino performed large-scale pullout tests to gain an understanding of the failure mechanism.[12,17] The pullout test geometry was scaled up from conventional tests by a factor of about 12 so that strain gauges could be embedded in the concrete to measure strain distributions during the test. Two specimens were tested having apex angles of 54° and 70°.

By analyzing the strain gauge readings as a function of pullout load, Stone and Carino concluded that the failure sequence comprised three phases:[12]

Phase I — Initiation of circumferential cracking at the edge of the insert head at about 1/3 of the ultimate load

Phase II — Completion of circumferential cracking, from the insert head to the bearing ring, at about 2/3 of the ultimate load

Phase III — Shear failure of the matrix (mortar) and degradation of aggregate interlock beginning at about 80% of the ultimate load

These phases are illustrated in Figure 13. Thus Stone and Carino proposed that aggregate interlock is the ultimate load-carrying mechanism and that the ultimate load is reached when the coarse aggregates bridging the failure surface are pulled out of the mortar.

Figure 14 is a simple model of concrete to illustrate how aggregate interlock permits load to be carried across a crack. The model consists of a round aggregate particle embedded in a matrix and a crack passing through the matrix and around the aggregate particle. In Figure 14 a, the relative displacement of the two halves of the model is perpendicular to the crack. In this case, there is no interference between the matrix and the aggregate particle. Therefore, force cannot be transmitted across the crack. In Figure 14b, the relative displacement of the two halves of the model is inclined to the crack. In this case, there is

* In 1988, the name was changed to the National Institute of Standards and Technology.

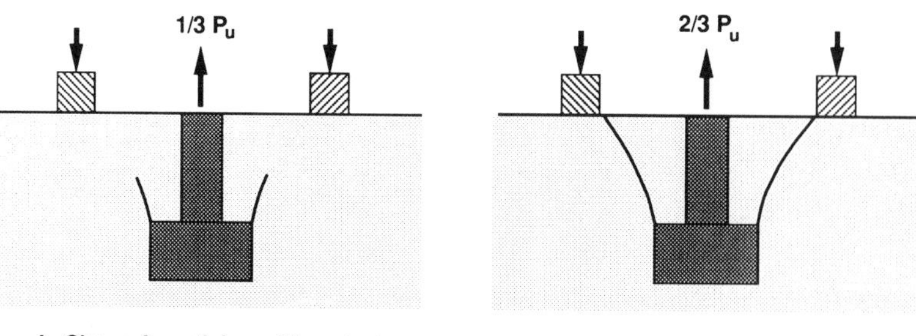

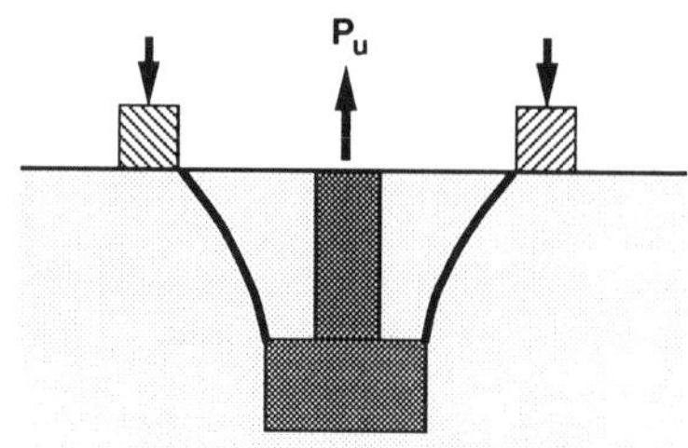

FIGURE 13. Failure sequence based on NBS large-scale pullout tests.[12]

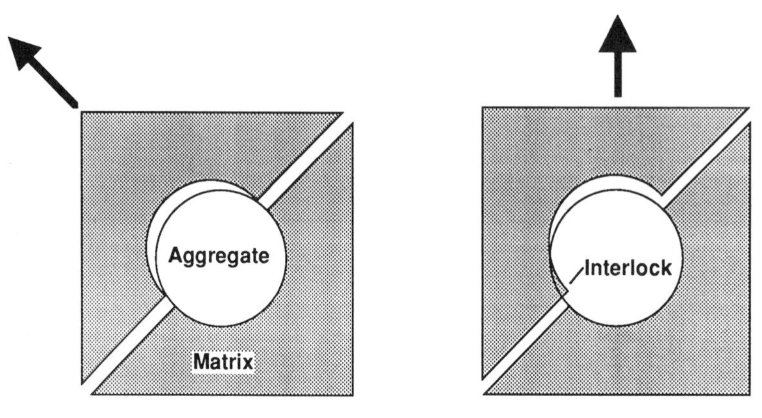

FIGURE 14. Model to illustrate aggregate interlock mechanism during pullout test.

interference between the matrix and the aggregate, and a force can be transmitted across the crack. In the pullout test, the displacement condition is similar to Figure 14b. This model justifies Stone and Carino's assertion[12] that additional load can be applied to the insert even though there is a circumferential crack from the insert head to the bearing ring.

If the aggregate interlock mechanism were to govern the ultimate pullout load, Stone and Carino argued that for a homogeneous matrix (no aggregate) the ultimate load would be reached when the circumferential crack extended from the insert head to the bearing ring. Verification of this hypothesis was provided by pullout tests in concrete and in mortar, in which the displacement of the insert head was measured as a function of the pullout load.[20] Figure 15 compares the load-displacement histories for pullout tests (apex angle = 62°) in

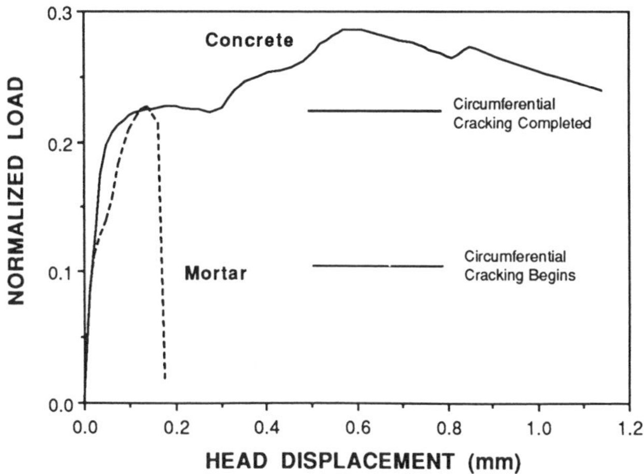

FIGURE 15. Load vs. insert head displacement for pullout tests in concrete and mortar specimens.[13]

concrete and in mortar. The pullout load has been normalized by dividing by the product of the compressive strength of companion cylindrical specimens and the area of the idealized conic frustum. At small loads, the load-displacement response is linear. At about 1/3 of the ultimate load, the response deviates from linearity as the circumferential crack propagates between the head and bearing ring. When the circumferential crack is formed completely, there is a marked difference between the responses of the concrete and mortar specimens. In the mortar, there is no aggregate interlock and the load drops abruptly. On the other hand, for the concrete there is a large increase in head displacement with no increase in load. Then the load begins to increase until the maximum load is reached. Beyond the ultimate load, the load decreases gradually with increasing head displacement. Thus the pullout test in the concrete behaves in a more "ductile" manner than the test in the mortar. The behavior in Figure 15 is consistent with the aggregate interlock mechanism.

Ottosen[15] stated that the observed ductile failure mode of the pullout test was because of compressive failure of the concrete. The aggregate interlock mechanism provides an alternative explanation for this ductile behavior. Peterson[13] noted that when a pullout cone is extracted "crushed materials are observed, aggregates as well as cement paste." In addition, for low strength concrete the failure surface of the cone consists of "fish scale layers." These observations were cited as further evidence that compressive failure occurs. However, these observations are also consistent with the aggregate interlock mechanism. The "crushed material" could result from pullout of aggregate particles bridging the circumferential crack. The "fish scale" could be due to cracking of the matrix as the aggregate particles are pulled out, rather than due to parallel cracks associated with compressive failure.

Microcracking analysis — Krenchel and Shah[21] loaded pullout inserts to predetermined fractions of the expected ultimate load. The specimens were unloaded, cut in half, and prepared for examination of microcracks. They also monitored acoustic emissions (AE) during the tests.

Based on the microcracking analysis, Krenchel and Shah concluded that the pullout test involves two circumferential crack systems. There is a stable primary crack which begins at the insert head at about 30% of the peak load and propagates to a point below the bearing ring at an apex angle between 140° and 160°. There is a secondary system which becomes fully developed after the peak load and defines the shape of the extracted cone. The existence of two crack systems agrees with the independent analytical results of Hellier et al.[18] However, there is disagreement as to when the secondary system develops. Hellier et al. claim that it develops before the peak load, while Krenchel and Shah state that it becomes fully

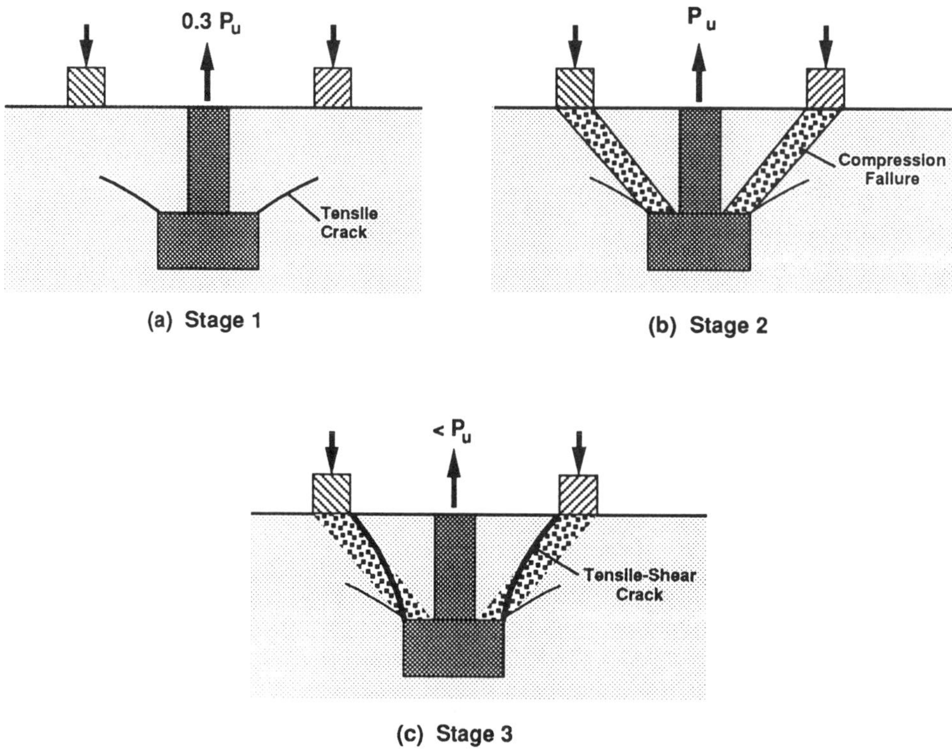

FIGURE 16. Failure mechanism of pullout tests according to Krenchel and Bickley.[22]

developed after the peak load. Krenchel and Shah reported that specimens unloaded from the peak load showed no surface cracking at the bearing ring. This would support their claim. However, some of the figures in Reference 21 appear to show a fully developed secondary crack at loads less than the peak.

Krenchel and Shah observed that the start of AE activity coincided with the limit of proportionality of the load vs. insert displacement curve, and this was interpreted as the initiation of the primary crack. AE activity increased gradually as the load approached the peak value and increased dramatically thereafter.

Upon review of the various analytical and experimental investigations that had been conducted, Krenchel and Bickley concluded that the failure mechanism of the pullout test involves the following stages[22] (see Figure 16):

> **Stage 1** — At a load of about 30 to 40% of the ultimate, ''tensile cracks'' originate at the corner of the insert head and propagate into the concrete for a distance of 15 to 20 mm (0.6 to 0.8 in.) forming an apex angle between 100° and 135° (Figure 16 a). This cracking concentrates subsequent straining of the concrete so that ''all load is taken up in the truncated zone'' between the insert head and the bottom of the bearing ring.
>
> **Stage 2** — A large number of stable microcracks develop in the truncated zone. These cracks run from the top of the insert head to the bottom of the bearing ring, forming an apex angle of about 84° (Figure 16b). This second stage cracking occurs as the load increases up to and just past the ultimate load. These stable microcracks are analogous to the vertical cracks observed during an ordinary uniaxial compression test of a cylinder or prism.

Stage 3 — Beyond the ultimate load, a circumferential "tensile/shear" crack develops which forms the final shape of the extracted cone (see Figure 16c).

Krenchel and Bickley emphasize that the progression of microcracking in Stage 2 is directly related to the ultimate load during the test, and that the fracture surface formed when the cone is completely extracted has little to do with the failure mechanism at ultimate load. As an analogy, they cite the case where a cylinder is compressed past the ultimate load point, and the resulting fracturing that is observed has little to do with the fracture mechanism at ultimate load.

SUMMARY

A series of independent analytical and experiment studies have been performed to gain an understanding of the fundamental failure mechanism of the pullout test. It is understood that the pullout test subjects the concrete to a nonuniform, three-dimensional state of stress. It also has been demonstrated that there are at least two circumferential crack systems involved: a stable primary system which initiates at the insert head at about 1/3 of the ultimate load and propagates into the concrete at a large apex angle; and a secondary system which defines the shape of the extracted cone. However, there is not a consensus on the failure mechanism at the ultimate load. Some believe that ultimate load occurs as a result of compressive failure of concrete along a line from the bottom of the bearing ring to the top face of the insert head. This could explain why good correlation exists between pullout strength and compressive strength. Others believe that the ultimate failure is governed by aggregate interlock across the secondary crack system, and the ultimate load is reached when sufficient aggregate particles have been pulled out of the matrix. In the latter case, it is argued that good correlation exists between pullout strength and compressive strength because both properties are controlled by the strength of the mortar.

Despite the lack of agreement on the exact failure mechanism, it has been shown that pullout strength has good correlation with compressive strength of concrete and that the test has good repeatability. These aspects are discussed in the next section.

STATISTICAL CHARACTERISTICS

Two important statistical characteristics of tests for in-place concrete strength, such as the pullout test, are the within-test variability and the correlation relationship between the test results and compressive strength. Within-test variability, also called "repeatability", refers to the scatter of results when the test is repeated on identical concrete using the same test equipment, procedures, and personnel. For a given concrete, the repeatability of a test affects the number of tests required to establish, with a desired degree of certainty, the average value of the property being measured by the test.[23] The correlation relationship is required to convert the test results to a compressive strength value. This section examines these two characteristics of the pullout test.

REPEATABILITY

If pullout tests are repeated on the same concrete at the same maturity, the ultimate pullout loads would be expected to be normally distributed about the average value and the standard deviation would be the measure of repeatability. If replicate tests were performed on the same concrete but at different maturities, so that there would be different average pullout strengths, would the standard deviation be independent of the average pullout strength? If the standard deviation were found to be proportional to the average pullout strength, the coefficient of variation (standard deviation divided by the average) would be the correct measure of repeatability.

In a review of about 4300 field pullout tests, Bickley[24] concluded that the standard

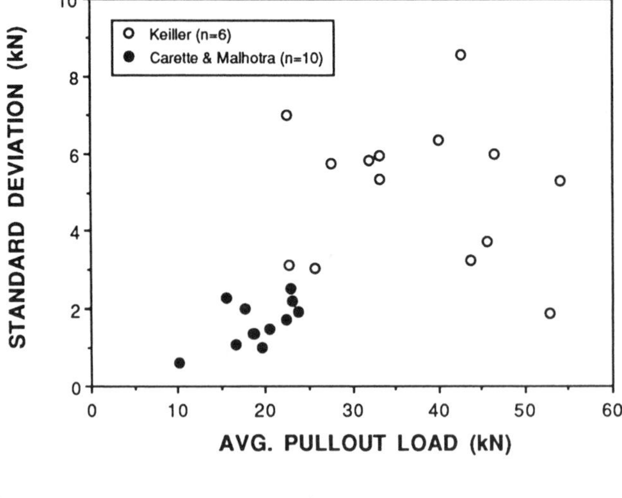

A

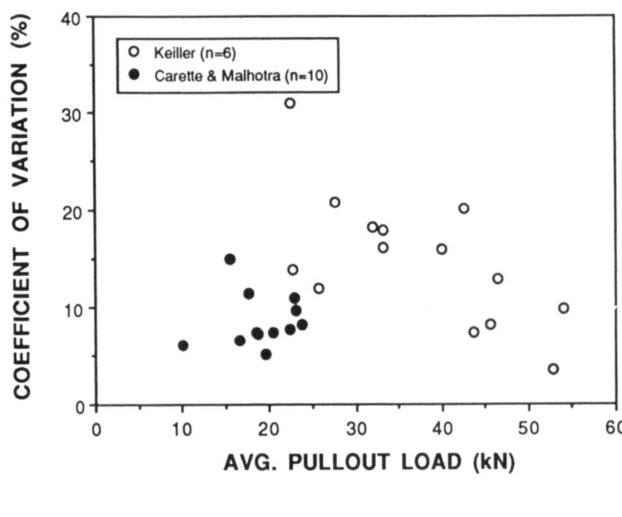

B

FIGURE 17. Repeatability of the pullout test as a function of average pullout strength:[25,26] (A) in terms of standard deviation and (B) in terms of coefficient of variation.

deviation of pullout strength was constant. Because average values were not given, it is not known whether this conclusion applies over a wide range of average pullout strength. The standard deviations reported by Carette and Malhotra[25] and by Keiller[26] offer some insight. Figure 17A shows the standard deviation of the ultimate pullout load plotted as a function of the average value. The same commercial test system (LOK-TEST) was used by the two research groups, but the number of replications were different. It is clear that the standard deviations reported by Carette and Malhotra are lower than those reported by Keiller, but so are the average ultimate pullout loads. For comparison, Figure 17B, shows the corresponding coefficients of variation as a function of average pullout load. The differences between the two groups of data are reduced and this suggests that the coefficient of variation may be the appropriate indicator of repeatability over a wide range of pullout load.

Additional evidence that the standard deviation is dependent on the level of pullout load

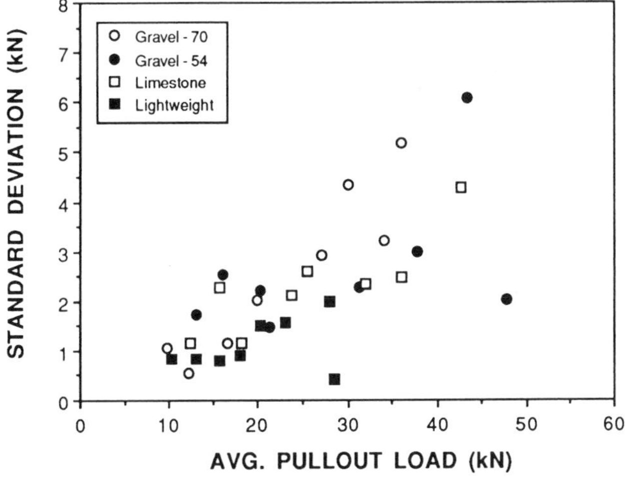

A

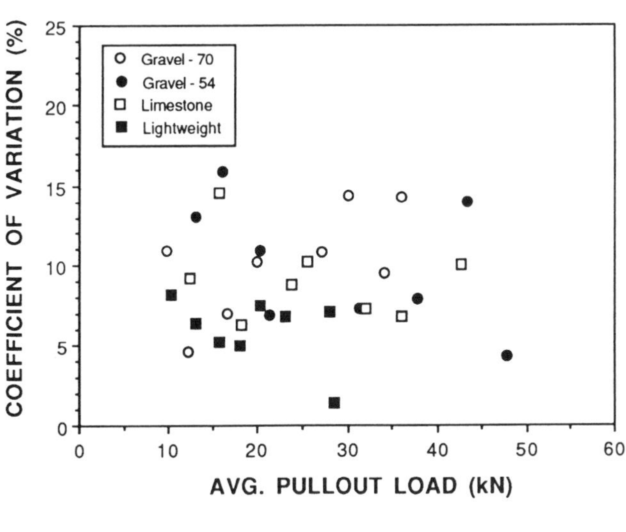

B

FIGURE 18. Repeatability of the pullout test for different test conditions:[27] (A) in terms of standard deviation and (B) in terms of coefficient of variation.

is provided by the results of Stone et. al[27] shown in Figure 18A. Three of the test series involved a pullout test configuration having a 70° apex angle and a 25-mm (1-in.) depth of embedment, but concretes with different types of coarse aggregate (river gravel, crushed limestone, and expanded shale) were used. The fourth series used a 54° apex angle and river gravel aggregate. Figure 18A shows that the standard deviation tends to increase with increasing average pullout load. In contrast, Figure 18B shows that the coefficient of variation appears to be independent of the average pullout load.

Based upon the above discussion, it is concluded that the coefficient of variation is the better statistic for quantifying the repeatability of the pullout test. Table 1, which was prepared by the author and included in Reference 28, lists some of the reported values of the coefficient of variation for replicate pullout tests. The average values vary from about 4 to 15%. In a

TABLE 1
Within-Test Coefficient of Variation of Pullout Tests

Ref.	Apex angle	Embedment depth (mm)	Max agg size (mm)	Type of aggregate	Sample size	Range of C.V. (%)	Average C.V. (%)
5	67	53	19	Limestone	3	2.3—6.3	3.9
6	67	53	25	R.Gravel	2	0.9—14.3	5.3
24	62	25	10	?	8	3.2—5.3	4.1
29	70	25	20	Granite	6	1.9—12.3	6.9
25	67	53	19	Limestone	4	1.9—11.8	7.1
25	62	25	19	Limestone	10	5.2—14.9	8.5
26	62	25	20	Limestone	6	7.4—31	14.8
27	70	25	19	R.Gravel	11	4.6—14.4	10.2
	70	25	19	Limestone	11	6.3—14.6	9.2
	70	25	19	Lightweight	11	1.4—8.2	6.0
	54	25	19	R.Gravel	11	4.3—15.9	10.0
30	67	30	12	Natural	24	2.8—6.1	4.3

summary of the experiences with the LOK-TEST system, Krenchel and Peterson[31] also reported coefficients of variation ranging between 4 and 15%, with an average of about 8%.

The data in Table 1 are for different test configurations (apex angle and embedment depth), aggregate types, aggregate sizes and sample sizes. Similarly, the results summarized by Krenchel and Peterson involved different types and sizes of aggregates as well as different sample sizes. The National Bureau of Standards performed a study to determine whether some of these variables have an effect on the repeatability of the pullout test. The following variables were considered: apex angle, embedment depth, maximum aggregate size, and aggregate type. The experimental details were reported by Stone and Giza.[20] Basically, 11 pullout inserts were embedded in 152 × 152 × 914 mm (6 × 6 × 36 in.) beam specimens. Mortar and concrete beams were cast, and the 11 inserts in each beam were extracted at a given test age. Concrete strength was not a variable, so the average companion cylinder strengths varied over a narrow range of 14 and 17 MPa (2000 to 2500 psi).

The author has used the techniques in ASTM E 178[32] to remove outliers from the data reported by Stone and Giza.[20] Usually, an erratic low pullout load resulted because of abnormal cracking of the beam outside of the bearing ring, and an erratic high load resulted when a large aggregate particle was situated across the fracture surface of the conical fragment. Table 2 summarizes the testing variables and the resulting coefficients of variation. To visualize the results, the data are also plotted. Figure 19A shows the effect of the apex angle. For pullout tests in concrete, a variation of the apex angle from 30° to 86° does not appear to have a strong effect on the repeatability, but there is a tendency for more scatter at lower apex angles. The tests in mortar were less variable, and, likewise, there was no dependence on the apex angle. The effects of embedment depth are summarized in Figure 19B. Because of the few data points, it is not possible to conclude whether embedment depth has an effect. However, it appears that when the embedment was equal to or less than the maximum aggregate size (19 mm (3/4 in.)), there was a tendency for more variability. Again, pullout tests in mortar have less variability. The effects of aggregate size are shown in Figure 19C. For an embedment depth of 25 mm (1 in.), the 19-mm (3/4-in.) maximum aggregate size resulted in greater variability than the smaller aggregate sizes. In addition, there were no differences between the tests in mortar and those in concrete with the 6-, 10-, and 13-mm (1/4-, 3/8-, and 1/2-in.) aggregates. Finally, Figure 19D shows the effects of coarse aggregate type. The tests in concretes made with normal weight aggregates had about the same coefficient of variation. However, the tests in lightweight concrete had a lower coefficient of variation of 5.7%, which is practically identical to the average value of 5.6% obtained from tests in mortar.

TABLE 2
Effects of Test Variables on the Within-Test Coefficient of Variation of the Pullout Test[20]

Test series	Apex angle	Embedment depth (mm)	Max agg size (mm)	Type of aggregate	No. of data sets[a]	Range of C.V. (%)	Average C.V. (%)
Apex angle	30	25	19	R.Gravel	2	9.1—11.4	10.3
			Mortar		1		4.6
	46	25	19	R.Gravel	4	5.6—13.3	8.9
			Mortar		2	4.5—6.5	5.5
	54	25	19	R.Gravel	2	6.3—6.7	6.5
			Mortar		1		4.3
	58	25	19	R.Gravel	2	7.3—8.6	8.0
			Mortar		1		4.9
	62	25	19	R.Gravel	2	7.5—9.6	8.6
			Mortar		1		4.1
	70	25	19	R.Gravel	4	8.0—10.1	8.8
			Mortar		6	4.0—6.9	5.6
	86	25	19	R.Gravel	2	7.5—9.0	8.3
			Mortar		1		2.8
Embedment depth	58	12	19	R.Gravel	1		12.9
			Mortar		1		5.3
	58	20	19	R.Gravel	2	7.7—10.9	9.3
			Mortar		1		6.4
	58	23	19	R.Gravel	2	6.5—6.7	6.6
			Mortar		1		4.7
	58	25	19	R.Gravel	2	7.3—8.6	8.0
			Mortar		1		4.9
	58	27	19	R.Gravel	2	8.0—8.8	8.4
			Mortar		1		2.8
	58	32	19	R.Gravel	2	8.1—9.1	8.6
			Mortar		1		3.1
	58	43	19	R.Gravel	2	7.9—9.4	8.7
			Mortar		1		4.1
Aggregate size	70	25	6	R.Gravel	2	4.1—7.0	5.6
	70	25	10	R.Gravel	5	3.5—6.5	4.9
	70	25	13	R.Gravel	5	3.3—10.6	5.5
	70	25	19	R.Gravel	4	8.0—10.1	8.8
Aggregate type	70	25	19	Lightweight	2	5.7—5.7	5.7
	70	25	19	R.Gravel	4	8.9—10.1	8.8
	70	25	19	Gneiss	2	7.2—7.5	7.4
	70	25	19	Porous Limest	2	7.7—10.9	9.3

[a] Number of data sets having various replicates after discarding outliers or bad tests:

n =	11	10	9	8
Concrete	25	18	2	2
Mortar	10	5	4	

The NBS study showed that, for a 25-mm (1-in.) embedment depth, tests in concrete made with 19-mm (3/4-in.) aggregate had higher variability, and tests in concrete made with lightweight aggregate had lower variability. These observations support the notion that aggregate interlock controls the ultimate load during the test. Stone and Giza[20] also found that the average pullout loads in the mortar specimens where consistently lower than the strengths in concrete specimens, and the difference tended to be higher for smaller apex angles. They argued that, at smaller apex angles, aggregate interlock effects are greater, and more load is needed to extract the aggregates from the mortar matrix. They reasoned that tests in concrete had greater variability than tests in mortar because of the random

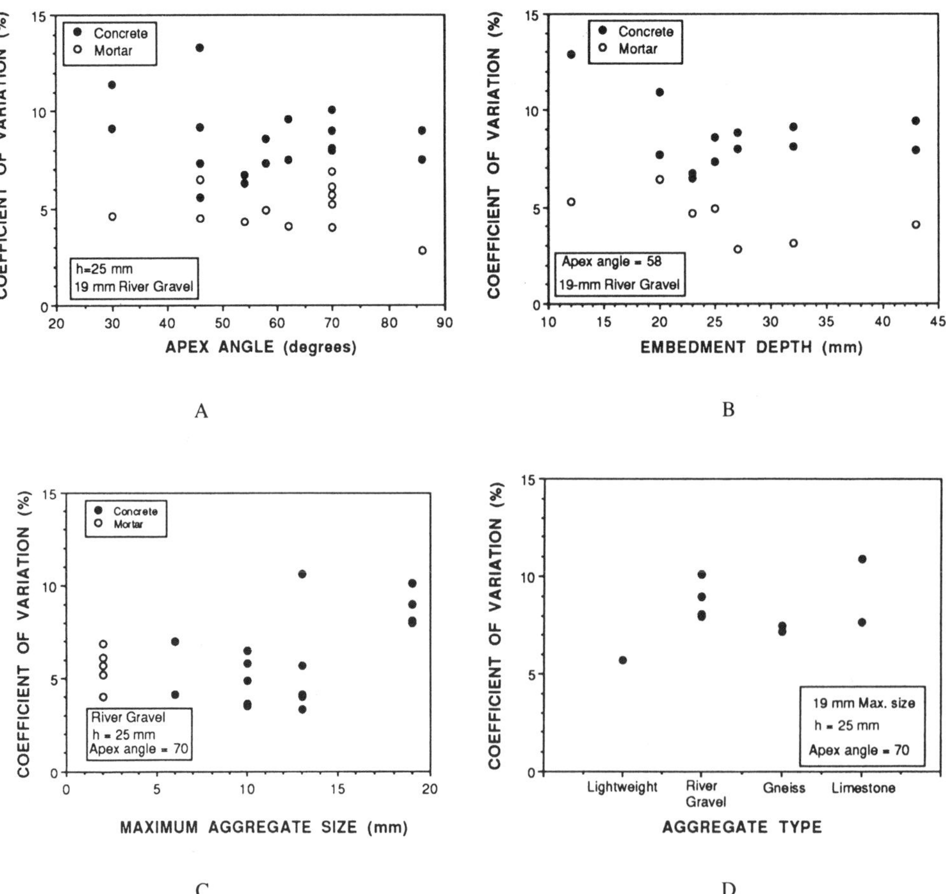

FIGURE 19. Coefficient of variation as a function of: (A) apex angle, (B) embedment depth, (C) maximum aggregate size, and (D) aggregate type (Data from Reference 20).

manner in which coarse aggregate particles cross the eventual failure surface. They further observed that, for the tests in concrete with lightweight aggregate, the fracture surface went through the aggregate particles rather than around them, as was observed in concretes made with the normal weight aggregates. Thus, for the weak lightweight aggregates, interlock effects were not significant, and the repeatability was similar to tests in mortar.

In summary, the repeatability of the pullout test is characterized by the coefficient of variation. Reported values of the coefficient of variation, for different aggregate materials and test configurations, have ranged from about 4 to 15% with an average value of about 8%. The maximum size of the aggregate in relation to the embedment depth appears to be a significant factor. Tests on concrete made with coarse aggregate having a maximum size less than the embedment depth tend to have lower variability. The type of coarse aggregate appears to be insignificant, except for lightweight aggregate which results in less variability.

CORRELATION RELATIONSHIP

In the review of the early pullout test developed in the Soviet Union, Skramtajew[1] reported that for concrete with cube strengths between 1.5 and 10.5 MPa (200 to 1500 psi) there was a constant ratio between pullout load and cube strength. On the other hand, Tremper[2] showed that, over a wide range of concrete strength, the relationship between pullout load and compressive strength was nonlinear and may be affected by the type of aggregate (Figure 4). Recall that these early tests did not involve a bearing ring.

TABLE 3
Linear Correlation Relationship by Bickley[24] with LOK-TEST System
(P = a + b C)

Strength range (MPa)	Intercept (a) (kN)	Slope (b) (kN/MPa)
7.1—38.3	−0.9	0.88
12.7—28.8	−2.0	1.05
9.7—44.4	2.4	0.85
5.9—32.5	1.7	0.81
13.7—34.4	−2.0	1.06
8.8—25.2	2.7	0.89

To improve the correlation between pullout strength and compressive strength, Kierkegaard-Hansen[3] introduced a bearing ring and concluded from his tests that: ''There is nothing to indicate that the relationship between the two strength measurements is nonlinear''. However, Kierkegaard-Hansen found that the relationship was linear but not proportional, i.e., the straight line had a nonzero intercept. In addition, he found that the relationship was dependent on the maximum size of the coarse aggregate. He suggested the following correlation relationships for his LOK-strength system:

$$P = 5.10 + 0.806\ C \quad \text{(16 mm maximum aggregate size)} \tag{5}$$

$$P = 9.48 + 0.829\ C \quad \text{(32 mm maximum aggregate size)} \tag{6}$$

where, P = ultimate pullout load (kN), and
C = cylinder compressive strength of concrete (MPa).

Thus, for equal cylinder compressive strength, concrete with a larger coarse aggregate will have a greater ultimate pullout load.

The manufacturer[33] of the widely used LOK-TEST system originally proposed the following correlation relationship for all concrete with aggregate sizes up to 32 mm ($1^1/_4$ in.):

$$P = 5 + 0.8\ C \tag{7}$$

However, Bickley[24] reported that correlation testing performed at six test sites, using the same LOK-TEST system, resulted in straight lines which differed from Equation 7. Table 3 summarizes the best-fit values of the slopes and intercepts obtained by Bickley. The table shows that there are positive and negative intercepts and that some of the slopes are significantly greater than the recommended value of 0.8 kN/MPa. However, it was shown that, in general, Equation 7 is a ''conservative'' relationship, i.e.; for a given pullout load Equation 7 estimates a lower compressive strength than the straight lines in Table 3.

The lack of agreement among the correlation relationships obtained with the same test system and the illogical result of a nonzero intercept, has caused skepticism among potential users of the pullout test method. This section demonstrates that, for a given test system, there is not a unique correlation relationship applicable to all concrete. Also, it will be shown that the correlation relationship for a particular combination of materials and test system is not necessarily linear. The discussion is limited to correlation relationships between ultimate pullout load and cylinder compressive strength.

First, the shape of the correlation relationship is investigated. Figure 20A shows correlation data obtained using concrete made with 19-mm (3/4-in.) crushed limestone.[27] The

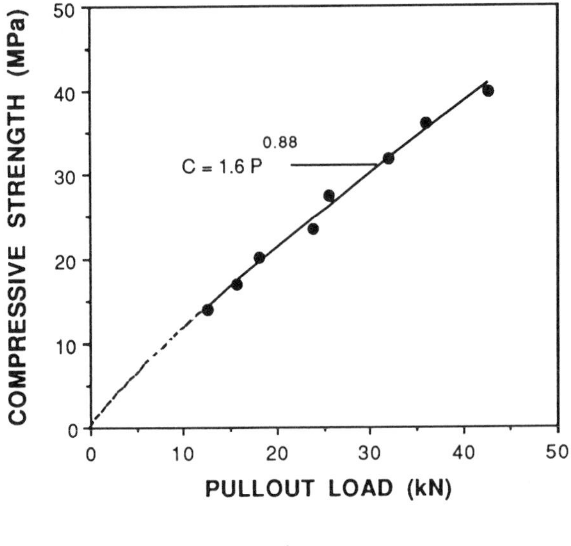

A

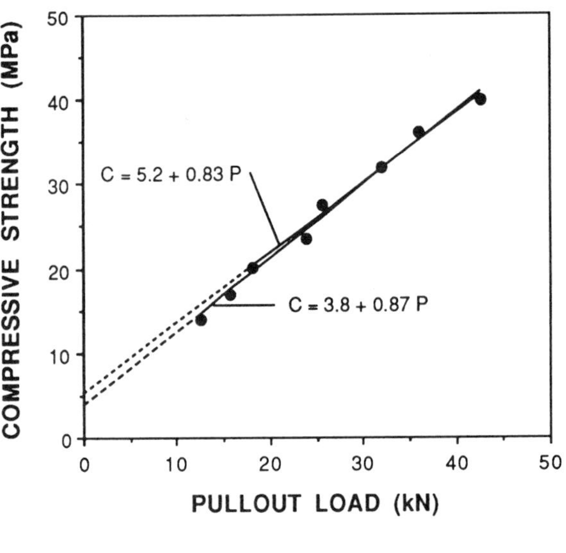

B

FIGURE 20. Correlation relationship for concrete made with crushed limestone:[27] (A) power function and (B) straight line relationships for different strength ranges.

pullout test system had a 70° apex angle and a 25-mm (1-in.) embedment depth. Rather than using a linear correlation relationship, consider a power function as follows:

$$C = \alpha P^{\beta} \tag{8}$$

By taking the logarithms of both sides, Equation 8 is transformed to

$$\log(C) = \log(\alpha) + \beta \log(P) \tag{9}$$

Thus, by plotting the logarithm of compressive strength vs. the logarithm of pullout load, the power function is transformed into a straight line. The best-fit values of α and β can be obtained by linear regression analysis using the transformed data.* The best-fit power function for the data in Figure 20A is

$$C = 1.6\ P^{0.86} \tag{10}$$

The power function fits the data quite well and the shape is nearly a straight line over the range covered by the data.

Now, examine what happens if a linear relationship is assumed having the equation:

$$C = a + b\ P \tag{11}$$

If all points are considered, the best-fit straight line is

$$C = 3.8 + 0.87\ P \tag{12}$$

However, considering only the six points for compressive strengths above 20 MPa (2900 psi), the best-fit straight line is

$$C = 5.2 + 0.83\ P \tag{13}$$

It is seen that the two straight lines in Figure 20C are practically the same for compressive strength above 20 MPa. The point of this exercise is to illustrate that, if the true correlation relationship is nonlinear and it is approximated with a straight line, the slope and intercept of the straight line depend on the strength range covered by the correlation data.

Next, consider correlation data for the same test system but for concrete made with 19-mm (3/4 in.) river gravel.[27] The data are shown in Figure 21A and the best-fit power function is

$$C = 1.07\ P^{1.02} \tag{14}$$

The power function looks very much like a straight line. In this case, the exponent is greater than 1, and the curvature of the correlation relationship is opposite to that shown in Figure 20A. Figure 21B, shows the best-fit straight lines for all the data and for only the six data points above 20 MPa (2900 psi). Again, the equations of the straight lines are different but the correlation relationships are similar for the data above 20 MPa. Also, note that because of the different curvature, the values of the intercepts are negative.

Thus, if the true correlation relationship is slightly nonlinear and if the curvature can depend on the type of aggregate, one can explain why Bickley[24] reported different linear correlation relationships for the same pullout test system.

The manufacturer[33] of the LOK-TEST system later proposed the following relationship, which differs from Equation 7, for concrete with compressive strengths between 3 and 25 MPa (400 and 3600 psi):

$$P = 1.0 + 0.96\ C \tag{15}$$

Since the correlation relationship is used to estimate compressive strength based upon the

* Linear regression analysis of the transformed data is preferred when the coefficient of variation of the dependent variable (concrete strength in this case) is constant.[27]

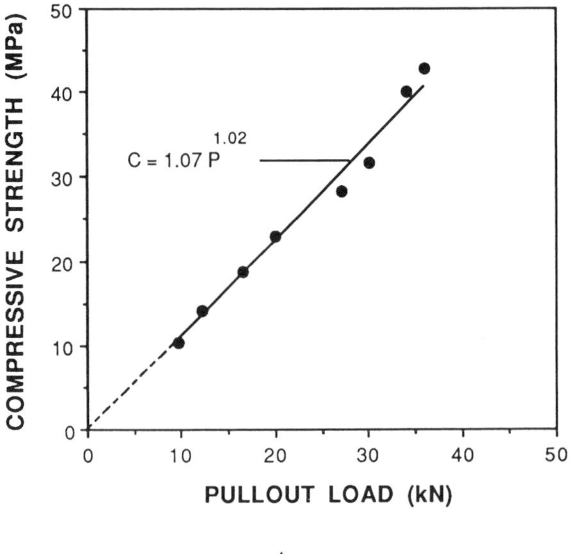

A

B

FIGURE 21. Correlation relationship for concrete made with river gravel:[27] (A) power function and (B) straight line relationships for different strength ranges.

measured pullout load, it is preferable to treat compressive strength as the dependent variable. Thus the relationships for the LOK-TEST system are as follows:

$$C = -1.0 + 1.04\ P \quad (\text{for } 3\ \text{MPa} < C < 25\ \text{MPa}) \tag{16}$$

$$C = -6.3 + 1.25\ P \quad (\text{for } C \geq 25\ \text{MPa}) \tag{17}$$

These two straight lines are shown in Figure 22A. For purposes of illustration, eight evenly spaced points were chosen along this bilinear correlation relationship, as shown in Figure

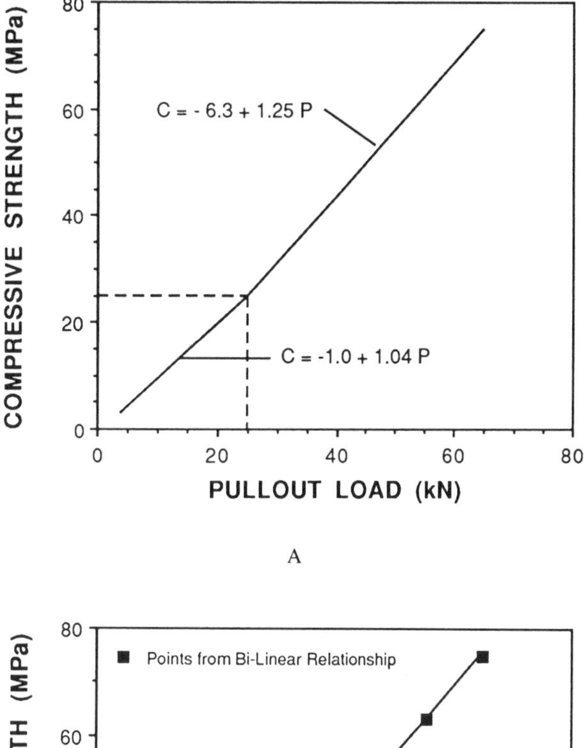

A

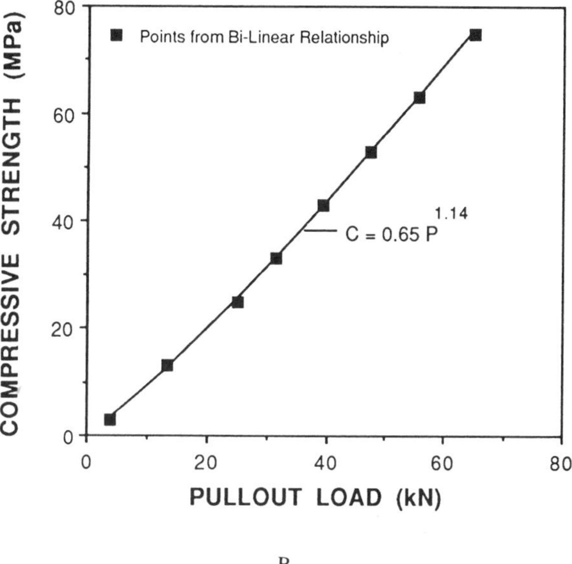

B

FIGURE 22. (A) Bi-linear correlation relationship proposed for LOK-TEST system;[33] (B) power function approximation of the two lines.

22B. A best-fit power function was fitted to the points, and the equation of the function is as follows:

$$C = 0.6\ P^{1.14} \tag{18}$$

It is seen that the power function is nearly identical to the bilinear function.

Figure 23 shows correlation data reported by Khoo[29] for pullout tests and tests of corresponding cores. The pullout configuration had an apex angle of 70° and the embedment depth was 25 mm (1 in.). The concrete was made with 20-mm (0.8-in.) maximum size

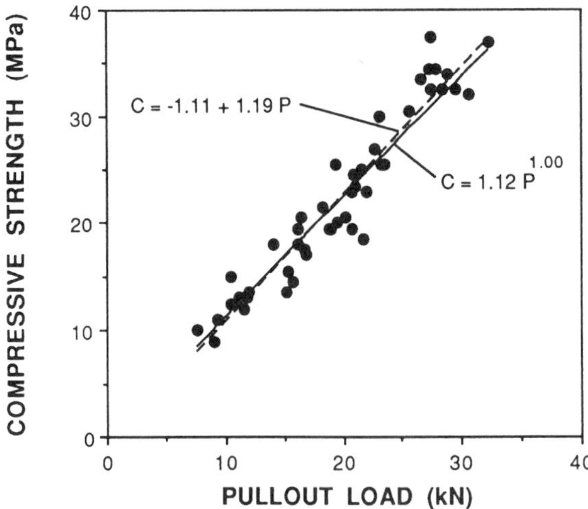

FIGURE 23. Correlation data by Khoo[29] and best-fit linear and power function relationships.

crushed granite, and the compressive strength ranged between 10 and 40 MPa (1500 and 5800 psi). The best-fit straight line correlation relationship for the data, as determined by this author, is

$$C = -1.11 + 1.19 P \tag{19}$$

While the best-fit power function is

$$C = 1.12 P^{1.00} \tag{20}$$

Since the exponent of the power function is equal to 1, the power function is actually a straight line passing through the origin. The intercept in Equation 19 is not statistically significant, and, for the range of strength considered, the compressive strength of the cores is proportional to the pullout load.

Finally, let us examine the effects of test geometry and aggregate type on the correlation relationship. Figure 24A shows the reported[27] power-function correlation relationships for two pullout test configurations: one had an apex angle of 54° and the other an angle of 70°. The embedment depth and insert head diameter were 25 mm (1 in.), and the concrete was made with 19-mm (3/4-in.) river gravel. The exponents for the two curves are close to one so both relationships are very close to linear. As shown in Table 1, the repeatability of the two test configurations were found to be similar. As was discussed by Stone and Giza,[20] because the slope of the relationship for the 54° pullout configuration is lower than for the 70° configuration, the relationship for the 54° configuration would result in slightly less uncertainty in the estimated compressive strength.

Figure 24B compares power-function correlation relationships for different types of aggregates using the 70° pullout test configuration.[27] The relationships were found to be statistically different. Note that for compressive strengths above 20 MPa (3000 psi), the concrete with crushed limestone resulted in much greater pullout loads. Thus there is evidence that the aggregate type can effect the correlation relationship.

SUMMARY

This section has reviewed available information on the within-test variability (repeatability) and the correlation relationship of the pullout test. Over a wide range of concrete

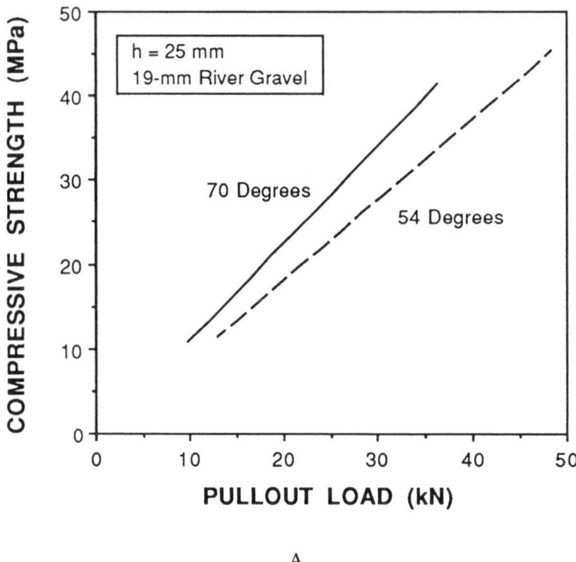

A

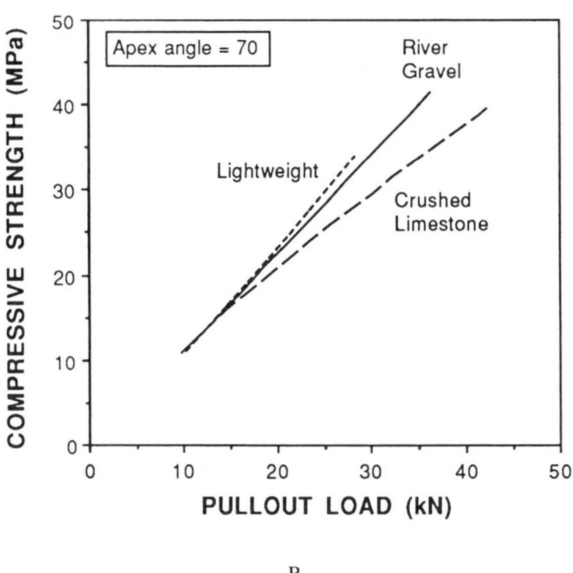

B

FIGURE 24. Correlation relationships as effected by: (A) apex angle and (B) aggregate type.[27]

strength, the standard deviation of the ultimate pullout loads, for repeated tests in the same concrete, increases with increasing strength. Hence the coefficient of variation is the appropriate statistic to describe repeatability. A significant amount of repeatability data have been published, and it appears that the average value of the coefficient of variation for the pullout test is about 8%. However, the size and type of the coarse aggregate affect the coefficient of variation, and the repeatability for a given concrete mixture can be higher or lower than 8%

Considerable correlation data have been published for the commercially available LOK-TEST system, which uses a 62° apex angle and a 25-mm embedment depth. The majority of the empirically determined relationships have been reported to be straight lines with nonzero intercepts. It has been explained that these linear relationships are approximations to inherently nonlinear relationships. For this reason, the slopes and intercepts depend on the strength ranges used in developing the correlation data. It has been proposed that a power function is a superior equation for analyzing correlation data. The coefficients of the power function are readily determined by linear regression analysis of the logarithms of pullout and compressive strength data. As demonstrated by examples, the power function can accommodate various degrees of nonlinearity.

Some of the early studies indicated that, for a given test system, the correlation relationship is influenced by the maximum aggregate size. More recent results show that the type of aggregate also has an effect. Thus, for the most reliable estimates of in-place strength, a correlation relationship should be developed for the specific concrete mixture to be used in construction.[28] The next section discusses methods for developing correlation data.

APPLICATIONS

The pullout test has been adopted as a standard test method in many parts of the world, including North America, and its successful use on large construction projects has been reported.[24,34,35] This section reviews the evolution of the current ASTM (American Society for Testing and Materials) standard governing the pullout test, and discusses some of the practical aspects for implementing the method and interpreting test results.

STANDARDS

The first standard for the pullout test was established in Denmark in 1977,[33] and the method is recognized for the acceptance of concrete in structures. In North America, ASTM adopted a tentative test method in 1978, which was subsequently revised and issued as a standard in 1982. The ASTM standard does not limit the test configuration to a specific geometry. The following compares some of the geometrical requirements in the 1978 tentative method with those in the 1982 ASTM standard:

	ASTM C 900-78T	ASTM C 900-82
Embedment depth	1.0 d to 1.2 d	1.0 d
Bearing ring	2.0 d to 2.4 d	2.0 d to 2.4 d
Apex angle	45° to 70°	53° to 70°

The 1982 standard set the embedment depth equal to the insert head diameter, d, thereby limiting the range of possible apex angles form 53° to 70°. The 1987 revision of the ASTM standard made no changes to the allowable test configurations.[36]

The current ASTM standard allows three procedures for placement of pullout inserts:

1. Attached to the surface of formwork prior to concrete placement
2. Attached to formwork with special hardware to enable testing deep within the concrete (refer to Figure 7a).
3. Placed into the surface of freshly placed concrete

In the third procedure, inserts are placed manually into the top surface of the fresh concrete. Special inserts with a ‘‘cup’’ or a plastic plate are used to provide a smooth surface for proper seating of the bearing ring. Manual placement requires care to assure that the concrete around the insert is properly consolidated and surface air voids are minimized. In general, manually placed inserts result in higher variability[24,37] and are not recommended unless absolutely necessary. In all cases, the clear spacing between the inserts and the edges of the member should be at least four times the insert head diameter. Also, each insert should be placed so that reinforcing steel does not interfere with the eventual fracture surface when the insert is pulled out.

The number of required pullout tests in the field was a controversial subject during the development of the ASTM standard. The 1978 tentative method had no requirement. The 1982 standard stated that a ''minimum of three pullout tests shall comprise a test result,'' and Note 6 stated: ''Often it will be desirable to provide more than three individual pullout inserts in a given placement''. In 1987, the section on the number of tests was expanded to the following:

> When pullout test results are used to assess the in-place strength in order to allow the start of critical operations, such as formwork removal or application of post tensioning, at lest five individual pullout tests shall be performed for a given placement for every 115 m^3 (150 yd^3), or fraction thereof, or for every 470 m^2 (5000 ft^2), or a fraction thereof, of the surface area of one face in the case of slabs or walls.

A note to this requirement stated: ''Inserts shall be located in those portions of the structure that are critical in terms of exposure conditions and structural requirements''. In addition, the following statement was also added to the 1987 standard:

> When planning pullout tests and analyzing test results, consideration should be given to the normally expected decrease of concrete strength with increasing height within a given concrete placement in a structural element.

This so-called ''top-to-bottom'' effect is well documented.[38-41] However, standards and codes have not addressed the significance of the effects, therefore, there are no guidelines on how to deal with such variability. The important point is that when high variability is obtained from pullout tests performed at different elevations within a structural element, it should not be interpreted to mean that the pullout test is unreliable. Engineering judgment is required in selecting the test locations and in interpreting the results.

The 1978 and 1982 versions of the ASTM test method allowed the option of reporting the pullout strength as a stress, obtained by dividing the ultimate pullout load by the area of the idealized conic frustum. There were criticisms that the calculation was not meaningful because the pullout force is inclined to the surface of the frustum. In 1987, the procedure was changed to allow the calculation of a nominal normal stress as given by the previous Equation 2. As was discussed, the normal stress distribution on the idealized conic frustum is nonuniform. Therefore, this calculated normal stress is fictitious and should be used only for comparing results of different pullout test configurations.

Finally, the 1987 ASTM standard has the following statements regarding correlation relationships between pullout strength and other strength tests:

> Such strength relationships depend on the configuration of the embedded insert, bearing ring dimensions, depth of embedment, and the level of strength development in that concrete. Prior to use, these relationships must be established for each system and new combination of concreting materials.

Aside from assuring a correlation relationship which is applicable to the particular combinations of equipment and materials, this requirement also forces testing agencies to become familiar with the details of pullout testing procedures prior to using the test equipment at the project site.

CORRELATION RELATIONSHIP

The development of the correlation relationship applicable to the specific construction project is a critical step in implementing the pullout test. Unfortunately, at the time of writing, no standard procedures exist to obtain correlation data.

Historically, various techniques have been used to acquire companion pullout strength and compressive strength data. Kierkegaard-Hansen[3] placed pullout inserts in the bottoms of standard cylindrical specimens. A pullout test was performed on the cylinder, and then the same cylinder was capped and tested for compressive strength. If the pullout test was ceased just beyond the point of maximum load, the pullout cone was not extracted, and it was shown that the cylinder could be tested in compression without significant effect on the compressive strength. Bickley[24] also provided data showing negligible effects of this procedure on compressive strength. Such a procedure is possible because, during a standard compression test, the ends of the cylinder are subjected to confining stresses,[8] which prevent premature failure of the cylinder due to the damage incurred during the pullout test. However, it was found that, as concrete strength increased, radial cracking occurred at the end of the cylinder outside of the bearing ring, and this reduced the ultimate pullout load.[3,24] Later studies confirmed that, for concrete with compressive strengths above 40 MPa (5800 psi), pullout tests in 150-mm (6-in.) diameter cylinders resulted in lower strengths than pullout tests on larger specimens which did not experience radial cracking. For this reason and because there is a limit to the size of the pullout test configuration than can be used on the bottom of a cylinder, this approach is not recommended.

An alternative to placing inserts in standard cylinders, is to place them in slabs and cast companion standard cylinders. At designated ages, replicate pullout tests are performed on the slab and replicate cylinders are tested in compression. A drawback to this approach is the need to assure that the pullout tests and compression tests are performed at the same maturity. Because of their different masses and shapes, the slab and cylinders are not likely to experience the same temperature history during the critical early ages, when strength changes rapidly with age and is strongly dependent on temperature history.[42] Failure to account for possible maturity differences can lead to inaccurate correlation relationships. Either maturity meters should be used to ensure companion testing at equal maturities, or compression tests should be performed on cores drilled from the slab. While the latter approach helps assure equal maturities, it is time consuming.

For the commercially available pullout systems, having embedment depths of 30 mm (1.2 in.) or less and apex angles of 70° or less, the preferred approach is to place inserts on the side faces of 200-mm (8-in.) cubes and cast companion standard cylinders. Because of similar surface-to-volume ratios, the early-age temperature histories of the two types of specimens will be similar. The cubes and cylinders should be compacted similarly, and the use of an internal vibrator or a vibrating table is recommended.

Committee 228 of the American Concrete Institute (ACI) recommends performing eight replicate pullout tests and two cylinder compression tests at each test age.[28] These numbers of tests assure that the average pullout strength and average compressive strength are determined with about the same degree of certainty. By placing four inserts in each cube, this recommendation requires two cubes and two cylinders at each test age. The specimens should be moist-cured until time of testing.

The next question is how many sets of pullout and compression tests should be performed to establish the correlation relationship. The chosen number should satisfy two needs: the tests should span as wide a range of strength as is possible, and there should be enough points to define the relationship with a reasonable degree of accuracy. Based on field experience, Bickley[24] suggested that the range of compressive strength should be at least 20 MPa (3000 psi) but preferably greater than this. ACI Committee 228[27] recommends performing companion tests at a minimum of six evenly spaced strength levels. Generally, if test ages are increased by a factor of two there will be about the same strength increase between successive tests. For example, tests at ages of 1, 2, 4, 8, 16, and 32 days should result in approximately evenly spaced test points. This of course assumes a constant temperature during the curing period. If the pullout test will be used to estimate in-place strengths at very low levels, the first test age should be reduced to 12 h. This will require care in handling the low-strength specimens.

Thus the recommended correlation testing program involves casting at least 12 cubes, with 4 inserts per cube, and 12 cylinder specimens. The inserts in two cubes and two cylinders are tested at different ages so as to produce evenly spaced points when the correlation data are plotted. The average of the pullout strengths and compressive strengths are used in a least-squares fit analysis to develop the correlation relationship.[27,43,44]

INTERPRETATION OF RESULTS

To estimate in-place strength, pullout tests are performed on a particular part of the structure and the correlation relationship is used to convert the test results to a compressive strength value. To judge whether sufficient strength has been attained, the estimated compressive strength is compared with the required strength in the Project Documents. However, to provide for a margin of safety, the pullout test results should be treated statistically rather than simply comparing the average estimated in-place strength with the required strength.

In assessing the safety of a structure, the "specified" concrete strength is used in the design equations to calculate member resistances. The specified strength is the strength that is expected to be exceeded by a large proportion of the concrete in the structure, and it is often called the "characteristic strength". In North American practice,[45] this proportion (or fraction) is about 90%. Alternatively, it is expected that 10% of the concrete in the structure will be weaker than the specified strength. Thus, in interpreting test results, the characteristic strength indicated by the pullout tests should be computed for comparison with the required strength.

One approach, which was developed in Denmark and has been used in North America,[24,46] uses the lower tolerance limit of the in-place strength as the characteristic strength. The lower tolerance limit is a statistical term which represents the value that is expected to be exceeded by a certain fraction of the population with certain degree of confidence (or probability level). It is calculated by subtracting the product of the standard deviation and the appropriate tolerance factor from the average value. In applying this approach to pullout tests, the following procedure is used:

- Perform the pullout tests
- Convert the test results to compressive strength values by means of the correlation relationship
- Compute the average and standard deviation of the compressive strength values
- Compute the characteristic strength as follows

$$C_{0.1} = C_a - K\,S \tag{21}$$

where,

$C_{0.1}$	=	the characteristic strength, i.e., the strength not expected to be exceeded by 10% of the concrete in the structure
C_a	=	average compressive strength based on the pullout test results
K	=	one-sided tolerance factor
S	=	the standard deviation of the compressive strength values

The tolerance factor value depends on the number of tests and the confidence level. A confidence level of 0.75 is usually used by proponents of this approach.[24,46] Table 4 lists the one-sided tolerance factors for an understrength fraction of 10% and a confidence level of 0.75.

In a strict sense, the tolerance limit approach is not intended for application to values (compressive strength) obtained through the use of an empirical correlation relationship. The approach has been criticized[27,43] because of the following assumptions implied in its use:

- The correlation relationship has no error, i.e., the coefficients describing the correlation relationship are known with absolute certainty.
- The standard deviation of the actual in-place compressive strength is assumed to equal the standard deviation of compressive strengths estimated from the correlation curve.

TABLE 4
One-Sided Tolerance Factor for 10% Understrength and 0.75 Confidence Level for Normal Distribution[47]

Number of tests	K	Number of tests	K
3	2.501	12	1.624
4	2.134	13	1.606
5	1.961	14	1.591
6	1.860	15	1.577
7	1.791	16	1.566
8	1.740	17	1.554
9	1.702	18	1.544
10	1.671	19	1.536
11	1.646	20	1.528

The first assumption can lead to unconservative estimates of the in-place strength. The second assumption is likely to be conservative because the variability of the in-place compressive strength is not expected to be as great as that of pullout strength. In addition to these assumptions, the correlation relationships have been determined using ordinary least squares analysis, which is not strictly applicable when the independent variable (pullout strength in this case) has measurement error.

To overcome the deficiencies of the tolerance limit approach, the National Bureau of Standards developed a rigorous statistical procedure to establish the correlation relationship and estimate the in-place characteristic strength to a desired confidence level.[43] Basically, the NBS procedure estimates the expected characteristic strength and its uncertainty. From these estimates, one determines the value the characteristic strength that is expected to be exceeded with a high level of confidence. The procedure is complex, but it is well suited for implementation on a personal computer. The NBS method assumes that the ratio of the standard deviation of cylinder strength to the standard deviation of pullout strength has the same value in the field as was obtained during the laboratory correlation testing. Characteristic strengths computed by using the rigorous approach were compared with the values computed using the tolerance limit method.[27] The comparison showed that the tolerance limit approach leads to overly conservative estimates of in-place characteristic strength, especially when the variability of the pullout test results is high. While this is acceptable for safety, it may lead to unnecessary delays in the construction schedule.

A simplified procedure was developed to interpret pullout test results and was implemented with spreadsheet software.[44] A spreadsheet template was prepared which contained the necessary equations to develop the correlation relationship and analyze subsequent in-place test results. To use the template, the user enters the test data from the correlation testing program, and the correlation relationship is automatically computed. The user then enters the in-place pullout test results, and the characteristic strength is automatically computed.

In arriving at a reliable estimate of in-place characteristic strength, the simplified method[44] considers the following sources of variability or uncertainty:

- The variability of the in-place concrete strength
- The uncertainty in the average value of the in-place pullout strength
- The uncertainty in the correlation relationship

Simulation studies[44] indicated that the simplified method and the rigorous approach resulted in similar estimates of in-place characteristic strength.

At the time of writing, a consensus had not been reached in North America on the recommended approach for analyzing in-place test results. The user is encouraged to study the cited references for additional guidance on the interpretation of pullout test results.

NUMBER OF TESTS

As was stated earlier, an area of contention during the development of the 1987 ASTM standard was the number of pullout tests that should be performed in a given placement to have a reliable estimate of the in-place strength. Before discussing this issue, consider a statistical principle. Suppose there are two test methods with values of repeatability, as expressed by the coefficient of variation, equal to V_1 and V_2. If the respective material properties are to be measured with the same degree of certainty, the number of replicate tests, n_1 and n_2, for each method should obey the following relationship:[23]

$$\frac{n_1}{n_2} = \left[\frac{V_1}{V_2}\right]^2$$

Thus, if the coefficient of variation of method 1 is twice the coefficient of variation of method 2, four times as many replicate measurements should be made with method 1. Of course, the degree of uncertainty in the average value decreases as the absolute number of tests increases.

Now consider the sampling requirements in ACI 318[48] for acceptance of concrete. The current requirement is that two cylinders should be tested for every 115 m^3 (150 yd^3) of concrete, or in the case of slabs and walls for every 470 m^2 (5000 ft^2) of surface area (of 1 face). The coefficient of variation of the standard cylinder compression test is about 4%.[28] If the same sampling frequency were used for pullout tests and if the coefficient of variation of the pullout test is taken as 8%, at least eight pullout tests should be performed for the above quantities of concrete. Note that this only assures that the pullout strength is known with the same degree of certainty as the standard cylinder strength. The estimated in-place compressive strength will have greater uncertainty because of the additional uncertainties associated with the correlation relationship.

Some have suggested that a minimum of ten pullout tests should be performed for a given concrete placement.[25,29,34] As a practical matter, Bickley[34] advocated the placement of 15 inserts per 100 m^3 (130 yd^3). When it is anticipated that the desired strength level has been reached, five inserts are randomly selected for testing. If the results indicate less than the required strength level, testing is discontinued and additional curing is provided. At a later age, the remaining ten inserts are tested. This procedure provides for a reserve in the event that testing is begun too soon. Bickley and others[49] also advocate the use of a maturity meter to determine the appropriate time to perform the pullout tests (see Chapter 5 for additional information).

Thus, while the 1987 ASTM standard requires a minimum of 5 pullout tests for every 115 m^3 (150 yd^3) of concrete, or in the case of slabs or walls for every 470 m^2 (5000 ft^2) of the surface area of one face, a greater number of inserts is recommended for added reliability and as a safety measure in the event testing is begun too soon.

DRILLED-IN TESTS

A drawback of the standard pullout test is that the locations of the inserts have to be planned in advance of concrete placement and the inserts have to be fastened to the formwork. This limits the applicability of the method to new construction. In an effort to extend the application of pullout testing to existing structures, various "drilled in" techniques have been investigated.

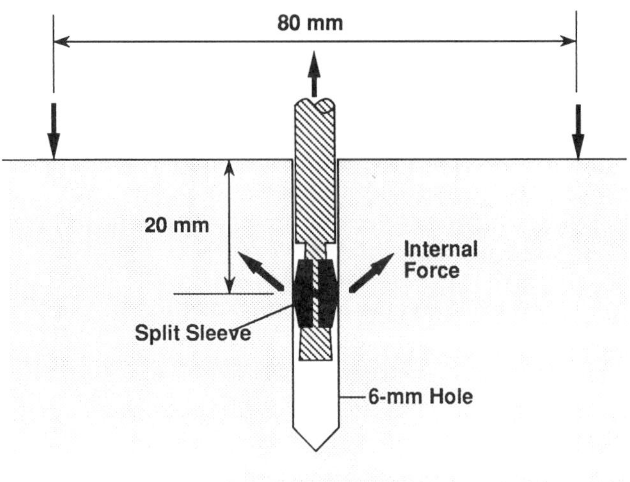

FIGURE 25. BRE internal fracture test.[50,51]

Spurred by the need to evaluate distress in structures made with high alumina cement, the Building Research Establishment (BRE)[50,51] developed a technique based on commercial anchor bolts with expanding sleeves, as shown in Figure 25. A 6-mm (1/4-in.) hole is drilled into the concrete, the hole is cleaned, and an anchor bolt is inserted into the hole so that the split-sleeve is at a depth of 20 mm (0.8 in.). After applying an initial load to expand and engage the sleeve, the bolt is loaded in tension and the maximum load during the extraction is recorded. Reaction is provided by three "feet" located along the perimeter of a 80-mm (3.1-in) diameter ring. The expanding sleeve applies to the concrete a force having vertical and horizontal components, as indicated by the inclined arrows in Figure 25. The concrete fracture differs from that in the standard pullout test, and the test is referred to as an "internal fracture test" rather than a "pullout test". The reported correlation relationship between ultimate load and compressive strength has a pronounced nonlinearity, indicating that the failure mechanism is probably related to the tensile strength of the concrete. The within-test variability was found to be greater than that of the standard pullout test, and the 95% confidence limits of the correlation relationship were found to range between ±30% of the mean curve.

In the BRE test system, the pullout force is applied by turning a nut on the end of the anchor bolt and measuring the maximum torque achieved during the test. Bungey[52] developed a mechanical loading system with the aim of reducing the scatter of test results compared with using the torque loading system. Using the mechanical loading system, the 95% confidence intervals for the estimated compressive strength were estimated to be ±20%, which is a significant improvement. The comparatively low precision of the internal fracture test has been attributed to two principal causes:[52] (1) the variability in the hole drilling and preparation; and (2) the influence of aggregate particles on the load transfer mechanism and on the failure initiation load.

Mailhot et al.[53] investigated the feasibility of several drilled-in pullout tests. One of these used a split-sleeve and tapered bolt. In this case, the bolt assembly was placed in a 19-mm (3/4 in.) hole drilled into the concrete. As shown in Figure 26A, this technique differs from the BRE method because the reaction to the pulling force acts through a specially designed high strength, split-sleeve assembly. Thus the force transmitted to the concrete is predominantly a lateral load due to the expansion of the sleeve. The developers claimed that failure occurred by shear. However, the author believes that failure is more likely to occur by splitting as in the standard splitting tension test of a cylinder. Similar to the BRE test, the variability of this test was reported to be rather high.[53] A second successful method

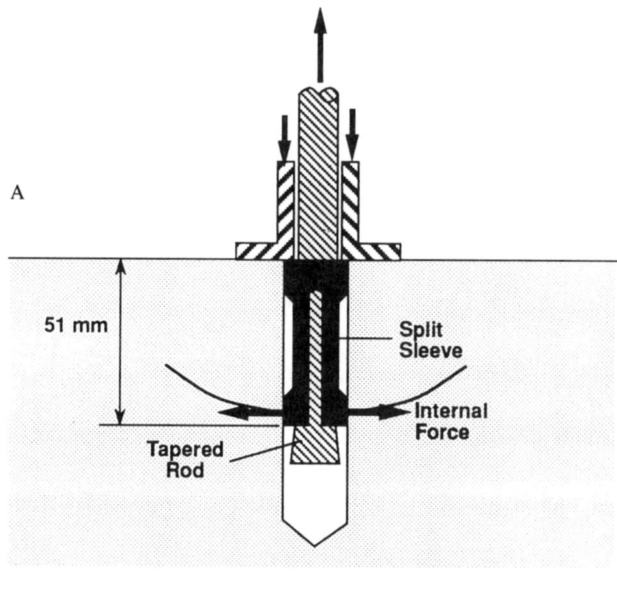

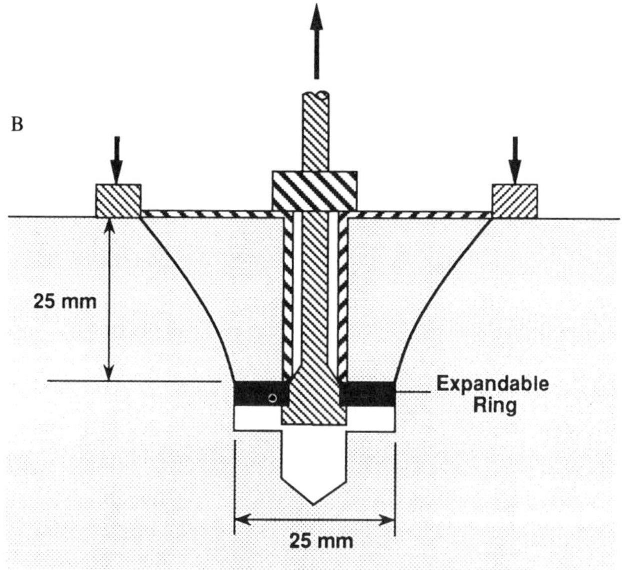

FIGURE 26. (A) Drilled-in test using spilt-sleeve and tapered bolt;[53] (B) the CAPO test using undercut and expandable ring.[33]

involved epoxy-grouting a 16-mm (5/8-in.) threaded rod to a depth of 38 mm ($1^1/_2$-in.) in a 19-mm (3/4-in.) diameter hole. After the epoxy had cured, the rod was pulled using a tension jack reacting against a bearing ring. This method was also reported to have high variability. The study concluded that these two methods have the potential for assessing the strength in existing construction. However, additional research was recommended to enhance their reliability.

Recently, Domone and Castro[54] developed a technique similar to that shown in Figure 26A, except that the load was applied by a torquemeter and the embedment was 20 mm (3/4 in.) as in the BRE method. Based on a limited number of comparative laboratory tests, it was concluded that the new method resulted in better correlation relationships than the BRE method.

Another method was developed by the manufacturer of the LOK-TEST system and is referred to as the CAPO test (for **C**ut **A**nd **P**ull**O**ut).[33] The method involves drilling an 18-mm (0.7-in.) hole into the concrete and using a special milling tool to undercut a 25-mm (1-in.) diameter slot at a depth of 25 mm (1 in.). An expandable ring is placed into the hole, and the ring is expanded using special hardware. Figure 26B shows the ring after expansion. The entire assembly used to expand the ring is then pulled out of the concrete using the same loading system as for an ordinary pullout test. Thus, unlike the other methods discussed above, the CAPO test subjects the concrete to the same type of loading as the standard pullout test. The performance of the CAPO test in laboratory evaluations has been reported to be similar to the LOK-TEST.[33] However, attempts at using the CAPO test in the field have indicated that the test is cumbersome and has high variability. The high variability is probably because of the need for a flat concrete surface that is perpendicular to the drilled hole. If these conditions are not achieved, the bearing ring will not seat properly and test results are erratic.

SUMMARY

This section has discussed some of the practical considerations in the application of the pullout test. Considerable information has been published on laboratory and field experiences involving the method, and standard test procedures and recommended practices have been established.

Prior to using the pullout test to estimate in-place strength, a correlation relationship between ultimate pullout load and compressive strength must be established for the particular test system and concrete materials. The maximum size and type of coarse aggregate can have a significant influence on the correlation relationship. While no standards currently exist, recommendations for establishing this relationship have been published. The preferred procedure is to perform the pullout tests on cube specimens. Correlation data should span as wide a strength range as is practicable, and the strength level that is expected to be measured in the field should fall within this range. At least six data sets should be used to develop the correlation relationship.

In implementing pullout testing in the field, the number of pullout tests and statistical analysis of the results are critical. The number of tests should be chosen so that the average value and the variability of the pullout strength are established with a reasonable degree of confidence. Most practitioners use more than the minimum number required by the current ASTM standard. The inserts should be located in critical portions of the structure. Test results should be subjected to statistical analysis so that the estimated in-place strength will be exceeded by a large fraction of the concrete in the structure. Several statistical methods have been proposed, but there is no consensus on which should be used. Danish practice has promoted the tolerance limit approach, but there are more rigorous methods that can be implemented using a personal computer.

Finally, there has been a brief discussion of recent developments related to drilled-in tests which do not require the installation of inserts prior to concrete placement. Some of these methods load the concrete in a different manner compared with the standard pullout test, and higher within-test variability have been reported. One of these methods loads the concrete in a manner similar to the standard test, and comparable performance has been reported in laboratory evaluations. These methods have not been standardized in North America.

CONCLUDING REMARKS

The pullout test measures the load required to extract a conical-shaped fragment of specified geometry from a concrete mass. The modern test is an outgrowth of earlier attempts which did not use a bearing ring to transmit the reaction of the tension load to the concrete mass. Danish research conducted in the late 1960s demonstrated that, by introducing the bearing ring, there was an approximately linear relationship between the ultimate pullout load and the compressive strength of concrete.

The pullout test subjects the concrete to a static load and, therefore, the test is amenable to theoretical analysis. Independent analytical and experimental investigations have been performed to gain an understanding of the failure mechanism. There has been agreement on some aspects of the failure process and divergent points of view on others. It is agreed that the concrete is subjected to highly nonuniform, triaxial stresses and that there is a stress concentration at the edge of the insert head. At about 1/3 of the ultimate load, circumferential cracking begins in the highly stressed region. This first crack propagates at a greater apex angle than that defining the extracted conical fragment, and the first crack stabilizes at less than the ultimate load. A second circumferential crack forms which defines the eventual shape of the extracted fragment. At about 70% of the ultimate load, this second crack has extended from the insert head to the bearing ring. The ultimate load carrying mechanism is a point of contention. Some believe that there is a compression strut between the insert head and the bearing ring, and others believe that additional load is carried by an aggregate interlock mechanism. Experimental evidence has been used to support both points of view. Despite the lack of agreement on the ultimate failure mechanisms, there is consensus that the ultimate pullout load is governed by the same strength properties that govern the compressive strength of concrete.

Some of the early proponents of the pullout test have claimed that the repeatability was similar to, and in some cases less than, the repeatability of standard compression tests. However, consideration of a wide variety of published data indicates that the within-test coefficient of variation of the pullout test is about twice that of the cylinder compression test. The size of the coarse aggregate in relation to the insert embedment appear to have the most significant effects on the scatter of pullout test results. In addition, the variability is lower in mortar and lightweight concrete than in normal weight concrete.

A correlation relationship between pullout strength and compressive strength is needed to estimate in-place strength. Some claim that for a given test system there is a unique relationship. However, there is evidence that the nature of the coarse aggregate influences the relationship. Therefore, the recommended practice is to develop the correlation relationship for the particular concrete to be used in construction. A large number of correlation studies have reported that compressive strength is a linear function of pullout strength. However, recent research suggests that the relationship may be nonlinear, and a power function is a more suitable equation for the correlation relationship.

An important step in implementing the method is choosing the locations and number of pullout tests in a given placement of concrete. The inserts should be located in the most critical portions of the structure and there should be a sufficient number of tests to provide statistically significant results. Additional inserts are recommended in the event that testing begins too soon, and the concrete has not yet attained the required strength. The use of maturity meters along with pullout tests is encouraged to assist in selecting the correct testing times and in interpreting possible low-strength results.

In-place pullout data should be statistically analyzed so that there is a high confidence in the estimated strength. Although a standard analysis procedure has not been adopted, several methods are available. The more rigorous approaches are well-suited for computer implementation.

Techniques have been developed which permit testing in existing construction by drilling a hole and inserting some type of expansion anchor. Some methods subject the concrete to different stress conditions and have different failure mechanisms than the standard pullout test. One method produces a failure surface which is similar to that of the standard test.

However, these methods have not found widespread acceptance due to their high variability.

In summary, the pullout test has been standardized and is recognized as a reliable method for assessing the in-place strength of concrete during construction so that critical activities may be performed safely. As with other in-place tests, the active involvement of a qualified individual in all aspects of the testing program, from the correlation testing to the analysis of in-place data, is recommended to realize the potential benefits of the method.

REFERENCES

1. **Skramtajew, B. G.,** Determining concrete strength for control of concrete in structures, *J. Am. Concr. Inst.,* 34, 285, 1938.
2. **Tremper, B.,** The measurement of concrete strength by embedded pull-out bars, *Proc. Am. Soc. Testing Mater.,* 44, 880, 1944.
3. **Kierkegaard-Hansen, P.,** Lok-strength, *Nordisk Betong,* No. 3, 1975, 19.
4. **Rutenbeck, T.,** New developments in in-place testing of concrete, in *Use of Shotcrete for Underground Structural Support,* ACI SP-45, American Concrete Institute, Detroit, MI, 1973, 246.
5. **Malhotra, V. M.,** Evaluation of the pull-out test to determine strength of in-situ concrete, *Materials and Structures* (RILEM), 8(43), 17, 1975.
6. **Malhotra, V. M. and Carette, G. G.,** Comparison of pullout strength of concrete with compression strength of cylinders and cores, pulse velocity, and rebound number, *ACI J.,* 77(3), 17, 1980.
7. **Richards, O.,** Pullout strength of concrete, *Reproducibility and Accuracy of Mechanical Tests,* ASTM SP 626, 1977, 32.
8. **Ottosen, N. S.,** Evaluation of concrete cylinder tests using finite elements, *ASCE J. Eng. Mech.,* 110(3), 465, 1984.
9. Mass concrete for dams and other massive structures, ACI 207.1R-70, Report of ACI Committee 207, American Concrete Institute, Detroit, MI.
10. Standard practice for determining the mechanical properties of hardened concrete under triaxial loads, ASTM C 801-81(86), 1988 Annual Book of ASTM Standards, Vol. 04.02, American Society for Testing and Materials, Philadelphia.
11. **Jensen, B. C. and Braestrup, M. W.,** Lok-tests determine the compressive strength of concrete, *Nordisk Betong,* No. 2, 1976, 9.
12. **Stone, W. C. and Carino, N. J.,** Deformation and failure in large-scale pullout tests, *ACI J.,* 80(6), 501, 1983.
13. Discussion of Reference 12, *ACI J.,* 81(5), 525, 1984.
14. **Yener, M. and Chen, W. F.,** On in-place strength of concrete and pullout tests, *ASTM J. Cement Concrete Aggregates,* Winter 1984, 90.
15. **Ottosen, N. S.,** Nonlinear finite element analysis of pullout test, *J. Struct. Div. ASCE,* 107(ST4), 591, 1981.
16. **Yener, M. and Vajarasathira, K.,** Plastic-fracture finite element analysis of pullout tests, Preprints of the 22nd Annual Tech. Meet. of the Society of Engineering Science, ESP22/8so38, Penn State Univ., October 7—9, 1985.
17. **Stone, W. C. and Carino, N. J.,** Comparison of analytical with experimental strain distribution for the pullout test, *ACI J.,* 81(1), 3, 1984.
18. **Hellier, A. K., Sansalone, M., Carino, N. J., Stone, W. C., and Ingraffea, A. R.,** Finite-element analysis of the pullout test using a nonlinear discrete cracking approach, *ASTM J. Cement, Concr. Aggregates,* 9(1), 20, 1987.
19. **Ballarini, R., Shah, S. P., and Keer, L. M.,** Failure characteristics of short anchor bolts embedded in a brittle material, *Proc. R. Soc. London,* A404, 1986, 35.
20. **Stone, W. C. and Giza, B. J.,** The effect of geometry and aggregate on the reliability of the pullout test, *Concr. Int.,* 7(2), 27, 1985.
21. **Krenchel, H. and Shah, S. P.,** Fracture analysis of the pullout test, *Materials and Structures* (RILEM), Vol. 18, No. 108, Nov.—Dec. 1985, 439.
22. **Krenchel, H. and Bickley, J. A.,** Pullout testing of concrete: historical background and scientific level today, *Nordisk Betong,* Publ. #6, Nordic Concrete Federation, 1987, 155.
23. Recommended practice for choice of sample size to estimate the average quality of a lot or process, E 122, 1988 ASTM Annual Book of Standards, Vol. 14.02.
24. **Bickley, J. A.,** The variability of pullout tests and in-place concrete strength, *Concr. Int.,* 4(4), 44, 1982.

25. **Carette, G. G. and Malhotra, V. M.,** In situ tests: variability and strength prediction at early ages, in ACI SP-82, *In Situ/Nondestructive Testing of Concrete,* Malhotra, V. M., Ed., American Concrete Institute, Detroit, MI, 1984, 111.
26. **Keiller, A. P.,** A preliminary investigation of test methods for the assessment of strength of in situ concrete, Technical Rep. 42.551, Cement and Concrete Association, Wexam Springs, U.K., 1982, 37.
27. **Stone, W. C., Carino, N. J., and Reeve, C. P.,** Statistical methods for inplace strength prediction by the pullout test, *ACI J.,* 83(5), 745, 1986.
28. In-place methods for determination of strength of concrete, Rep. of ACI Committee 228, *ACI J.,* 85(5), 446, 1988.
29. **Khoo, L. M.,** Pullout techniques — an additional tool for in-situ concrete strength determination, in ACI SP-82, *In Situ/Nondestructive Testing of Concrete,* Malhotra, V. M., Ed., American Concrete Institute, Detroit, MI, 1984, 143.
30. **Bocca, P.,** The application of pull out test to high strength concrete strength estimation, *Materials and Structure* (RILEM), 17(99), 211, 1984.
31. **Krenchel, H. and Peterson, C. G.,** In-place testing with LOK-Test: ten years' experience, paper presented at Int. Conf. on In Situ/Non-Destructive Testing of Concrete, Ottawa, October 2—5, 1984.
32. Recommended practice for dealing with outlying observations, E 178, 1988 ASTM Annual Book of Standards, Vol. 14.02.
33. **Peterson, C. G.,** LOK-test and CAPO-test development and their applications, *Proc. Inst. Civ. Engrs.,* Part I, 76, May 1984, 539.
34. **Bickley, J. A.,** Evaluation and acceptance of concrete quality by in-place testing, in *In Situ/Nondestructive Testing of Concrete,* Malhotra, V. M., Ed., ACI SP-82, American Concrete Institute, Detroit, MI, 1984, 95.
35. **Bickley, J. A.,** Concrete optimization, *Concr. Int.,* 4(6), 38, 1982.
36. Standard test method for pullout strength of concrete, ASTM C900-87, Annual Book of ASTM Standards, Vol. 04.02, 1988.
37. **Vogt, W. L., Beizai, V., and Dilly, R. L.,** In situ pullout strength of concrete with inserts embedded by 'finger placing', in ACI SP-82, *In Situ/Nondestructive Testing of Concrete,* Malhotra, V. M., Ed., American Concrete Institute, Detroit, MI, 1984, 161.
38. **Dilly, R. L. and Ledbetter, W. B.,** Concrete strength based on maturity and pullout, *ASCE J. Struct. Eng.,* 110(2), 354, 1984.
39. **Murphy, W. E.,** The interpretation of tests on the strength of concrete in structures, in ACI SP-82, *In Situ/Nondestructive Testing of Concrete,* Malhotra, V. M., Ed., American Concrete Institute, Detroit, MI, 1984, 377.
40. **Munday, J. G. L. and Dhir, R. K.,** Assessment of in situ concrete quality by core testing, in ACI SP-82, *In Situ/Nondestructive Testing of Concrete,* Malhotra, V. M., Ed., American Concrete Institute, Detroit, MI, 1984, 393.
41. **Haque, M. N., Day, R. L., and Langan, B. W.,** Realistic strength of air-entrained concrete with and without fly ash, *ACI J.,* 85(4), 241, 1988.
42. **Parsons, T. J. and Naik, T. R.,** Early age concrete strength determination by pullout testing and maturity, in ACI SP-82, *In Situ/Nondestructive Testing of Concrete,* Malhotra, V. M., Ed., American Concrete Institute, Detroit, MI, 1984, 177.
43. **Stone, W. C. and Reeve, C. P.,** A new statistical method for prediction of concrete strength from in-place tests, *ASTM J. Cement Concr. Aggregates,* 8(1), 3, 1986.
44. **Carino, N. J. and Stone, W. C.,** Analysis of in-place test data with spreadsheet software, in *Computer Use of Statistical Analysis of Concrete Tests,* ACI SP-101, Balaguru, P. and Ramakrishnan, V., Eds., American Concrete Institute, Detroit, MI, 1987, 1.
45. Recommended Practice for Evaluation of Strength Test Results of Concrete, ACI 214-77(Reapproved 1983), in *ACI Manual of Concrete Practice,* American Concrete Institute, Detroit, MI.
46. **Hindo, K. R. and Bergstrom, W. R.,** Statistical evaluation of the in-place compressive strength of concrete, *Concr. Int.,* 7(2), 44, 1985.
47. **Natrella, M.,** *Experimental Statistics,* Handbook No. 91, National Bureau of Standards, U.S. Govt. Printing Office, Washington, D.C., October 1966.
48. Building code requirements for reinforced concrete, ACI 318-89, American Concrete Institute, Detroit, MI.
49. **Peterson, C. G. and Hansen, A. J.,** Timing of loading determined by pull-out and maturity tests, RILEM International Conference on Concrete at Early Ages, Paris 1982, Ecole Nationale des Ponts et Chausses, Vol. I, 173.
50. **Chabowski, A. J. and Bryden-Smith, D. W.,** A simple pull-out test to assess the strength of in-situ concrete, *Precast Concr.,* 8(5), 243, 1977.
51. **Chabowski, A. J. and Bryden-Smith, D. W.,** Assessing the strength of concrete of in situ portland cement concrete by internal fracture tests, *Mag. Concr. Res.,* 32(112), 164, 1980.

52. **Bungey, J. H.,** Concrete strength determination by pull-out tests on wedge anchor bolts, *Proc. Inst. Civ. Engrs.,* Part 2, 71, June 1981, 379.
53. **Mailhot, G., Bisaillon, G., Carette, G. G., and Malhotra, V. M.,** In-place concrete strength: new pullout methods, *ACI J.,* 76(12), 1267, 1979.
54. **Domone, P. L. and Castro, P. F.,** An expanding sleeve test for in-situ concrete and mortar strength evaluation, *Proc. Structural Faults and Repairs 87,* Engineering Technics Press, Edinburgh, 1987, 149.

Chapter 4

THE BREAK-OFF TEST METHOD

Tarun R. Naik

ABSTRACT

In-place concrete strength is not the same as the cylinder concrete strength because the in-place concrete is placed, compacted and cured in a different manner than the cylinder specimen concrete. Determination of accurate in-place strength is critical in form removal and prestress or post-tension force release operations. Also, fast construction techniques and recent construction failures emphasize the need for adopting methods for determining in-place concrete strength. Presently several such methods exist and a considerable amount of information is available. Out of many of these currently available NDT methods, only the Break-Off and the Pull-Out tests measure a direct strength parameter. The Break-Off (B.O.) test consists of breaking off an in-place cylindrical concrete specimen at a failure plane parallel to the finished surface of the concrete element. The Break-Off stress at failure can then be related to the compressive or flexural strength of the concrete using a predetermined relationship which relates the concrete strength to the Break-Off strength for a particular source of concrete. The Break-Off test was developed in Norway by Johansen in 1976; and it was introduced recently in North America, initially by Malhotra in Canada and later by Naik in the U.S.

This chapter provides a more complete and detailed recent information regarding the theory behind the B.O. method, factors affecting this method, and the practical use of this method for laboratory and site investigations. Selected case histories and lab investigations are also included as an attempt to introduce the B.O. method to North America's concrete industry.

INTRODUCTION

For many years questions have been raised regarding concrete quality assurance test methods based upon standard cylinder tests, which measure the potential strength of a concrete batch. In-place concrete strength is not the same as the cylinder concrete strength because the in-place concrete is placed, compacted, and cured in a different manner than the cylinder specimen concrete. Determination of accurate in-place strength is critical in form removal and prestress or post-tension force release operations. Also, fast construction techniques and recent construction failures emphasize the need for adopting methods for determining in-place concrete strength. Presently several such methods exist and a considerable amount of information is available.[1-5] Out of many of these currently available NDT methods, only the Break-Off and the Pull-Out tests measure a direct strength parameter. The Break-Off (B.O.) test consists of breaking off an in-place cylindrical concrete specimen at a failure plane parallel to the finished surface of the concrete element. The Break-Off stress at failure can then be related to the compressive or flexural strength of the concrete using a predetermined relationship which relates the concrete strength to the Break-Off strength for a particular source of concrete. The Break-Off test was developed in Norway by Johansen in 1976.[6] The B.O. test is still not very widely used in North America. The primary factor in limiting the widespread use of this method being the lack of necessary technical data and experience in North America. Initial work at CANMET in the early 1980s had indicated a lack of reproducibility in results of this test method.[21] Several papers have been published in Europe about the B.O. method. This chapter provides more complete

FIGURE 1. Tubular plastic sleeves for inserting in fresh concrete for the B.O. test.

and detailed recent information regarding the theory behind the B.O. method, factors affecting this method, and the practical use of this method for laboratory and site investigations. Selected case histories and lab investigations are also included as an attempt to introduce the B.O. method to North America's concrete industry.

THEORETICAL CONSIDERATIONS

The B.O. method is based upon breaking off a cylindrical specimen of in-place concrete. The test specimen has a 55 mm (2.17 in.) diameter and 70 mm (2.76 in.) height. The test specimen is created in the concrete by means of a disposable tubular plastic sleeve, which is cast into the fresh concrete and then removed at the planned time of testing, or by drilling the hardened concrete at the time of the B.O. test. Figures 1 and 2 show tubular plastic sleeves and a drill bit, respectively. Both the sleeve and the drill bit are capable of producing a 9.5 mm ($^3/_8$ in.) wide groove (counter bore) at the top of the test specimen, Figure 3, for seating the load cell (see section on Break-off Test Equipment). A force is applied through the load cell by means of a manual hydraulic pump. Figure 3 is a schematic of a B.O. concrete cylindrical specimen obtained by inserting a sleeve or drilling a core. The figure also shows location of the applied load at the top of the B.O. test specimen. In essence, the load configuration is the same as a cantilever beam with circular cross section, subjected to a concentrated load at its free end. The force required to Break-Off a test specimen is measured by a mechanical manometer. The Break-Off stress can then be calculated as:

$$B.O. = M/S$$

where:

M = PB.O. *h
PB.O = Break off force at the top
h = 65.3 mm

S = (d)/32
d = 55 mm

FIGURE 2. Core drill bit for drilling a core for B.O. testing of existing concrete element.

In the above simplified formula the manufacturer uses the elementary theory of strength of materials and does not apply the concept of deep beam analysis even though the diameter to length ratio is 1:1.3. This point is being explored.[13-15] The Break-Off method assumes that the ultimate flexural strength of the concrete is reached at the extreme outside fiber at the base of the B.O. test specimen. In this case, the circular cross-section area would restrict the ultimate fiber stress theoretically to a point, and a crack is initiated at this point. The exact location of the rupture is determined by the loading arrangement, Figure 3, at a distance of 55 mm from the concrete surface. Away from the extreme outside fiber at the base, the stresses successively change in the direction of the neutral axis from tension to compression. The B.O. method is presently the only available test method for directly determining flexural strength of in-place concrete; and, there is a linear relationship between the B.O. flexural strength and modulus of rupture as determined by a beam test.[7-11]

BREAK-OFF TEST EQUIPMENT

The Break-Off tester, Figure 4, consists of a load cell, a manometer, and a manual hydraulic pump capable of breaking a cylindrical concrete specimen having the specified dimensions given in Theoretical Considerations. The load cell has two measuring ranges: low range setting for low strength concrete up to approximately 20 MPa (3000 psi) and high range setting for higher strength concrete up to about 60 MPa (9000 psi), Figure 5. A tubular plastic sleeve, with internal diameter of 55 mm (2.17 in.) and geometry shown in Figure 1, is used for forming cylindrical specimen in fresh concrete. A sleeve remover, Figure 6, is used for removing the plastic sleeve from the hardened concrete. A diamond tipped drilling bit is used for drilling cores for the B.O. test in hardened concrete, Figure 2. The bit is capable of producing a cylindrical core, along with a reamed ring (counter bore) in the hardened concrete at the top with dimensions similar to that produced by using a plastic

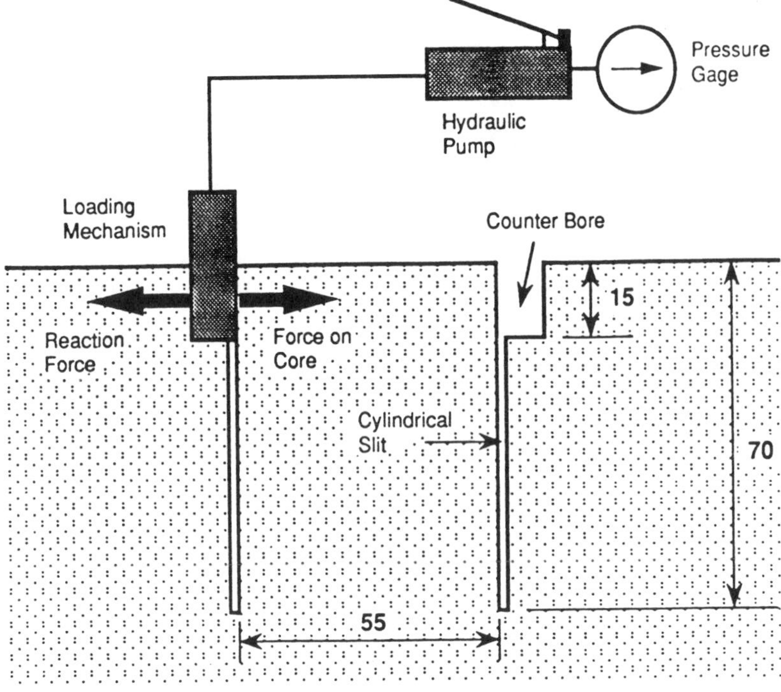

Dimensions in millimeters

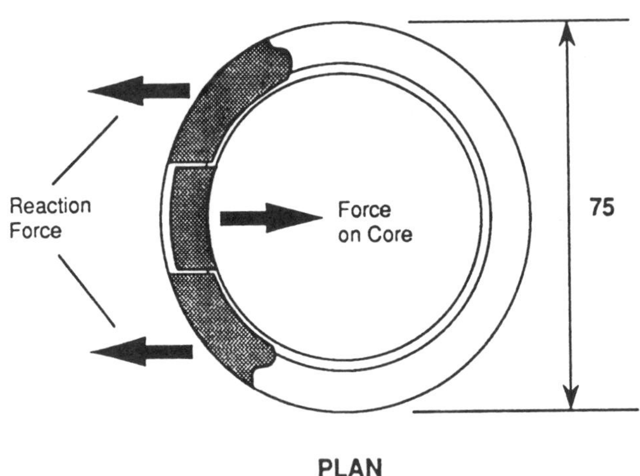

FIGURE 3. Schematic of concrete cylindrical specimen obtained by inserting a sleeve or drilling a core, and location of applied load.

sleeve. The manufacturer also provides a calibrator for calibration and adjustment of the B.O. tester, Figure 7. The procedure of calibrating a B.O. tester is discussed later.

HISTORICAL BACKGROUND

The B.O. method is a relatively new nondestructive test. The first paper was published by Johansen in 1976[6] and the research work was done at the Norwegian Technical University

FIGURE 4. The B.O. test equipment: (1) load cell, (2) manometer, and (3) hydraulic hand pump.

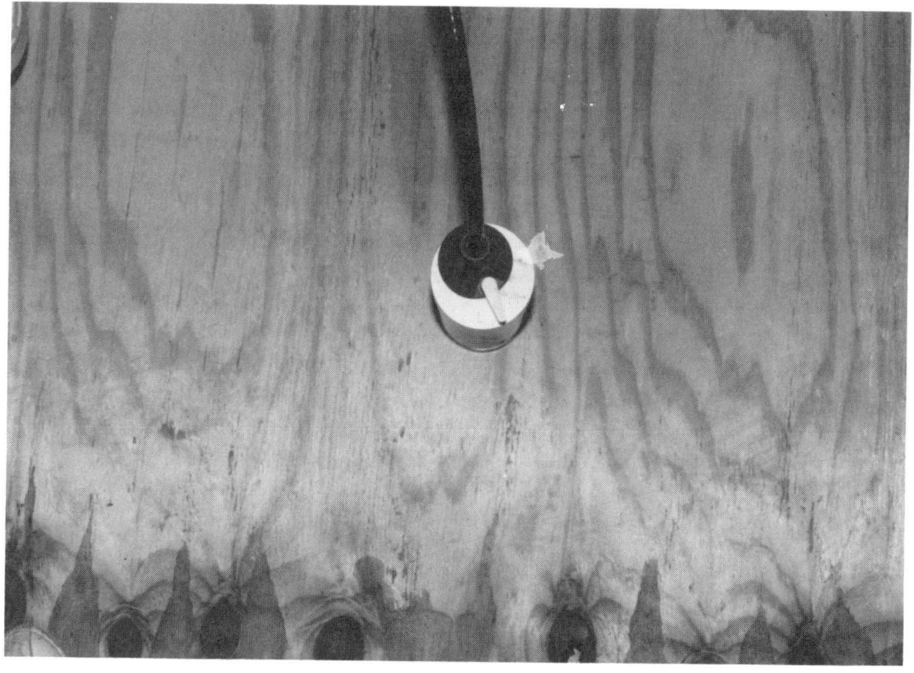

FIGURE 5. Low and high range settings of the B.O. tester load cell.

FIGURE 6. Sleeve remover.

FIGURE 7. The B.O. tester calibrator.

(NTH). In 1977 researchers at the NTH and the Research Institute for Cement & Concrete in Norway developed and patented the Break-Off tester as a method for determination of compressive strength of in-place concrete.

Johansen,[6] in his first paper, indicated the main use of this method as a very efficient way of determining the in-place concrete strength for form removal. In 1979, Johansen and Dahl-Jorgensen published a paper on the use of the B.O. method to detect variation in the concrete strength and curing conditions.[12] In their research a comparison was made between the B.O. method and the Pull-Out test method. The compressive strength of cores obtained from the B.O. tests and the standard cube compressive strength was also compared. They concluded that the Pull-Out test method and the core compressive strength values obtained from the B.O. tests have a better ability to differentiate between concrete qualities than the cube test. On the other hand, the B.O. test results and the core compressive strength results demonstrated their ability in detecting variation in curing conditions, while the Pull-Out test method did not register some of the curing differences demonstrated by the B.O. and the core results.

Also, in 1979, Johansen published another paper[9] on the use of the B.O. method, with a particular reference to its application to airport pavements made of vacuum concrete. The author concluded that the variation of the concrete strength detected by the B.O. method is of the same order of magnitude as the variation detected by conventional flexural beam test. Furthermore, the B.O. strength was about 30% higher than the conventional modulus of rupture because of deviations in the load configurations and geometric parameters between the two test methods. A high sensitivity of the B.O. method to sense the influence of the ambient air temperature on early strength was also indicated in this paper. A good relationship was obtained between B.O. test reading and the compressive strength of the concrete obtained by standard cube testing.

In 1980 Byfors tested concrete at early ages using the Break-Off method.[7] In his research Byfors tested concretes with different water to cement ratios, and different aggregate sizes 8, 16, and 32 mm ($^5/_{16}$", $^5/_8$" and $1^1/_4$"). The conclusion was that the B.O. method is well suited to detecting low strength concrete made with different sizes of coarse aggregates.

After modifications to the B.O. tester, Dahl-Jorgensen published two reports on the use of the B.O. method.[12,16] In this study, he investigated the use of the new equipment in testing epoxy to concrete bond strength, and compared the results of the B.O. and the Pull-Out methods. He concluded that the B.O. test provided results with smaller variations between individual tests than the Pull-Out method. Also, fewer tests were rejected for the B.O. method compared to the Pull-Out method.

Nishikawa published his work in 1983[10] after conducting very limited laboratory research on the use of the B.O. method for determining flexural strength of concrete. He concluded that the relationship between the B.O. test results and cylinder compressive strength tests was complex and of little practical value. Therefore, no attempt was made to correlate these two test results. This is, of course, contrary to all other published information about the B.O. method. He further concluded that the change in the shape of the aggregates was not sensed by the B.O. method. These two conclusions are further discussed in the Evaluation of Test Specimens, Nishikawa indicated a relatively high within test variation for the inserted sleeve B.O. tests as compared to cylinder and beam tests. With respect to other variables, he found that the B.O. test results were affected by water-cement ratio, age, curing conditions, and cement type.

In 1984 Carlsson et al. published a paper on field experiences with the use of the B.O. tester.[17] Six case histories were discussed. The authors concluded a trend towards greater acceptance of the B.O. test method in the field.

In 1987 another very limited exploratory investigation was done by Barker and Ramirez.[18] The Scancem version of the tester was used. They investigated the effects of the water to

TABLE 1
Results of Standard Cylinder Compressive Strength and Break-Off Tests for Mix 1 (30 MPa)

	Cylinder compressive strength (psi)		Break-off readings (bar)							
			Inserted sleeve slab thickness (in.)				Drilled core slab thickness (in.)			
			5		7		5		7	
Test age (days)	Actual	Average	Actual	Ave	Actual	Ave	Actual	Ave	Actual	Ave
	2120		41,50		59,59					
1	2210	2180	59,57	55	57,56	59	—	—	—	—
	2210		57,64		62,61					
	3360		76,70		76,72					
3	3380	3315	70,76	73	73,70	74	—	—	—	—
	3200		70,77		77,74					
	4085		78,75		70,86		104,94		91,87	
5	3925	3935	80,83	79	77,79	79	91,98	97	89,94	88
	3800		80		81		100		80	
	4155		67,75		86,82		95,84		93,84	
7	4070	4100	73,80	73	79,79	82	80,88	87	80,90	86
	4070		72		85		89		83	
	5040		104,92		102,90		123,113		118,120	
28	4950	4955	83,96	94	91,98	96	105,105	112	118,112	118
	4880		96		98		112		120	

cement ratio, maximum aggregate size, and aggregate shape. They indicated a relatively low within test variation of the method of 6.1%, while that of the cylinder and beam tests were 7.6 and 4.6%, respectively. A regression analysis was performed between B.O. test results and the cylinder test results. Good correlation was obtained and a detailed statistical study was performed on the effects of different parameters on the B.O. compressive and flexural strengths relation. Also in 1987, Naik et al. conducted a comprehensive laboratory investigation at the University of Wisconsin-Milwaukee.[13-15] They were the first to study the effects of the method of obtaining the B.O. test specimen, either by inserting a plastic sleeve in fresh concrete or by drilling a core after the concrete had hardened. Furthermore, they investigated the applicability of the B.O. test for high strength concrete. The strengths investigated were 45 and 55 MPa (6000 and 8000 psi), along with 30 MPa (4000 psi). They also studied the effect of aggregate shape and slab thickness on the B.O. test results. A total of 524 B.O. tests were performed. They concluded that B.O. readings for crushed aggregates were on the average 10% higher than that for the rounded aggregates. Also, the B.O. test is less variable for crushed aggregate concrete. They also stated that the drilled core B.O. test results were on the average about 9% higher than the inserted sleeve B.O. test results. According to Naik et al. the drilled core test method is preferable, although both methods of obtaining the B.O. test specimen showed good correlation with the compressive strength of the in-place concrete, and also yielded equally consistent B.O. test results, see Tables 1, 2, and 3, and Figures 8 and 9. They have discussed some of the difficulties encountered in inserting sleeves in harsh concrete or concrete with high amounts of bleeding. Additional statistical analysis is being performed on the large amount of data that this investigation has yielded.

TEST PROCEDURE

INSERTING SLEEVES IN FRESH CONCRETE

Sleeves should be at center to center and edge distance of minimum 150 mm (6 in.).

TABLE 2
Results of Standard Cylinder Compressive Strength and Break-Off Tests for Mix 2 (45 MPa)

	Cylinder compressive strength (psi)		Break-off readings (bar)							
			Inserted sleeve slab thickness (in.)				Drilled core slab thickness (in.)			
			5		7		5		7	
Test age (days)	Actual	Average	Actual	Ave	Actual	Ave	Actual	Ave	Actual	Ave
	2650		57,58		61,59					
1	2760	2705	68,59	60	52,68	61	—	—	—	—
	2705		60,60		60,65					
	4880		85,82		71,81					
3	4740	4820	79,85	81	87,70	76	—	—	—	—
	4845		72		71					
	5525		85,87		89,89		89,87		89,89	
5	5465	5560	89,79	84	78,69	83	85,95	91	84,83	87
	5685		78		89		100		92	
	5850		90,88		92,95		103,96		94,92	
7	5215	5735	83,87	87	90,97	91	106,97	100	104,107	95
	6135		85		80		97		81	
	6490		122,103		95,90		142,125		105,109	
28	6100	6320	113,109	110	103,95	97	119,143	129	108,95	103
	6365		103		101		115		100	

TABLE 3
Results of Standard Cylinder Compressive Strength and Break-Off Tests for Mix 3 (55 MPa)

	Cylinder compressive strength (psi)		Break-off readings (bar)							
			Inserted sleeve slab thickness (in.)				Drilled core slab thickness (in.)			
			5		7		5		7	
Test age (days)	Actual	Average	Actual	Ave	Actual	Ave	Actual	Ave	Actual	Ave
	2775		92,85		56,59					
1	2760	2745	74,80	85	74,68	62	—	—	—	—
	2795		88,91		67,56					
	5020		108,90		82,97		98,108		119,108	
3	4635	4865	90,78	93	106,84	95	108,94	97	92,91	110
	4935		82,92		97,106		94,79		90,98	
	5555		107,108		110,95		110,106		118,111	
5	5215	5335	108,114	106	95,110	103	105,116	106	115,119	115
	5235		95		103		94		110	
	5820		117,110		109,112		117,110		103,106	
7	6225	5920	125,109	112	110,107	106	99,108	108	113,115	109
	5710		97		92		106		110	
	8310		119,137		125,120		113,140		123,124	
28	8455	8150	123,110	124	109,108	118	117,131	130	137,124	129
	7690		132		129		149		136	

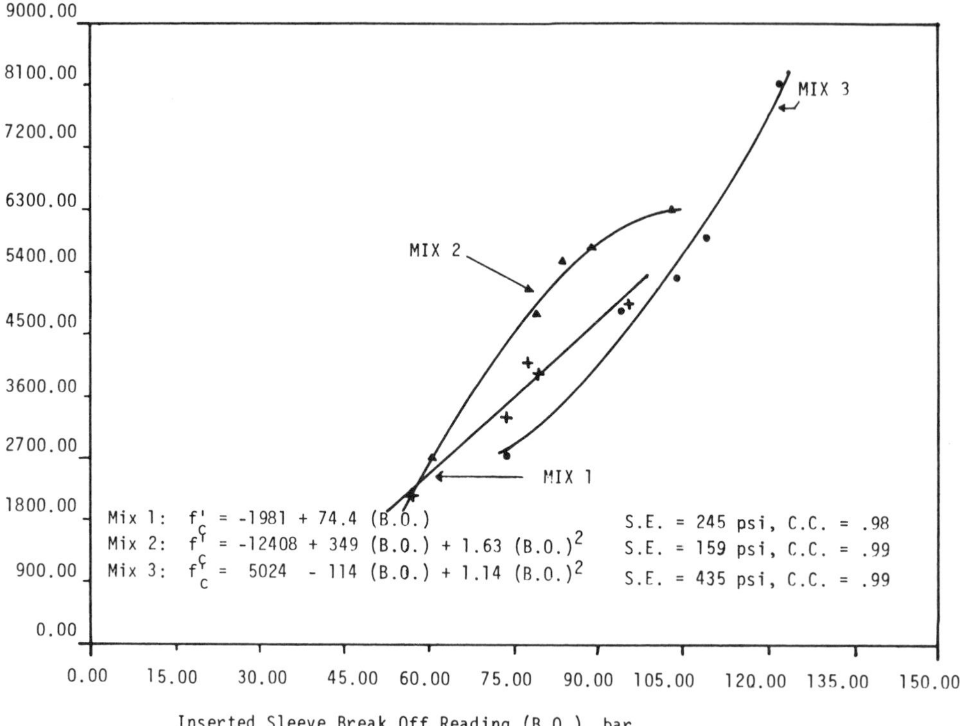

FIGURE 8. Plots of regression equations for inserted sleeve specimens for Mix 1, 2, and 3.

They are best pushed in-place by a rocking and twisting action, Figure 10. Concrete inside the sleeve and the top of plastic sleeve itself should then be tapped by fingers to insure good compaction for the B.O. specimen. Sleeves should then be moved gently up and down in-place and brought to the same level as the concrete surface at its final position. For stiff mixes (i.e., low slump concrete) a depression may occur within the confines of the sleeve during the insertion process. In such cases the sleeve should be filled with additional concrete, tapped with fingers, and slightly jiggled from side to side. On the other hand, for wet, high slump mixes, the sleeve may move upward due to bleeding. For such cases, sleeves should be gently pushed back in-place, as necessary, to the level of the finished concrete surface. Sometimes this process may have to be repeated until the uplift movement stops after the initial setting has occurred. A small weight may be placed on the sleeve in order to prevent its upward movement. Heavy grease, or other similar material, should be used to lubricate the plastic sleeves for easier removal after the concrete hardens.

PREPARATION FOR CORE DRILLING FROM HARDENED CONCRETE

The finished concrete surface should be evaluated for sufficient smoothness in order to fix the vacuum plate of the core drilling machine. The core barrel should be perpendicular to the concrete surface at all times. The drilling process should be continued to the full depth required to produce a cantilever cylindrical core of 70 mm (2.76 in.) length, with a groove at the top of the core for setting the B.O. tester load cell. A slightly longer drilled core will not affect the B.O. reading, while a slightly shorter drilled core will affect the B.O. reading, Figure 11.

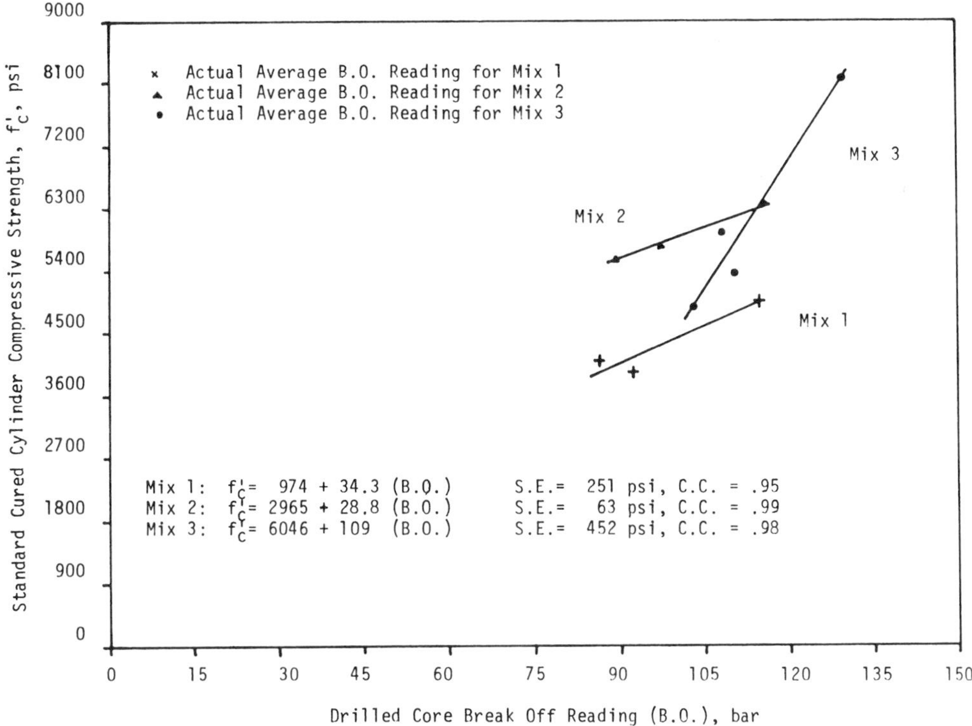

FIGURE 9. Plots of regression equations for drilled core specimens for Mix 1, 2, and 3.

FIGURE 10. Inserting sleeve by rocking action.

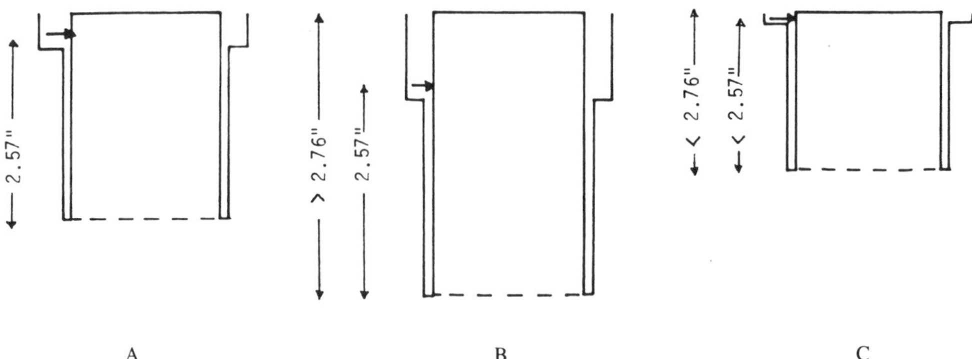

FIGURE 11. The B.O. drilled specimen dimensions: (A) The B.O. drilled specimen with the exact depth. (B) The B.O. drilled specimen with larger depth than 2.76''. (C) The B.O. drilled specimen with shorter depth than 2.76''.

CONDUCTING THE B.O. TEST

At the time of the B.O. test, remove the inserted plastic sleeve by means of the key supplied with the tester, Figure 6. Leave the plastic ring in-place. Remove loose debris from around the cylindrical slit and the top groove, see Figure 3. Select the desired range setting and place the load cell in the groove on the top of the concrete surface so that the load is applied according to Figure 3. The load should be applied to the test specimen at a rate of approximately one stroke of the hand pump per second. This rate is equivalent to about 0.5 MPa (70 psi) of hydraulic pressure per second. After breaking off the test specimen, record the B.O. manometer reading. This manometer reading can then be translated to the concrete strength using curves relating the B.O. reading to the desired concrete strength (i.e., flexural and/or compressive).

THE B.O. TESTER CALIBRATION PROCEDURE

The B.O. tester should be calibrated preferably each time before use, otherwise periodically. To calibrate the tester, follow the following steps:

1. Set the calibrator gauge to zero.
2. Place the calibrator in the load cell (Figure 12).
3. Set the load cell on the high setting.
4. Apply the load to the calibrator by pumping the handle until the load cell manometer reading is 100.
5. Record the dial gauge reading and compare it with the expected value obtained from the manufacturers calibration chart (Figure 13). The dial gauge value should be within 4% of the manufacturer's chart value.
6. Repeat the above procedure for the low range setting.

Adjustment of the B.O. tester is necessary if error in the reading obtained is greater than $\pm 4\%$ of the expected value from the chart. For a well calibrated tester, the needle on the manometer should move five bars per one hand stroke, while the first and/or second strokes might not move it that much. A good rate of applying the load would be one stroke per second.

DEVELOPING A CORRELATION CURVE

The B.O. manufacturer provides correlation curves relating the B.O. reading and the compressive strength of the standard 150 × 300 mm cylinders and 150 mm cubes. Figure 14 is the manufacturer's curves for the high range setting of the load cell, while Figure 15

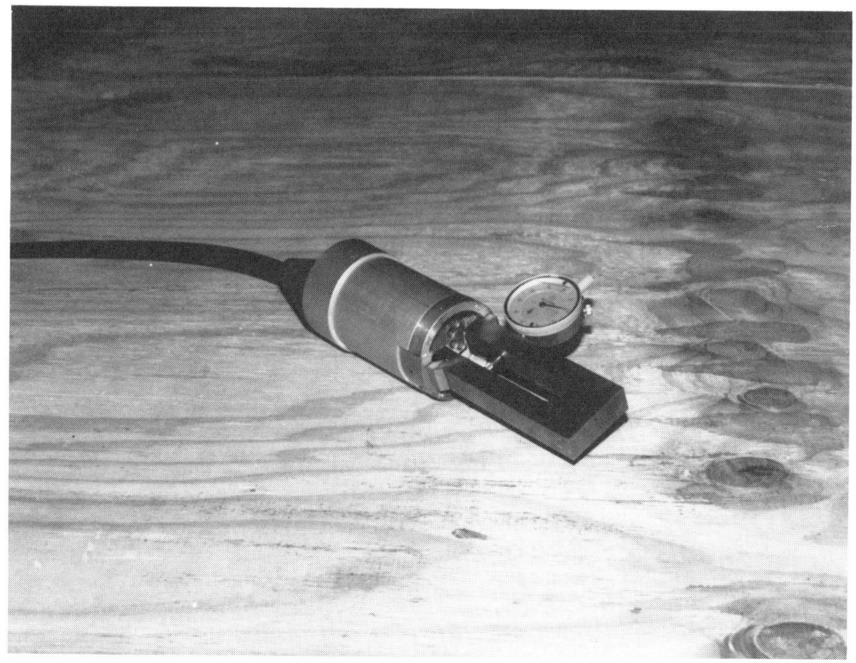

FIGURE 12. Calibrator placed in the load cell.

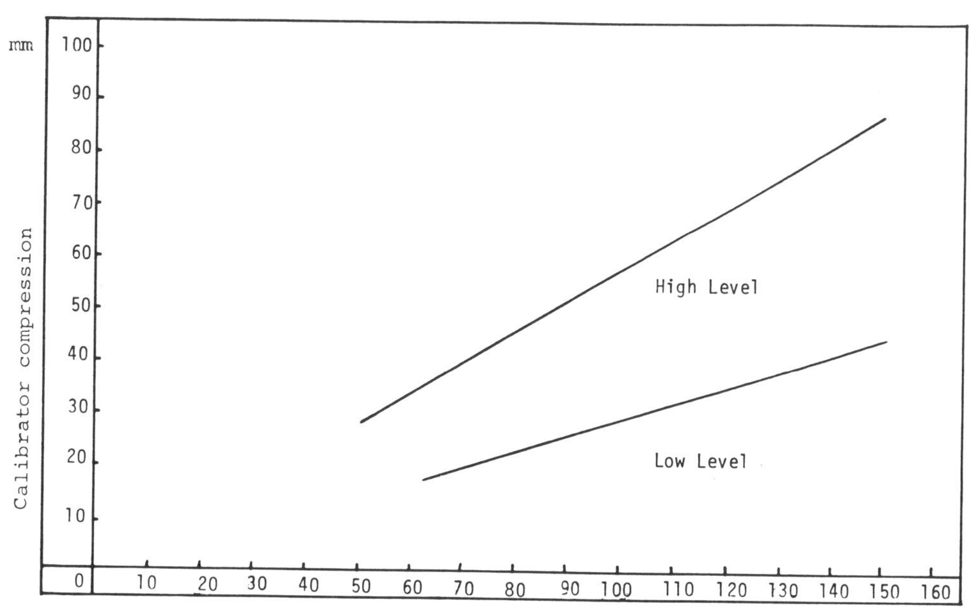

FIGURE 13. Calibrator chart as provided by the manufacturer.

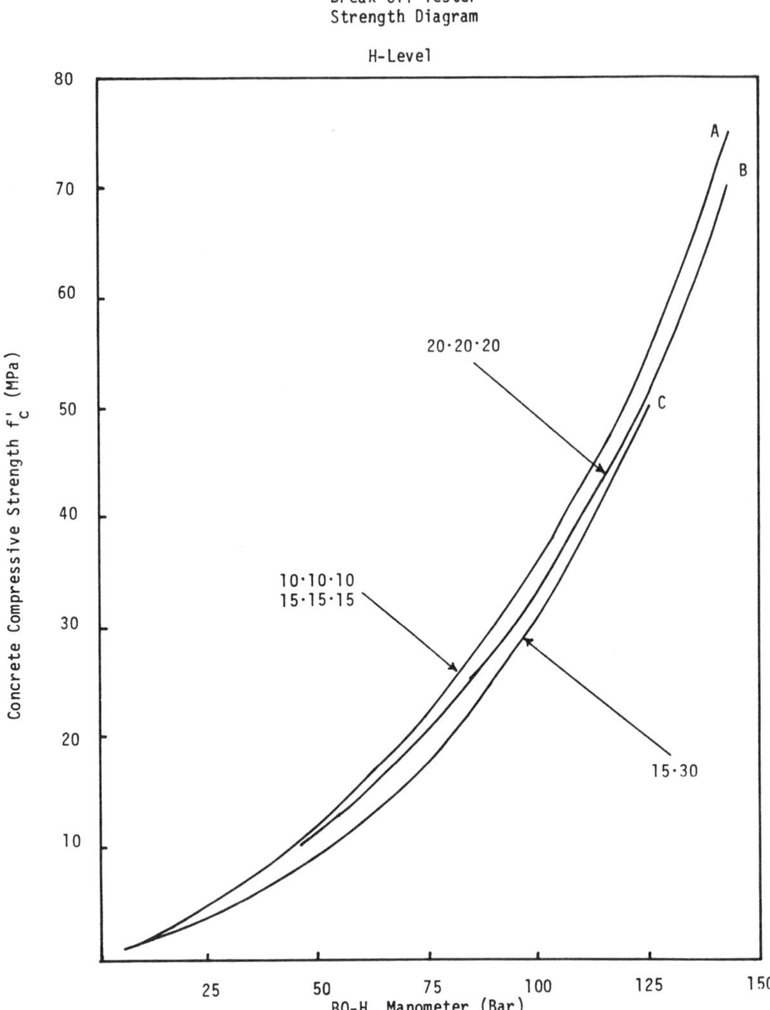

FIGURE 14. B.O. manometer reading vs. concrete compressive strength as provided by the manufacturer, for the high range setting.

is for the low range setting (see Historical Background). This correlation is nonlinear and was empirically derived. The curve relating the B.O. reading and the compressive strength is concaved upward. This seems to indicate that the B.O. tester is less sensitive for higher concrete strengths. It should be noted that the manufacturer's curves consider many variables. However, knowing that concrete itself has inherent variability, a user should develop his own correlation curves for a particular concrete batch. Developing correlation curves for different types of concrete would efficiently increase the accuracy and dependability of the method in predicting the in-place strength.

The following precautions should be taken when developing data for correlations:

1. Keep the center-to-center and the edge distances of at least 150 mm (6 in.) in the process of inserting sleeves or drilling B.O. cores.
2. Obtain a minimum of five B.O. readings and three corresponding standard strength test specimens values, i.e., cylinders for compressive strength, and beams for flexural strength, for each test age.

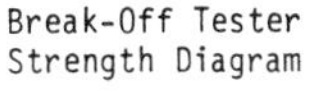

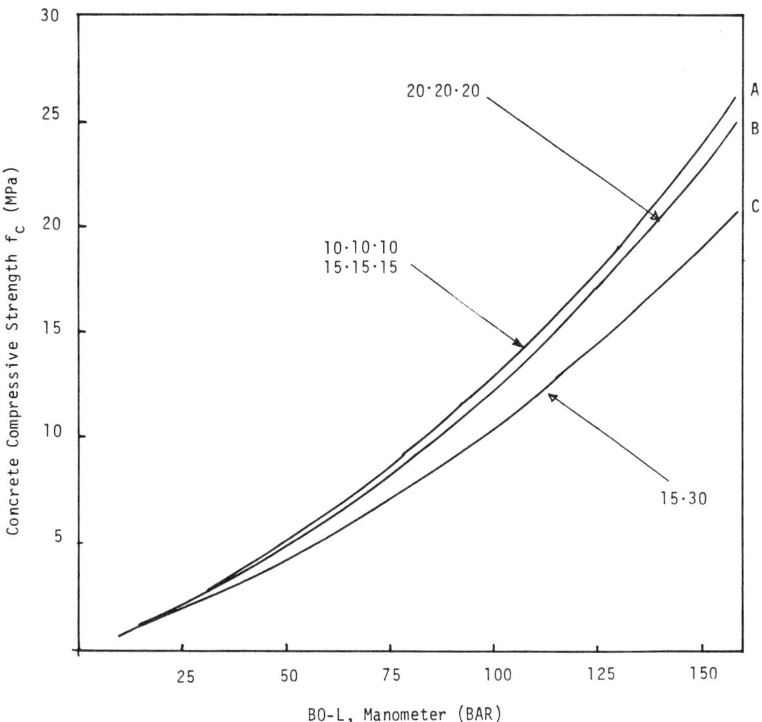

FIGURE 15. B.O. manometer reading vs. concrete compressive strength as provided by the manufacturer, for the low range setting.

3. An average of the five B.O. readings and the average of the three standard cylinder test results represent one point on the graph relating the B.O. reading to the desired standard strength of the concrete.
4. Cover the range of concrete strengths expected in the project, at early as well as at later ages, such as 1, 3, 5, 7, 14, and 28 days.

Regression equations between the mean values of the B.O. tests and standard strength tests should then be developed after sufficient data points are obtained for the correlation curve. The least square method can be applied to fit the best curve between the data points. The manufacturer's correlation curve could be used as a preliminary estimate only if no correlation chart has been obtained for the particular concrete under consideration. The strength value obtained from the manufacturer's correlation curves is considered only an approximation of the true in-place strength because it does not consider the combination of concrete making materials for the specific concrete under investigation.

EVALUATION OF TEST SPECIMENS

Before accepting a particular B.O. reading, the B.O. specimen tested should be examined to insure a ''good'' test. The B.O. test specimen must be perpendicular to the concrete surface. A minimum center to center and edge distance of 150 mm (6 in.) should be maintained. The failure plane should be approximately parallel to the concrete surface. It must be at a depth of 70 mm (2.76 in.) from the finished surface. The presence of honey-

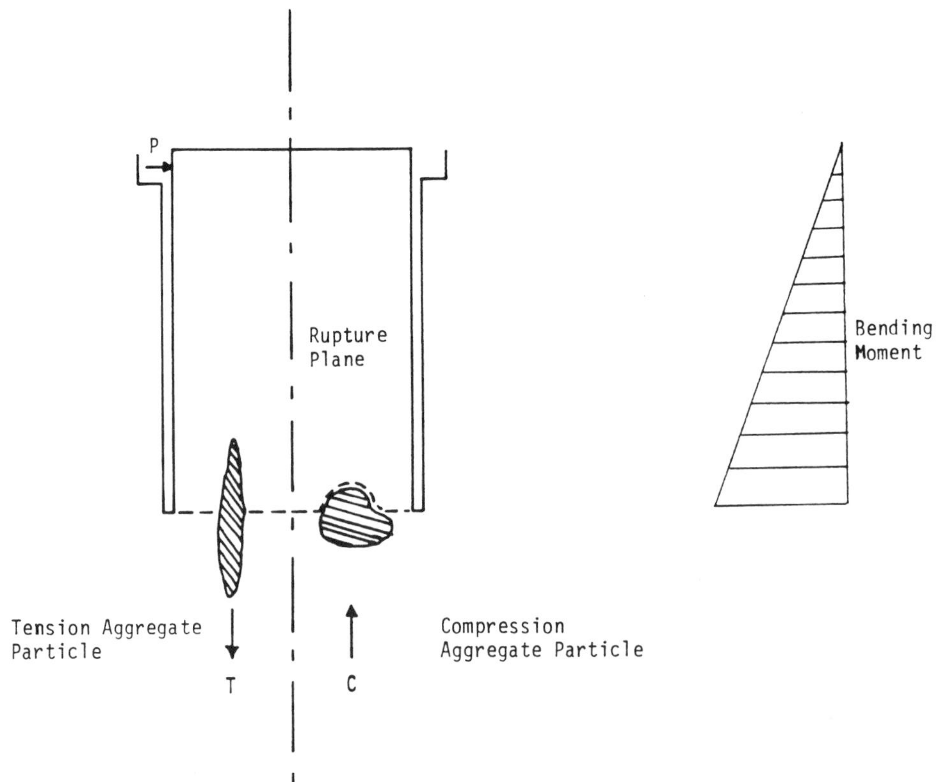

FIGURE 16. Rejected B.O. specimen due to presence of aggregate(s) in the failure plane.

combed concrete, excessive air voids, and/or reinforcement at the failure plane of test specimens could shift the rupture plane from its intended place. Such test specimen results should be rejected. The rejection criteria is somewhat dependent on the engineering judgement of the user. Figure 16 illustrates an example of irregular resistance mechanism to the applied B.O. force. In this special case, two fairly large aggregates exist in the rupture plane in such a way that the combination of the two particles could create a resistance couple. The rupture plane is forced to pass through the tensile particle, as illustrated in Figure 16, because the aggregate has a large aggregate-mortar bond area. Such cases, for example, could lead to a greater flexural resistance than that of other B.O. tests. Nishikawa and others[10,18] report rejection of very few tests. Dahl-Jorgensen[12] stated that less tests were rejected for the B.O. method than that for the Pull-Out method. None of the published research reports on a statistical rejection technique that would provide a procedure for excluding bad test specimens, results, or personal judgement.

It is important to note that the inserted sleeve B.O. specimen tends to be trapezoidal in shape rather than cylindrical (the top diameter is 4 mm (0.16 in.) less than the bottom diameter), while it is exactly cylindrical in the case of the B.O. drilled core specimen. However, inserted sleeve B.O. specimen reading is not affected by the trapezoidal shape because the bottom diameter at the failure plane always remains 55 mm (2.17 in.).[13-15] In evaluating inserted sleeve specimens, it is also reported that the drilled core specimens give higher readings than the inserted sleeve specimens.[13-15] This is because in the case of the inserted sleeve specimens, the accumulation of bleed water under the bottom edge of the sleeve would tend to create a weaker zone of concrete exactly where the failure plane for the B.O. inserted sleeve test occurs.

APPLICATIONS

The potential of the B.O. test is promising. This method can be used both for quality control and quality assurance. The most practical use of the B.O. test method is for determining the time for safe form removal, and the release time for transferring the force in prestressed or post-tension members. The B.O. method can also be used to evaluate existing structures. It has been reported that the B.O. test provides a more effective way in detecting curing conditions of concrete than the Pull-Out and the standard cylinder tests.[7-9,12,16]

In 1982 the B.O. tester was used to control the time for safe form removal for a new "Bank of Norway" building and an apartment building in Oslo. In 1983 the B.O. tester was used in England. Recently the B.O. tester has been used by the Norwegian Contractors Company which is responsible for building off-shore platforms for the oil fields in the North Sea. The B.O. method can also be used to measure the bonding strength of overlays or the bonding between concrete and epoxy, but this usage has not been applied in the field.[12,16]

ADVANTAGES AND LIMITATIONS

The main advantage of the B.O. test is that it measures the in-place concrete (flexural) strength. The equipment is safe and simple; and the test is fast to perform, requiring only one exposed surface. The B.O. test does not need to be planned in advance of placing the concrete because drilled B.O. test specimens can be obtained. The test is reproducible to an acceptable degree of accuracy and does correlate well with the compressive strength of concrete.

Two limitations for the B.O. test equipment are worth noting (1) the maximum aggregate size; and (2) the minimum member thickness for which it can be used. The maximum aggregate size is 19 mm ($^3/_4$ in.) and the minimum member thickness is 100 mm (4 in.). However, the principle of the method can be applied to accommodate larger aggregate sizes or smaller members. The major disadvantage of the B.O. test is that the damage to the concrete member must be repaired if the member is going to be visible. However, this test is nondestructive since the tested member need not be discarded.

STANDARDIZATION OF THE B.O. METHOD

Recently the B.O. method was standardized in England,[3] Norway,[19] and Sweden.[20] It is also undergoing the process of standardization in the U.S.

REFERENCES

1. **Malhotra, V.,** Testing hardened concrete: nondestructive methods, *ACI Monog.,* No. 9, 1976.
2. **Jones, R.,** A review of nondestructive testing of concrete, *Proc. Symp. Nondestructive Testing of Concrete and Timber,* Institute of Civil Engineers, London, June 1969, 1.
3. British Standard, B.S. 1881, Part 201, 1986, 17.
4. Methods of Mechanical Nondestructive Determination of Probable Compressive Strength of Concrete, Bulgarian National Standard 3816 - 65.
5. **Vossitch, P.,** Outline of the various possibilities of nondestructive mechanical tests on concrete, *Proc. Int. Symp. Nondestructive Testing of Materials and Structures,* Vol. 2, Paris, 1959, 30.
6. **Johansen, R.,** A New Method for Determination of In-Place Concrete Strength of Form Removal, 1st Eur. Colloq. on Construction Quality Control, Madrid, Spain, March 1976.
7. **Byfors, J.,** Plain Concrete at Early Ages, Swedish Cement and Concrete Research Institute, Rep. No. Facks-10044, Stockholm, 1980.

8. **Dahl-Jorgensen, E.,** In-situ Strength of Concrete, Laboratory and Field Tests, Cement and Concrete Research Institute, the Norwegian Institute of Technology, Rep. No. STF 65A, Trondheim, Norway, June 1982.
9. **Johansen, R.,** In-situ strength of concrete, the break-off method, *Concr. Int.*, p. 45, 1979.
10. **Nishikawa, A.,** A Nondestructive Testing Procedure for In-Place Evaluation of Flexural Strength of Concrete, Rep. No. JHRP 83-10, Joint Highway Research Project, School of Civil Engineering, Purdue University, West Lafayette, IN, 1983.
11. **Smith, L.,** Evaluation of the Scancem Break-Off Tester, Ministry of Works and Development, Central Laboratories, Rep. No. 86/6, New Zealand, 1986.
12. **Dahl-Jorgensen, E. and Johansen, R.,** General and Specialized use of the Break-Off Concrete Strength Test Method,'' ACI, SP 82-15, 1984, 294.
13. **Hassaballah, A.,** Evaluation of In-Place Crushed Aggregates Concrete by the Break-Off Method, Civil Engineering Department, University of Wisconsin-Milwaukee, December 1987.
14. **Salameh, Z.,** Evaluation of In-Place Rounded Aggregates Concrete by the Break-Off Method, Civil Engineering Department, University of Wisconsin-Milwaukee, December 1987.
15. **Naik, T. R., Salameh, Z., and Hassaballah, A.,** Evaluation of In-Place Strength of Concrete by the Break-Off Method, Department of Civil Engineering and Mechanics, The University of Wisconsin-Milwaukee, Milwaukee, WI, March 1988.
16. **Dahl-Jorgensen, E.,** Break-Off and Pull-Out Methods for Testing Epoxy-Concrete Bonding Strength, Project No. 160382, The Foundation of Scientific and Industrial Research of the Norwegian Institute of Technology, Trondheim, Norway, September 1982.
17. **Carlsson, M., Eeg, I., and Jahren, P.,** Field Experience in the Use of Break-Off Tester, ACI, SP 82-14, 1984.
18. **Barker, M. and Ramirez, J.,** Determination of Concrete Strengths Using Break-Off Tester, School of Civil Engineering, Purdue University, West Lafayette, IN, 1987.
19. Nordtest Method NTBUILD 212, Edition 2, Approved 1984-05.
20. Swedish Standard, SS 137239, Approved 1983-06.
21. Personal communication from V. M. Malhotra, CANMET, Ottawa, Ontario, Canada, 1984.

Chapter 5

THE MATURITY METHOD*

Nicholas J. Carino

ABSTRACT

This chapter reviews the history and technical basis of the maturity method, a technique for estimating the strength gain of concrete based upon the measured temperature history during curing. The combined effects of time and temperature on strength gain are quantified by means of a maturity function. The widely used maturity functions are reviewed critically. It is shown that the traditional maturity function is inadequate compared with the function based upon the Arrhenius equation. The concept of equivalent age, which is the most convenient measure of maturity, is explained. The strength gain of a specific concrete mixture is estimated using the measured maturity and the strength vs. maturity relationship for that mixture. Various proposed strength-maturity relationships are reviewed. It is also explained why the maturity method can only be used reliably to estimate relative strength. Examples are presented to illustrate how this technique can be used in combination with other in-place tests of concrete strength. The ASTM standard dealing with the method is also summarized.

INTRODUCTION

As is well known, the strength of a given concrete mixture, which has been properly place, consolidated and cured, is a function of its age and temperature history. At early ages, temperature has a dramatic effect on strength development. This temperature dependence presents problems in trying to estimate the in-place strength based on strength development data obtained under standard laboratory conditions.

Around 1950, an approach was proposed to account for the combined effects of time and temperature on strength development of concrete. The motivation for the development of this approach was the need for a method to predict the effects of steam curing treatments on strength development. Subsequently, application of the method was extended to ordinary curing conditions. It was proposed that the measured temperature history during the curing period could be used to compute a single number which would be indicative of the concrete strength. Saul called this single factor "maturity" and he proposed the well-known "maturity rule" for estimating the strength of concrete.[1]

Following the publication of the maturity rule, there were reports of its validity.[2,3] However, there were also reports of cases where the method failed.[4-6] From the time of the initial proposal, extensive research has been performed and modifications have been proposed for improving the accuracy of strength estimated from temperature history. Today, the maturity method is viewed as a useful and simple means for accounting approximately for the complex effects of time and temperature on strength development. The method is used during the curing period and is not applicable to existing concrete structures. Various standards and recommended practices dealing with curing, cold weather protection, and formwork removal refer to the maturity method.[7-10]

The objective of this chapter is to provide the reader with knowledge needed to correctly use the maturity method to estimate the strength of concrete in a structure while it is curing.

* Contribution of the National Institute of Standards and Technology, not subject to copyright in the U.S.

The chapter begins with a brief review of the historical development leading to current practice. This is followed by an analytical treatment of concrete strength gain to establish the basis for the maturity method. After the theoretical development, the practical aspects in applying the technique are discussed. Finally, a standard practice for implementing the maturity method is summarized.

HISTORICAL BACKGROUND

A comprehensive review of the maturity method up to 1971 was published by Malhotra.[12] Therefore, only a brief history is presented in this section. First, the various maturity functions which have been proposed to account for the effects of temperature and time on strength development are discussed. This is followed by a discussion of the relationships which have been proposed to represent strength development as a function of maturity.

MATURITY FUNCTIONS

Maturity functions are used to convert the actual temperature history of the concrete to a factor which is indicative of how much strength has developed. The roots of the earliest maturity function are found in a series of papers dealing with accelerated curing methods for concrete. In 1949, McIntosh[13] reported on experiments to develop procedures to predict the strength development of concrete during electric curing. He suggested that the product of time and concrete temperature above a datum temperature could be used to summarize the effects of the curing history. A datum temperature of 1.1°C (30°F) was suggested, and the product of time and temperature above the datum temperature was called the "basic age". However, it was found that when strength development at different temperatures was plotted as a function of basic age, there was not a unique relationship. McIntosh concluded that the strength development of concrete was governed by more complex factors than a simple product of temperature and time.

Soon after publication of McIntosh's work, Nurse[14] wrote on the principles of low-pressure steam curing. He also suggested that the product of time and temperature could be used to summarize the effects of different steam curing cycles. Nurse did not suggest using a datum temperature, and his calculations involved the curing chamber temperatures, not the actual concrete temperatures. Nevertheless, he showed that when the *relative strength* development was plotted as a function of the product of time and temperature, the data for different concretes and curing cycles fell reasonably close to a single nonlinear curve. This was the first evidence to show that the product of time and temperature could be used to approximate the combined effects of these two factors on strength development.

In 1951, Saul[1] summarized the conclusions drawn from research on the principles of steam curing performed at the Cement and Concrete Association in England. The term "maturity" was for the first time linked to the product of time and temperature. Saul suggested that maturity should be calculated with respect to a "datum temperature," which is the lowest temperature at which strength gain is observed. Thus maturity is computed from the temperature history using the following:

$$M = \sum_{0}^{t} (T - T_o)\,\Delta t \tag{1}$$

where

M = maturity* at age t,
T = average temperature of the concrete during time interval Δt, and
T_0 = datum temperature.

* The term "temperature-time factor" is also used for this quantity.

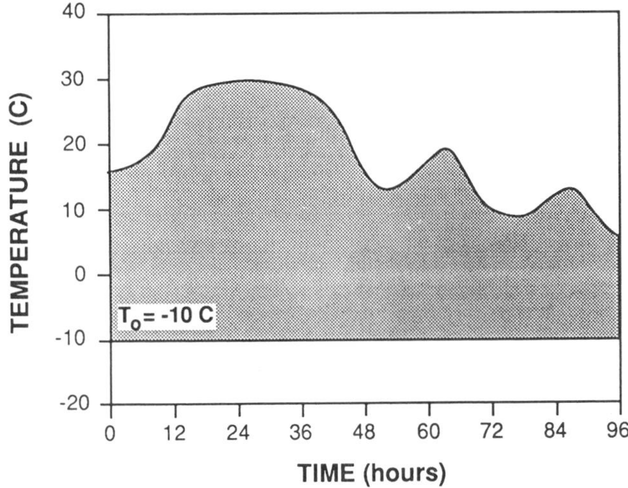

FIGURE 1. The Nurse-Saul maturity function.

This equation has become known as the Nurse-Saul function. In using Equation 1, only those time intervals in which the concrete temperature is greater than T_0 would be considered. If the concrete temperature is plotted vs. age, Equation 1 is simply equal to the area between the datum temperature and the temperature curve. This concept is demonstrated in Figure 1, which shows the temperature history at a particular point in a concrete member; the shaded area represents the maturity after 96 h.

Saul recognized that once concrete has set it will continue to harden (gain strength) at temperatures below 0°C (32°F). Thus, Saul recommended a datum temperature of − 10.5°C (13°F) for use in Equation 1. In 1956, Plowman[3] reported the results of a study designed to determine the temperature at which concrete, which has previously undergone setting, ceases to gain strength with time. Based on test data, Plowman suggested a value of − 12°C (11°F) for the datum temperature. Generally, a value of − 10°C (14°F) has been used in subsequent applications of the Nurse-Saul function.

In the 1951 paper, Saul presented the principle which has become known as the "maturity rule"[1]

> Concrete of the same mix at the same maturity (reckoned in temperature-time) has approximately the same strength whatever combination of temperature and time go to make up that maturity.

However, in applying the rule to steam-cured concrete, Saul noted that this principle was valid provided the concrete temperature did not reach about 50°C within the first 2 h or about 100°C within the first 6 h after the start of mixing. If the early-age temperature rise was excessive, the Nurse-Saul maturity function underestimated strength during the first few hours of treatment and the strength at later ages was adversely affected. Thus Saul recognized important limitations of the maturity rule and the Nurse-Saul maturity function. These limitations are discussed further in the next section.

The Nurse-Saul function can be used to convert a given temperature-time curing history to an equivalent age of curing at a reference temperature as follows:

$$t_e = \frac{\sum (T - T_0)}{(T_r - T_0)} \Delta t \tag{2}$$

where

t_e = the equivalent age at the reference temperature and
T_r = the reference temperature.

In this case equivalent age represents the duration of the curing period at the reference temperature which would result in the same value of maturity as the curing period at other temperatures. The equivalent age concept, originally introduced by Rastrup,[15] is a convenient method for using other functions besides Equation 1 to account for the combined effects of time and temperature on strength development.

Equation 2 can be written as follows:

$$t_e = \sum \alpha\ \Delta t \tag{3a}$$

where

$$\alpha = \frac{(T - T_0)}{(T_r - T_0)} \tag{3b}$$

The ratio α, which is called the "age conversion factor," has a simple interpretation: it converts a curing interval Δt to the equivalent curing interval at the standard reference temperature. For example, if $T_0 = -10°C$ (14°F), a curing period of 2 h at 43°C (109°F) is equivalent to $1.6 \times 2 = 3.2$ h at 23°C (73°F). Note that according to Equation 3b the age conversion factor is a linear function of the curing temperature. This is an important characteristic of the Nurse-Saul function and it helps explain some of its observed deficiencies.

Saul's introduction of the maturity rule was an outgrowth of studies dealing with accelerated curing. In 1953, Bergstrom[2] demonstrated that the maturity method was equally applicable for curing at normal temperatures. He used the maturity method to analyze previously published data on the effects of temperature on strength development. To calculate maturity, Bergstrom assumed that the temperatures of the concrete specimens were the same as the ambient curing temperatures. He found that, for a given concrete, there was generally little deviation of the strength data from a common strength vs. maturity curve.

In 1954, Rastrup[15] proposed the following function for equivalent age, which was based on a "...well-known axiom from physical chemistry which states: the reaction velocity is doubled if the temperature... is increased by 10°C."

$$t_e = \sum 2^{(T - T_r)/10} \Delta t \tag{4}$$

For example, a curing period of 2 h at 43°C (109°F) is equivalent to 8 h at 23°C (73°F). However, Wastlund[16] reported that subsequent studies at the Swedish Cement and Concrete Research Institute showed that over a wide temperature range the Rastrup function was not as accurate as the Nurse-Saul function in representing the effects of time and temperature.

In 1956, Plowman[3] presented a landmark paper on the maturity method. He made standard concrete cubes which were cured at temperatures varying between −11.5 and 18°C (11 to 64°F). Cubes were tested at regular intervals, and he demonstrated that for each mixture there was a unique relationship between strength and maturity. An important detail in Plowman's procedure was that all specimens were initially cured at a normal curing temperature (16 to 19°C [61 to 66°F]) for 24 h before being exposed to the different curing temperatures. Thus, the early-age temperature histories of all specimens were identical and their long-term strengths were approximately equal. It was for this reason that Plowman was able to obtain unique strength-maturity relationships for each mixture.

In 1956, a series of papers dealing with the maturity method were presented at the first

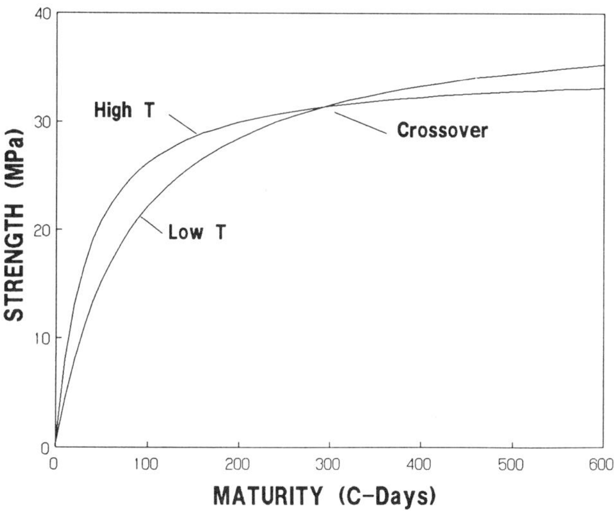

FIGURE 2. The effect of early-age curing temperature on the strength-maturity relationship (maturity based on Nurse-Saul function).

RILEM Symposium on Winter Concreting. McIntosh[4] reported on the results of a study in which specimens were exposed to different early-age temperatures. For equal maturities (based on the Nurse-Saul function), it was shown that specimens exposed to low early-age temperature were weaker at early maturities and stronger at later maturities than specimens exposed to a higher early-age temperature. It was concluded that a maturity function based on the product of time and temperature above a datum value cannot account for the "quality of cure" as affected by initial curing temperature. Two years later, Klieger[5] also noted that initial curing temperature influenced the shape of the strength-maturity relationship.

In 1962, Alexander and Taplin[6] reported the results of a study to determine whether strength gain of concrete and cement paste obeyed the maturity rule for curing at different temperatures (5, 21, and 42°C [41, 70, and 108°F]). Maturity was calculated using the Nurse-Saul function. In agreement with the previous observations of McIntosh and Klieger, they found that the curing temperature had systematic effects on the strength-maturity relationships of the pastes and concretes. The nature of the effects are summarized in a schematic fashion in Figure 2. At the same value of low maturity, a high curing temperature results in greater strength than a low curing temperature. Conversely, at later maturities, a high curing temperature results in lower strength. If Saul's maturity rule were correct, there should be a single strength-maturity curve.

In 1968, Verbeck and Helmuth[17] presented a qualitative explanation for the "cross-over effect" illustrated in Figure 2. They suggested that a higher initial temperature results in more than a proportional increase in the initial rate of hydration. Therefore, during the early stage of curing, when there is rapid strength development, the strength of concrete cured at the high temperature is greater than that of concrete cured at a lower temperature despite having the same maturity according to the Nurse-Saul function. However, with rapid hydration, reaction products do not have time to become uniformly distributed within the pores of the hardening paste. In addition, "shells" made up of low permeability hydration products build up around the cement grains. The nonuniform distribution of hydration products leads to more large pores which reduce strength and the shell impedes hydration of the unreacted portion of the grains at later ages. Thus the lower long-term strength under high curing temperature may be a result of the inability of the cement grains to continue hydrating due to the "shell" of low permeability reaction products.

In the late 1960s, there surfaced a new interest in the maturity method. Swenson[18] used the method in a failure investigation to estimate the concrete strength at the time of the collapse. In the U.S., Hudson and Steele[19,20] proposed the maturity method to predict the 28-day strength of concrete based on tests at early ages. Their results were later incorporated into an ASTM Standard.[21]

Weaver and Sadgrove[22] used the equivalent age concept to develop a manual for formwork removal times under various temperature conditions. They suggested the following new expression to calculate the equivalent age at 20°C (68°F):

$$t_e = \frac{\sum (T + 16)^2 \, \Delta t}{1296} \tag{5}$$

It was reported that Equation 5 gave better strength predictions at low maturity values than th Nurse-Saul function, Equation 2. However, in a later report,[23] Sadgrove noted that, at later maturities, the Nurse-Saul function was more accurate than Equation 5.

In the mid-1970s several reports from Canada appeared dealing with application of the maturity method under field conditions. Bickley[24] reported using it during slipforming of the C.N. Tower in Toronto, Mukherjee[25] reported its use for predicting the in-place strength of slabs, and Nisbett and Maitland[26] used the method for in-place strength predictions on a canal bypass project.

In 1960, Verbeck[27] suggested that the effects of temperature on the early rate of hydration of cement could be described by the Arrhenius equation. In 1977, Freiesleben Hansen and Pederson suggested the following expression for equivalent age based upon the Arrhenius equation:[28]

$$t_e = \sum_{0}^{t} e^{\frac{-E}{R}\left[\frac{1}{273 + T} - \frac{1}{273 + T_r}\right]} \Delta t \tag{6}$$

where

t_e = equivalent age at the reference curing temperature,
T = average temperature of concrete during time interval Δt, C,
T_r = reference temperature, C,
E = activation energy, J/mol, and
R = universal gas constant, 8.3144 J/(mol K).

In Equation 6, the exponential function is the age conversion factor and is a function of the absolute temperature. The exact shape of the curve describing the variation of the age conversion factor with temperature depends on the value of E, which according to Freiesleben Hansen and Pederson had the following values:

$$\text{for } T \geq 20°\text{C:} \quad E = 33500 \text{ J/mol} \tag{6a}$$

$$\text{for } T < 20°\text{C:} \quad E = 33500 + 1470\,(20 - T) \text{ J/mol} \tag{6b}$$

At this point it is helpful to compare the various age conversion factors which have been proposed for computing equivalent age. Figure 3 shows the age conversion factors given by Equations 3 through 6 for a reference temperature of 20°C (68°F). There are several notable observations. For temperatures below 20°C (68°F), the three nonlinear equations give similar results and the age conversion factors are lower than that obtained from the linear Nurse-Saul function. For temperatures above 20°C (68°F), the linear function yields

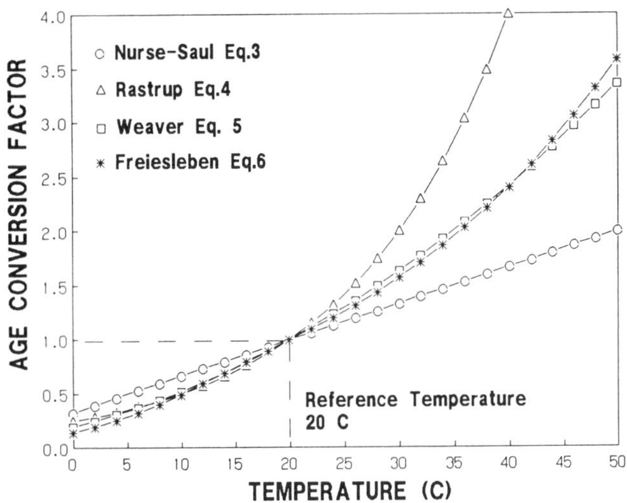

FIGURE 3. Various proposed functions to represent temperature dependence of the age conversion factor used to compute equivalent age.

the lowest age conversion factors, and the nonlinear functions diverge. Equations 5 and 6 give similar results which are significantly lower than the Rastrup function, Equation 4. Which of these functions is the most accurate? In independent studies of various maturity functions, Byfors[29] and Naik[30] demonstrated that, over a wide temperature range, the function based on the Arrhenius equation, Equation 6, was best able to account for the effects of temperature on strength gain. The next section of this chapter discusses the basis for the maturity method and will explain why Equation 6 has been found to be superior to other maturity functions.

As a result of its investigations of construction failures involving concrete, the U.S. National Bureau of Standards* (NBS) undertook research on the application of the maturity method as a tool for estimating in-place strength of concrete at early ages. Studies at the NBS demonstrated that the Nurse-Saul function was applicable under certain conditions, but they also showed that it had deficiencies. Lew and Reichard[31] reported that the Nurse-Saul function could be used to estimate the development of other mechanical properties of concrete besides compressive strength. The development of indirect tensile strength, modulus of elasticity and pullout bond strength of steel bars under different curing temperatures could be related to maturity. The initial concrete temperature was constant for all specimens. Specimens were then moved into the different temperature chambers soon after molding had been completed.

In a later study at NBS,[32] the applicability of the maturity method under simulated field conditions was investigated. Three different concrete mixtures were used to fabricate slabs containing push-out cylinder molds.** In addition, push-out cylinder molds were filled with concrete and stored in a moist curing room. The slabs were cured outdoors (during the spring). The objective was to determine whether the strength-maturity relationships for the field-cured push-out cylinders were the same as those for the companion laboratory-cured cylinders. The results of this study were perplexing: for one mixture there was good agreement between the strengths at equal maturities of field-cured and lab-cured specimens. For the other two mixtures there were significant differences. Examination of the temperature his-

* Name has been changed to the National Institute of Standards and Technology.

** See ASTM C 873-85, Test Method for Compressive Strength of Concrete Cylinders Cast in Place in Cylindrical Molds, ASTM Annual Book of Standards, Vol. 04.02.

tories of all specimens revealed that, for those two mixtures, the outdoor-cured specimens experienced different early-age temperatures than the lab-cured specimens. Specimens with higher early-age temperatures resulted in lower long-term strength. Therefore, a unique strength-maturity relationship did not exist for a single concrete mixture. Thus the "crossover" effect reported in 1956 by McIntosh[4] resurfaced. Subsequent tests at the NBS confirmed the importance of early-age temperature on the resulting strength-maturity relationship, and it appeared that "early age" could be as early as the first 6 h.

NBS later embarked on a fundamental study of the maturity method to gain a basic understanding of the cause of the "crossover" effect and to develop alternative procedures to eliminate the problem.[33-36] The findings of this study are reviewed in the section dealing with the theoretical basis of the maturity method, where the inherent deficiencies of the maturity rule and the Nurse-Saul function are explained.

STRENGTH-MATURITY RELATIONSHIPS

The previous sub-section has reviewed functions proposed to estimate maturity based on the temperature history of the concrete. Proposed functions to relate concrete strength to the maturity value are considered next. In this discussion the term "maturity" refers to either a temperature-time factor computed using the Nurse-Saul function or an equivalent age computed using any of the maturity functions.

In 1956, Nykanen[37] proposed an exponential strength-maturity relationship as follows:

$$S = S_\infty (1 - e^{-kM}) \tag{7}$$

where

S	=	compressive strength
S_∞	=	limiting compressive strength,
M	=	maturity, and
k	=	a constant

The limiting compressive strength would be a function of the water-cement ratio. The constant k is related to the initial rate of strength development. According to Nykanen, the value of k would be expected to depend on the water-cement ratio and the type of cement.

Plowman[3] observed that when strength was plotted as a function of the logarithm of maturity (based on the Nurse-Saul function) the data fell very close to a straight line. Therefore, he suggested the following empirical strength-maturity relationship:

$$S = a + b \log(M) \tag{8}$$

The constants *a* and *b* are also related to the water-cement ratio of the concrete and the type of cement. The publication of this relationship led to controversial discussions.[38] The following major points were raised, indicating the limitations of Plowman's proposal:

1. The relationship predicts ever increasing strength with increasing maturity.
2. The linear relationship is not valid at very early maturities and only intermediate maturity values result in an approximately linear relationship between strength and the logarithm of maturity.

In 1956, Bernhardt[39] proposed a hyperbolic strength-maturity relationship. A similar function was independently proposed by Goral[40] to describe the development of strength with age. Later, Committee 229 of the American Concrete Institute proposed the same function to estimate concrete strength at different ages.[41] In 1971, Chin[42] proposed the same function and described a procedure to evaluate the function for given data. The hyperbolic strength-maturity function can be expressed in the following form:

$$S = \frac{M}{\frac{1}{A} + \frac{M}{S_\infty}} \tag{9}$$

where

M	=	maturity,
S_∞	=	limiting strength, and
A	=	initial slope of strength-maturity curve.

The shape of this curve is controlled by the value of the initial slope. Figure 4A shows two curves which obey Equation 9, but have different initial slopes. As the initial slope increases, strength approaches the limiting strength S_∞ more rapidly as maturity increases.

The hyperbolic equation can be transformed into the following linear equation:

$$\frac{1}{S} = \frac{1}{S_\infty} + \frac{1}{A}\frac{1}{M} \tag{10}$$

Thus, if test data obey the hyperbolic function, the data would lie on a straight line when the inverse of strength is plotted vs. the inverse of maturity. The intercept of the line equals the inverse of the limiting strength and the slope of the line equals the inverse of the initial slope of the hyperbolic curve. Figure 4B shows the straight lines representing the two hyperbolas in Figure 4A. Because the hyperbolas have the same limiting strength, the straight lines have the same intercept. Note that a steeper straight line corresponds to a lower rate of initial strength gain.

Chin demonstrated that the hyperbolic function fitted his data well,[43] and a similar good fit was obtained in one of the NBS studies.[32] However, the author[44] pointed out that Equation 9 is not a very good representation of the strength-maturity function at low values of maturity. The reason is because Equation 9 assumes that strength development begins at $M = 0$. In reality, strength development does not begin until after setting has occurred. Therefore, following an idea first suggested by McIntosh,[4] the author introduced an ''offset'' maturity, M_0, to account for the fact that strength development does not begin until a finite value of maturity has been reached.[33] Thus Equation 9 is modified into the following form:

$$S = \frac{(M - M_0)}{\frac{1}{A} + \frac{(M - M_0)}{S_\infty}} \tag{11}$$

Figure 4C is a plot of Equation 11. The hyperbola in Figure 4C is similar to curve 1 in Figure 4A, except that the curve has been shifted to the right by a value equal to M_0. The linear transformation of Equation 11 is similar to Equation 10 except that the term $M - M_0$ replaces M. If the offset maturity is not used, actual strength-maturity data at low maturities (generally corresponding to strengths below one-half the limiting strength) will not follow a straight line when the inverse plot is used.[44]

Another strength-maturity relationship was proposed by Lew and Reichard.[45] Based on their test data and previously published data, they recommended the following empirical strength-maturity relationship:

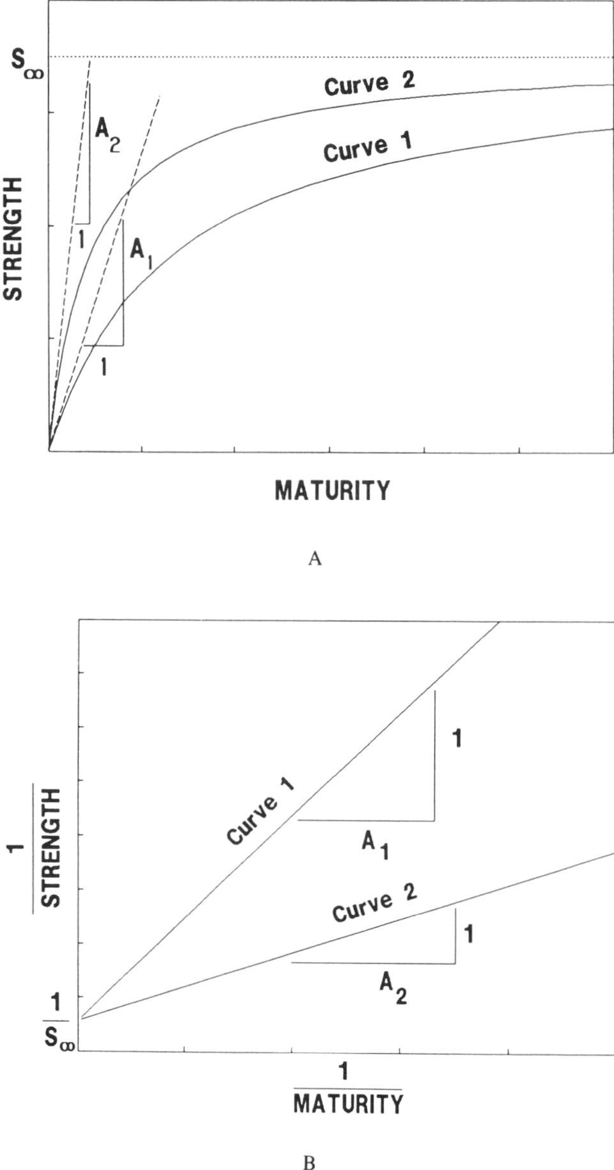

FIGURE 4. (A) Hyperbolic strength-maturity relationships having different initial slopes but the same limiting strength. (B) Plotting the reciprocal of strength vs. the reciprocal of maturity transforms hyperbolic curves in Figure 4A into straight lines. (C) Hyperbolic strength-maturity relationship using an offset maturity.

$$S = \frac{K}{1 + D\,[\log\,(M - 16.7)]^{b}} \tag{12}$$

This function was proposed as an improvement of Plowman's equation (Equation 8) in that strength approached a limiting value with increasing maturity. The coefficient b was found to vary between -1.5 and -4.3, depending on the water-cement ratio and type of cement. The coefficient D, which is related to the rate of strength gain, was also found to depend on water-cement ratio and type of cement. The coefficient K is the limiting strength, which

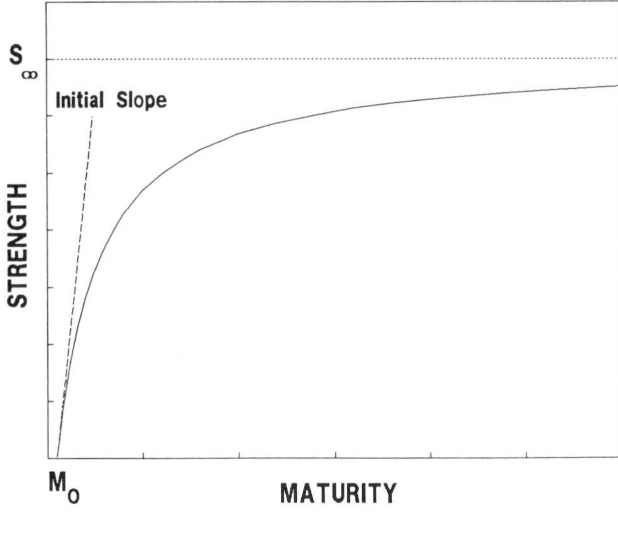

FIGURE 4C

was of course found to be strongly dependent on the water-cement ratio and to a lesser degree upon the type of cement. The three coefficients for Equation 12 were based upon using the Nurse-Saul maturity function with $T_0 = -12.2°C$ (10°F). An offset maturity, M_0, of 16.7°C-days (30°F-days) represented the maturity below which compressive strength was effectively zero. Plots were provided showing the variation of the coefficients for different water-cement ratios and cement types (ASTM Type I and III). It was suggested that a user could determine an approximate strength-maturity relationship for a particular concrete mixture by choosing the appropriate values of K, D and b from the published figures.

Freiesleben Hansen and Pederson[46] suggested that the strength-maturity relationship should be similar to the relationship between heat of hydration and maturity. They proposed the following empirical strength-maturity relationship:

$$S = S_\infty e^{-\left[\frac{\tau}{M}\right]^a} \tag{13}$$

where

S_∞	=	limiting strength,
M	=	maturity,
τ	=	characteristic time constant, and
a	=	shape parameter.

This relationship attempts to model the early part of the strength gain curve of concrete in that rapid strength development begins after a certain maturity has been attained. Figure 5 shows three curves which obey Equation 13, but have different values of the time constant and shape parameter. To accentuate the differences among the curves, strength is plotted as a function of the logarithm of maturity. Curve 2 has the same value of a as curve 1 but has a higher value of τ. Curve 3 has the same value of τ as curve 1 but has a higher value of a. The time constant has the following significance: when the maturity equals the value of τ, the strength $= S_\infty/e \approx 0.37\ S_\infty$. It is seen that changing the value of the time constant preserves the same general shape of the curve while shifting it to the left or right. Changing the value of the shape parameter alters the shape of the curve. As the value of a increases, the curve has a more pronounced S-shape, as shown by curve 3.

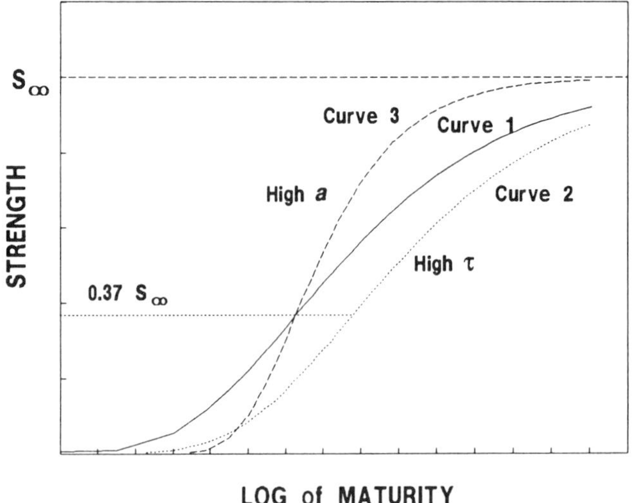

FIGURE 5. Effect of the time constant *(a)* and shape parameter (τ) on the strength-maturity relationship given by Equation 13.

TABLE 1
Compressive Strength vs. Age at 23°C (W/C = 0.45, Type I Cement)

Age (days)	Compressive strength[a] (MPa)	(psi)
0.45	4.10	595
0.98	13.52	1960
2.06	19.55	2835
4.33	27.24	3950
7.11	31.86	4620
14.11	36.48	5290
28.06	41.93	6080

[a] 100 × 200 mm (4 × 8 in.) cylinders.

Which of these various strength-maturity relationships best represents actual strength gain data? Table 1 lists compressive strength vs. age data for concrete cylinders cured in a water bath maintained at 23°C (73°F). The mixture had a water-cement ratio of 0.45 and was made with Type I cement. These data were used to obtain the best-fit curves for the strength maturity relationships that have been discussed. In using the relationship proposed by Lew and Reichard, the value of the offset maturity was chosen to produce the best least-squares fit. Maturity was defined to be the equivalent age at 23°C (73°F). Therefore, the ages in Table 1 are equivalent ages at 23°C (73°F). The strength data and the various best-fit curves are plotted in Figure 6. It is seen that for strength below about 50% of the limiting strength, all the curves accurately describe the strength development. However, at higher strength levels the relationships proposed by Nykanen, Equation 7, and by Plowman, Equation 8, do not model strength gain as accurately as the others. For the data in Table 1, Equations 11, 12, and 13 are accurate representations of the actual strength gain, and the only differences among them are the values of the limiting strength.

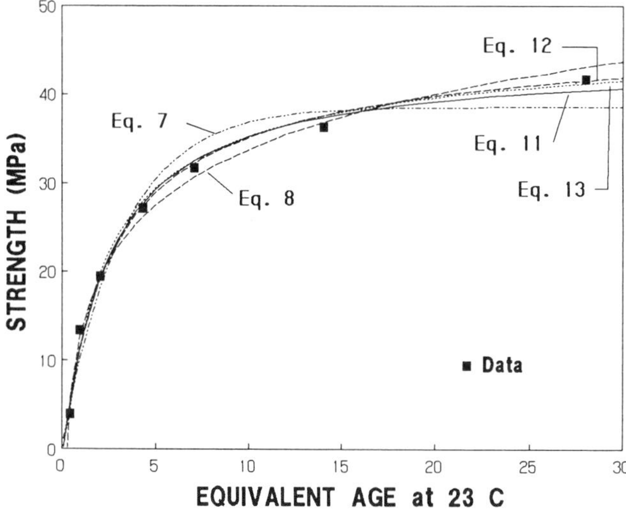

FIGURE 6. Comparison of various strength-maturity relationships with measured compressive strength of cylinders.

SUMMARY

This section has presented a brief review of the history of the maturity method. Various maturity functions have been proposed to account for the effects of time and temperature on the strength development of concrete. It is now generally recognized that the traditional Nurse-Saul function does not accurately represent time-temperature effects. Other maturity functions have been proposed which are more accurate than the traditional function. These functions are convenient to apply if the equivalent age approach is used to represent maturity.

This section has also reviewed some of the proposed functions to represent the relationship between maturity and strength development. These relationships are useful in computer applications of the maturity method, where the functions are used to estimate the strength based on the measured maturity of the in-place concrete. Regardless of which relationship is chosen, the key point is that the coefficients which define the exact shape of the curves depend on the particular concrete mixture.

The next section presents a theoretical basis for the maturity method. At the same time, it will explain why some maturity functions are superior to others.

THEORETICAL BASIS

The Nurse-Saul function was proposed as an approximate method to account for the effects of time and temperature on concrete strength, but its theoretical basis was not explained. Its applicability was based upon empirical evidence. In this section, the general form of the maturity function is derived, and it is shown that Equation 1 arises by assuming a specific type of behavior. It is also shown why other maturity functions are superior to the Nurse-Saul function.

STRENGTH GAIN OF CONCRETE

The development of a mathematical expression to describe the compressive strength development of concrete is discussed first. The derivation follows ideas originally presented in 1956 by Bernhardt.[39] The rate of strength gain (dS/dt) at any age (t) is assumed to be a function of the current strength (S) and the temperature (T), i.e.,

$$\frac{dS}{dt} = f(S) \cdot k(T) \tag{14}$$

where

f(S)	=	a function of strength and
k(T)	=	a function of temperature.

Based on empirical evidence, Bernhardt proposed that

$$f(S) = S_\infty \left[1 - \frac{S}{S_\infty} \right]^2 \tag{15}$$

where

S_∞	=	the limiting strength at infinite age.

Note that when strength begins to develop f(S) = S_∞. Therefore, the initial rate of strength development is

$$\left. \frac{dS}{dt} \right|_{S=0} = S_\infty \, k(T)$$

The temperature function k(T) is called the *rate constant* because it affects the initial rate of strength development.

If S_∞ is assumed to be independent of curing temperature and Equation 14 is combined with Equation 15, the following integral equation is obtained:

$$\int_0^S \frac{dS}{\left[(1 - \frac{S}{S_\infty}) \right]^2} = S_\infty \int_{t0}^{t} k(T) \, dt \tag{16}$$

Equation 16 departs from Bernhardt's original derivation by introducing the condition that strength gain does not begin until some time, t_0, after mixing. This approximation is introduced to account for the induction period between initial mixing and the start of strength gain described by Equation 14.

Note that the integral on the right side of Equation 16 involves the product of temperature and time. This integral is the general form of the maturity function, and it will be denoted as M(t,T):

$$M(t, T) = \int_{t_0}^{t} k(T) \, dt \tag{17}$$

Integrating the left side of Equation 16 and rearranging terms, one obtains the following general strength-maturity relationship:

$$S = S_\infty \frac{M(t,T)}{1 + M(t,T)} \tag{18}$$

Comparing Equation 18 with Equation 9, one sees that they have similar forms. Hence there is a basis for the hyperbolic strength-maturity function proposed by Chin.[42]

Let us now take a closer look at the nature of the maturity function by considering strength development under constant and variable temperature curing.

MATURITY FUNCTIONS

Isothermal conditions — When the curing temperature is constant, the temperature function k(T) has a constant value. Therefore, the maturity function becomes:

$$M(t,T) = k_T\,(t - t_0) \tag{19}$$

where k_T = value of the rate constant at the curing temperature.

The strength-maturity relationship, Equation 18, becomes the following strength-age relationship:

$$S = S_\infty \frac{k_T(t - t_0)}{1 + k_T(t - t_0)} \tag{20}$$

A plot of Equation 19 would be identical to the hyperbolic curve shown in Figure 4C except that the units for the X-axis would be age rather than maturity. An interesting feature of the hyperbolic curve is that the inverse of the rate constant ($1/k_T$) equals the time beyond t_0 needed to reach 50% of the limiting strength.

An expression similar to Equation 20 was elegantly derived by Knudsen[47,48] using a completely different approach from that given here. He worked with the degree of hydration of cement rather than concrete strength. He considered the reaction kinetics of the individual cement grains and the particle size distribution of the grains. The key assumptions in Knudsen's derivation are as follows:

1. All cement particles are chemically similar and need be classified only according to their size.
2. The cement particles react independently.
3. The particle size distribution is described by an exponential equation.
4. The kinetic equation for hydration of each particle is also described by an exponential equation.

Knudsen called the results of his derivation the "dispersion model" because the particle size distribution of the cement grains was found to play a dominant role in the overall hydration behavior. According to Knudsen,[47-49] Equation 20 is valid when the individual particles obey linear kinetics.*

Knudsen demonstrated[47,48] that the hyperbolic equation given by Equation 20 is valid for strength development and any other property of concrete that is directly related to the extent of cement hydration. In addition, he showed that the rate constant, k_T, depends upon the particle size distribution of the cement and the rate constant for each particle (which is temperature dependent). The importance of the particle size distribution on the rate constant was also discussed by Bezjak and Jelenic.[50]

Knudsen's assumption that cement particles react independently is significant. As is well known, hydration products form in the water-filled spaces between cement particles. As the water-cement ratio is lowered, the distance between cement particles is reduced. Therefore,

* This ratio has also been called the "affinity" ratio.[50]

particle interference increases and one would expect the hydration rate to decrease. Thus, Knudsen[47] noted that the assumption of particle independence would be expected to be violated at very low water-cement ratios. However, Copeland and Kantro[51] showed that for a low water-cement ratio, the effects of particle interference on hydration are not significant at early ages. Knudsen concluded that the assumption of independent particle reaction was not seriously violated at a water-cement ratio as low as 0.4.[47] Thus it is expected that the rate constant should be independent of the water-cement ratio during the early stages of hydration.

In summary, the hyperbolic curve given by Equation 20 appears to be a soundly based approximation of the strength development of concrete under constant curing temperature. As was shown in Equation 17, the integral of the rate constant with respect to time is the general form of the maturity function. The following section examines how the temperature dependence of the rate constant affects the specific form of the maturity function when the curing temperature is not constant.

Variable temperature conditions — When the curing temperature is not constant, the temperature function k(T) is not constant. Therefore, the temperature dependence of the rate constant must be considered. The simplest case is to assume a linear relationship, that is,

$$k(T) = K(T - T_0) \tag{21}$$

where

T_0 = temperature corresponding to k(T) = 0, and
K = slope of the straight line.

Substituting Equation 21 into Equation 17, the maturity function can now be expressed as

$$M(t,T) = \int_{t_0}^{t} K(T - T_0)\,dt = K \int_{0}^{t} (T - T_0)\,dt - K \int_{0}^{t_0} (T - T_0)\,dt \tag{22}$$

The two terms on the right hand side of Equation 22 are time-temperature factors based on the Nurse-Saul function, Equation 1. If these terms are called M and M_0, the maturity function is

$$M(t,T) = K\,(M - M_0) \tag{23}$$

Substituting Equation 23 into Equation 18, we obtain the following strength-maturity relationship

$$S = S_\infty \frac{K\,(M - M_0)}{1 + K\,(M - M_0)} \tag{24}$$

It can be shown that Equation 24 is identical to Equation 11 by making the substitution $A = K \cdot S_\infty$

The above derivation shows that if the rate constant, k(T), is assumed to vary linearly with the curing temperature, the resulting maturity function equals the traditional Nurse-Saul function. Thus the fundamental assumption of the Nurse-Saul function is revealed.

At this point, it is helpful to examine the relationship between equivalent age and the general maturity function. Equivalent age represents the age at a reference curing temperature which results in the same maturity as under the actual curing temperature. Using Equation 17 and performing the integration from t = 0, we obtain the following:

$$t_e\, k_r = \int_0^t k(T)\, dt \tag{25a}$$

$$t_e = \int_0^t \left[\frac{k(T)}{k_r}\right] dt \tag{25b}$$

where

t_e = equivalent age at the reference temperature, T_r, and
k_r = value of the rate constant at the reference temperature.

Hence, the ratio of value of the rate constant at any temperature to the value at the reference temperature equals α, the age conversion factor* previously mentioned. When the rate constant is a linear function of temperature, Equation 25b is similar to the previous Equation 2, which was based on the Nurse-Saul function.

However, there is no technical basis for the linear temperature dependence of the rate constant. Since hydration is an exothermic chemical reaction, it is reasonable to assume that the rate constant should vary with temperature according to the Arrhenius equation, i.e.,

$$k(T) = \beta\, e^{\left[\frac{-E}{R\,(273\, +\, T)}\right]} \tag{26}$$

where

β = a constant,
E = activation energy, and
R = universal gas constant

The age conversion factor obtained using Equation 26 is identical to the relationship proposed by Freiesleben Hansen and Pederson[28] which was previously given as Equation 6.

In summary, the mathematical form of the age conversion factor is dependent on the function used to represent the variation of the rate constant with temperature. Different rate constant vs. temperature functions account for the different curves in Figure 3. The next step is to examine how the actual variation of the rate constant with temperature compares with these proposed functions.

EXPERIMENTAL RESULTS

Strength-age data for mortar cubes cured at constant temperatures of approximately 5, 12, 23, 32, and 43°C (41, 54, 73, 90, and 109°F) are used to compare the temperature dependence of the rate constant with the linear and Arrhenius functions. Two mortar mixtures, having water-cement ratios of 0.43 and 0.56, were used to prepare 51-mm (2-in.) cubes. The specimens were prepared in an environmental chamber at the above curing temperatures, and the molded cubes were stored in water baths maintained within ±1°C (2°F) of the nominal values. For each curing temperature, three cubes were tested in compression at each of seven test ages. Complete test results are given elsewhere.[33,36]

The strength-age data for each curing temperature were fitted by the hyperbolic relationship given by Equation 20, which can be transformed into the following linear equation:

* Parabolic kinetics applies when hydration is controlled by diffusion. For parabolic kinetics, Equation 20 is modified by substituting the term $(t\text{-}t_0)^{1/2}$ in place of $(t\text{-}t_0)$.

$$\frac{1}{S} = \frac{1}{S_\infty} + \frac{1}{k_T S_\infty} \frac{1}{(t - t_0)} \tag{27}$$

There are several approaches to estimate the values of the three parameters (S_∞, k_T, and t_0). A general least-squares fit computer program can be used to evaluate all three parameters.[53] This procedure was used in Reference 33. Another approach involves trial and error. A value t_0 is assumed and the valued of S_∞ and k_T are evaluated using linear regression analysis of the transformed data. The procedure is repeated for different values of t_0 until the best fit is obtained. A third alternative was suggested by Knudsen[47] and, because of its practicality, it is discussed here.

First, the limiting strength, S_∞, is evaluated by considering the strength-age data at later ages. For later age data, we can make the approximation $(t - t_0) \approx t$, and Equation 27 can be rewritten as follows:

$$\frac{1}{S} = \frac{1}{S_\infty} + \frac{1}{k_T S_\infty} \frac{1}{t} \tag{28}$$

Thus a plot of 1/S vs. 1/t is a straight line. The intercept of the line is obtained from linear regression analysis, and the inverse of the intercept equals the limiting strength. It was found[36] that strength data for equivalent ages beyond about 7 days should be used for this step.

Using the estimated the value of S_∞, Equation 20 can be written in the following form to estimate k_T and t_0:

$$\frac{S}{S_\infty - S} = -k_T t_0 + k_T\, t \tag{29}$$

Thus a plot of $S/(S_\infty - S)$ vs. age is a straight line having a slope k_T, which can be obtained from regression analysis. The value of t_0 is equal to the negative value of the intercept divided by the slope. It was found[36] that strength data for equivalent ages up to about 3 to 4 days should be used for this step to obtain the best estimates of k_T and t_0.

The complete results of the above two-step procedure are given in Reference 36, and the resulting values of k_T are listed in Table 2. Figure 7 shows the values of k_T in Table 2 plotted as a function of the curing temperature. It is seen that over the temperature range 5 to 43°C (41 to 109°F) the rate constant is a nonlinear function of temperature. These particular data show that water-cement ratio does not have a significant effect on the rate constant, which is consistent with Knudsen's findings.[47]

First, consider whether the Arrhenius equation, Equation 26, describes the variation of the rate constant with curing temperature. By taking the natural logarithm of Equation 26, the following is obtained:

$$\ell n\, k(T) = \ell n\, \beta - \frac{E}{R} \frac{1}{(T + 273)} \tag{30}$$

Thus the Arrhenius equation is transformed into a linear relationship. The negative of the slope of the straight line is equal to the activation energy divided by the gas constant. Figure 8 shows the natural logarithms of the rate constant values in Table 2 plotted as a function of the reciprocal of the absolute temperature. Also shown is the best-fit straight line having the following equation:

$$\ell n\, k(T) = 16.65 - 5255 \frac{1}{273 + T} \tag{31}$$

TABLE 2
Values of Rate Constant Based on Compression Tests of Mortar Cubes

W/C = 0.43		W/C = 0.56	
Curing temperature (°C)	k_T (day^{-1})	Curing temperature (°C)	k_T (day^{-1})
5.5	0.116	5.5	0.087
13	0.184	12.5	0.203
23.5	0.341	23	0.313
32	0.571	32	0.577
43	1.031	43	0.944

From Carino, N. J., *ASTM J. Cem. Congr. Aggr.*, 6, 61, 1984. With permission.

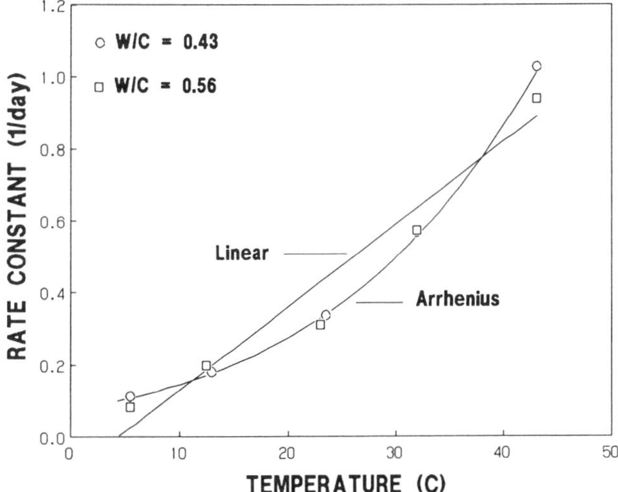

FIGURE 7. Variation of rate constant with temperature; results from tests compared with linear and Arrhenius equations.

The straight line fits the data quite well, so it is concluded that the Arrhenius equation provides a good representation of k(T) over the temperature range 5 to 43°C (41 to 109°F). Figure 7 shows the best-fit Arrhenius equation compared with the rate constant values.

Next, a straight line was fitted to the rate constant values. The best fit straight line was found to have the following equation:

$$k(T) = 0.023\ (T - 4.4°C) \tag{32}$$

This straight line is plotted in Figure 7. There are two significant observations. First, the straight line does not accurately represent the observed variation of the rate constant with curing temperature. Therefore, the fundamental assumption in the Nurse-Saul function is not valid for this particular data. Second, the best-fit straight line crosses the temperature axis at 4.4°C (40°F). Therefore, if the Nurse-Saul function is used to approximate the effects of temperature over the range 5 to 43°C (41 to 109°F), the best results would be obtained by using a datum temperature of 4.4°C (40°F) instead of −10°C (14°F) as is commonly used.

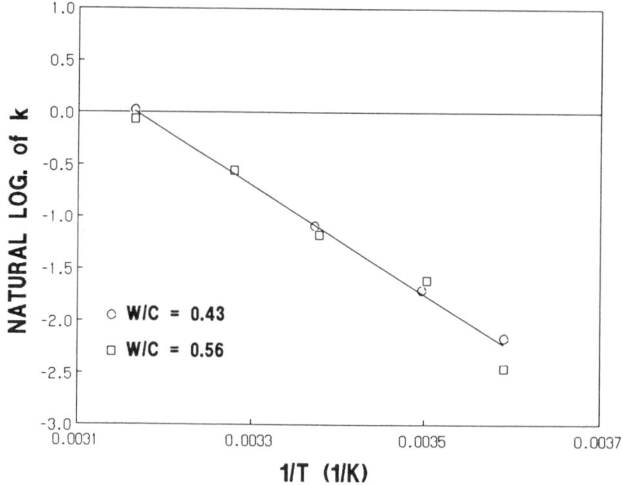

FIGURE 8. Logarithm of rate constant vs. the inverse of the absolute temperature (Arrhenius plot).

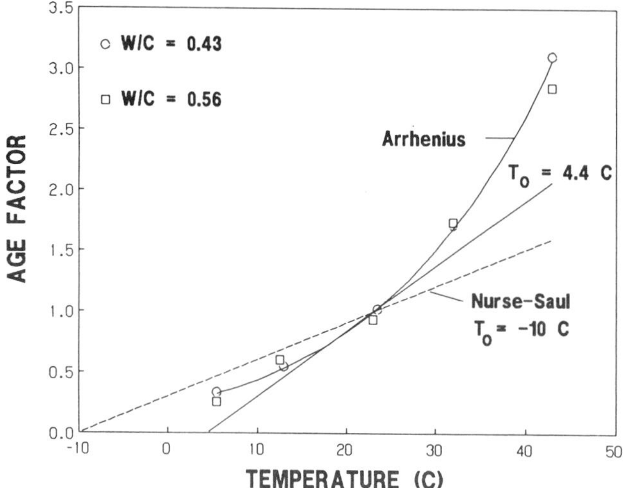

FIGURE 9. Variation of age factor with temperature.

The rate constant values in Table 2 can be converted to age conversion factors by dividing them by the value of the rate constant at the reference temperature. In this discussion the reference temperature is taken to be 23°C (73°F). From Equation 31, the rate constant is approximately 0.33 day^{-1} at 23°C (73°F). The values of the age factors are plotted vs. temperature in Figure 9. The age factor vs. temperature relationships based on various functions are also shown in Figure 9. The equation of the relationship based on the best-fit Arrhenius equation, Equation 31, is as follows:

$$\alpha = e^{-5255\left[\frac{1}{273 + T} - \frac{1}{296}\right]} \tag{33}$$

The solid straight line is the relationship based on the best fit straight line, Equation 32, and has the following equation:

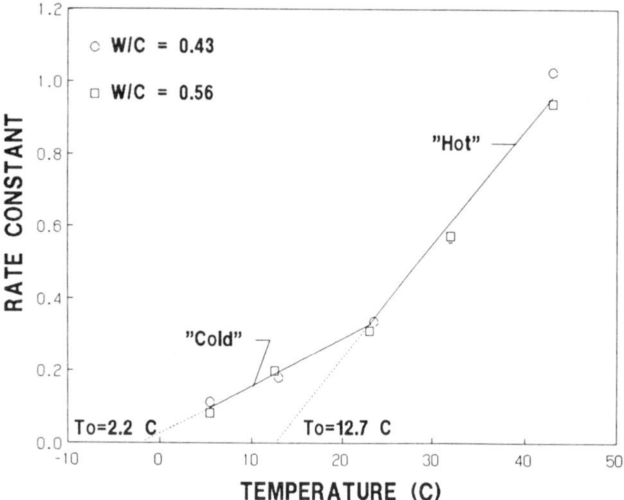

FIGURE 10. A bi-linear approximation of age factor vs. temperature relationship.

$$\alpha = \frac{(T - 4.4)}{18.6} \tag{34}$$

The dashed line is the linear age factor vs. temperature relationship for a datum temperature of -10°C, as commonly used with the traditional Nurse-Saul maturity function, and has the following equation:

$$\alpha = \frac{(T + 10)}{33} \tag{35}$$

Examination of Figure 9 shows that neither straight line is an accurate representation of the age factor vs. temperature relationship over the indicated temperature range. The line representing the Nurse-Saul function with $T_0 = -10$ C°(14°F) overestimates the age factor at temperatures below the reference temperature of 23°C (73°F). Thus using the traditional value of the datum temperature would overestimate the equivalent age. At temperatures above 23°C, the opposite is true, that is, the equivalent age would be underestimated. On the other hand, using the straight line with $T_0 = 4.4$°C always underestimates the equivalent age. The amount of underestimation increases as the temperature is further away from the reference temperature.

The results in Figure 9 show that a maturity function which assumes that the rate constant is a linear function of temperature will not accurately account for temperature effects when the curing temperature is significantly different from the reference temperature. If one wishes to use a linear maturity function because of its simplicity, accuracy can be improved greatly by using two values of the datum temperature. One value would be applicable when the curing temperature is below the reference temperature and the other value would be used above the reference temperature. Figure 10 illustrates this approach. It is seen that the variation of the rate constant for temperatures between 5 and 23°C (41 and 73°F) can be represented accurately by a straight line with a datum temperature of 2.2°C (36°F). Likewise, for temperatures between 23 and 43°C (73 and 109°F), a straight line with a datum temperature of 12.7°C (55°F) is a good approximation. These straight lines were forced to fit through a rate constant value of 0.33 day^{-1} at 23°C (73°F). The best values for the datum temperature

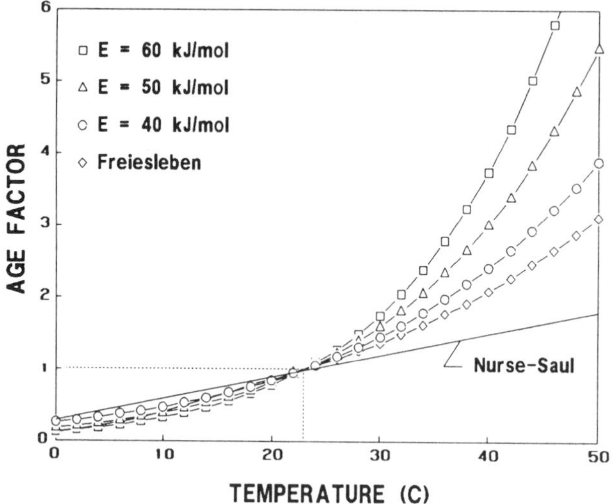

FIGURE 11. Effect of activation energy on the age factor vs. temperature relationship using the Arrhenius function.

are dependent on the actual shape of the rate constant vs. temperature function and the temperature range over which the linear approximation is made.[36]

VALUES OF ACTIVATION ENERGY

A good maturity function must accurately represent the effect of temperature on the age factor. In the previous section, it was shown that the Arrhenius function fitted experimental data well. Next, we examine the shape of the age factor vs. temperature relationship in more detail.

The value of the activation energy, E, in the Arrhenius equation, Equation 26, determines the shape of the age factor vs. temperature function. Figure 11 shows the variation of the age factor with temperature for activation energy values of 40, 50, and 60 kJ/mol and using the values suggested by Freiesleben Hansen and Pederson,[28] which were given in Equations 6a and 6b. The reference temperature is again taken as 23°C (73°F). The variation based on the Nurse-Saul function with $T_0 = -10$°C (14°F) is also shown. It is seen that with increasing value of activation energy, the age factor vs. temperature variation becomes more nonlinear. As a result, attempts to use a linear function as an approximation lead to greater errors for increasing values of activation energy. With this in mind, let us examine the values of activation energy which have been reported by others.

The value of the activation energy for a particular concrete can be determined in several ways. One approach is to make and cure concrete specimens at several different temperatures and analyze the strength-age data in the manner described above. However, there are alternative possibilities. It has been firmly established that the degree of hydration of cement correlates with the mechanical strength of concrete.[54,55] Thus it is possible to determine the activation energy from hydration studies of cement pastes. This approach is supported by the work of others[52,56,57] who have shown that the activation energies based upon heats of hydration are the same as those based upon the mechanical strength of mortars. In addition, Bresson[58] and Gauthier and Regourd[56] report that the same value of activation energy is obtained from strength tests of mortar specimens as from tests of concrete specimens. This was confirmed in a limited study by the author[36] and in an extensive study by Tank.[59] Another approach is to measure the chemical shrinkage of cement pastes.[60,61] Thus there is significant evidence showing that the activation energy for strength development can be determined without testing bulky, concrete specimens.

TABLE 3
Activation Energy Values for Strength Development

Cement type	Type of test	Activation energy, kJ/mol	Ref.
Type I (mortar)	Strength	42	33
Type I (mortar)	Strength	44	36
Type I (concrete)	Strength	41	
PC[a] (paste)	Heat of hydration	42—47	56, 57
PC + 70% BFS[a] (paste)	Heat of hydration	56	
PC (paste)	Chemical shrinkage	61	60
RHC[a] (paste)	Chemical shrinkage	57	
PC (paste)	Chemical shrinkage	67	61
Type I/II (paste)	Heat of hydration	44	62
Type I/II + 50% BFS (paste)	Heat of hydration	49	62

[a] PC = ordinary portland cement. BFS = blast furnace slag. RHC = rapid hardening cement.

Presently there is not much published data on the activation energy for strength development of concrete. Table 3 lists some of the activation energies that have been reported for different types of cement and using different test methods. The results for ordinary portland cement (ASTM Type I) are in reasonable agreement except for the higher values reported by Geiker.[60,61] Except for Freiesleben Hansen and Pederson,[28] no other researchers mentioned in Table 3 have suggested that the activation energy is a function of temperature.

Tank[59] conducted an extensive study of the isothermal strength development in concrete and mortar specimens made with different cementitious systems and having two water-cement ratios. The study had several objectives:

1. Verify the applicability of the hyperbolic strength-age model, Equation 20, for strength gain under constant temperature curing.
2. Confirm whether tests of mortar specimens result in the same activation energy values as tests of concrete.
3. Determine whether water-cement ratio affects the values of activation energy.
4. Determine the effects of admixtures and cement replacement on the activation energies.

Specimens were cured in constant temperature water baths at 10, 23, and 40°C (50, 73, and 104°F), and strength tests were performed at regular age intervals. The mortars had cement to sand proportions which were the same as the cement to coarse aggregate proportions of the corresponding concrete.

Tank found that the test data for concrete and mortar specimens obeyed the hyperbolic strength-age model, and the model was used to evaluate the rate constants. Several rate constant vs. temperature functions were examined, including the Arrhenius equation. Table 4 summarizes the values of activation energies for the Arrhenius equation. In Figure 12A, the activation energy values obtained from mortar tests are compared with the values obtained from concrete tests. The straight line is the line of equality. It is seen that the results are uniformly distributed about the line. Thus Tank verified that the activation energy for a concrete mixture can be obtained from the strength-gain data of mortar cubes.

Figure 12B compares the activation energies for the two water-cement ratios. Again, the straight line is the line of equality. In this case, the results are not so clear. For some mixtures, there was no effect of water-cement ratio, which is in agreement with previous findings.[36,47] However, with Type I and Type II cement, the low water-cement ratio mixtures

TABLE 4
Activation Energy Values Based on Compressive Strength Tests of Concrete Cylinders and Mortar Cubes

	Activation energy, kJ/mol			
	W/C = 0.45		W/C = 0.60	
Cement type	**Concrete**	**Mortar**	**Concrete**	**Mortar**
I	61	62	46	44
II	51	55	43	42
III	44	40	43	42
I + 20% Fly ash	30	32	31	36
I + 50% Slag	46	44	55	51
I + Accelerator	46	54	49	51
I + Retarder	39	42	39	34

From Tank, R. C., Ph.D. dissertation. With permission.

had significantly higher activation energies. On the other hand, the mixtures with Type I cement plus 50% slag (by mass) had higher values for the high water-cement ratio. Clearly, more information is required to establish how water-cement ratio relates to the activation energy.

Comparing the activation energies in Table 4 for the mixtures containing Type I cement it is seen that the addition of admixtures or cement replacements alters the activation energy of the concrete. Thus, when admixtures and cementitious additions (fly ash, slag, or silica fume) are used, it is necessary to determine the activation energies for the particular combinations of ingredients in the concrete to be used in construction.

Tank[59] examined the following alternative relationship to the Arrhenius equation for describing the variation of the rate constant with temperature:

$$k(T) = A\, e^{BT} \tag{36}$$

where

A = a constant, day^{-1}, and
B = temperature sensitivity factor, $°C^{-1}$

The constant A represents the value of the rate constant at T = 0. The value of B indicates the sensitivity of the rate constant to changes in curing temperature. For a temperature increase equal to 1/B, the rate constant increases by a value of approximately 2.72.

Tank found that Equation 36 fitted his results as well as the Arrhenius equation. He also found an empirical relationship between the activation energy and the temperature sensitivity factor:

$$B = 0.00135\, E \tag{37}$$

In Equation 37, E is expressed in units of kJ/mol. For example, if the activation energy is 45 kJ/mol, the corresponding value of B is $0.061°C^{-1}$, and the rate constant increases by a factor of 2.72 for an increase of 16.4°C in the curing temperature.

Using Equation 36, the expression for the age factor becomes the following:

$$\alpha = e^{B(T - T_r)} \tag{38}$$

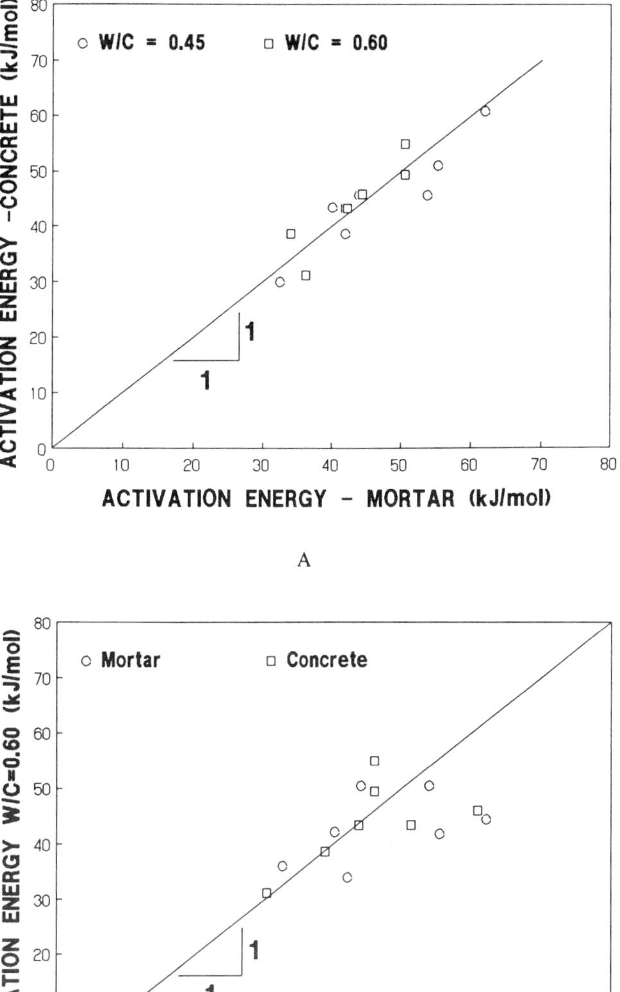

FIGURE 12. (A) Activation energy obtained from testing concrete cylinders compared with values obtained from testing mortar cubes. (B) Effect of water-cement ratio on activation energy.

Equation 38 is simpler to use than Equation 6 because one does not have to deal with the inverse of the absolute temperature.

This section has reviewed our understanding of values of activation energy which describe the age factor vs. temperature relationship based upon the Arrhenius equation. Due to the scarcity of data, there does not yet exist a thorough understanding of the factors which affect the activation energy. As more data are accumulated, it may be possible to predict the activation energy of a particular cement based upon its chemical composition. When admixtures or cementitious additions are used, their effects on the activation energy must also be determined. This can probably only be done by testing the combinations of cement and admixtures that will be used.

When the accuracy of strength predictions is not crucial, typical published values of activation energy can be used, such as those given in Tables 3 and 4. When maximum accuracy is desired, the value of activation energy for the particular combination of ingredients should be determined from testing. The applicable procedures are given in a subsequent section of this chapter.

RELATIVE STRENGTH GAIN

The maturity rule, as proposed by Saul,[1] states that samples of the same concrete will have equal strengths if they have equal maturities, irrespective of their temperature histories. In other words, there exists a unique strength-maturity relationship for a given concrete mixture. This section discusses why the traditional maturity rule is an approximation.

Up to now, it has been assumed that the limiting strength of a concrete mixture subjected to continuous curing is not affected by the early-age temperature history. However, it has been shown that higher early-age temperatures lower the limiting strength.[32,63-67] To illustrate the nature of this problem, Table 5 represents strength-gain data for concrete cylinders cured at three different temperatures. The complete analysis of the data was reported in Reference 36. The data are plotted in Figure 13A along with the best-fit hyperbolic curves. It is clear that the curing temperatures affect not only the initial rate of strength development but also the limiting strength. Table 5 shows the limiting strengths and rate constants obtained by regression analysis of the data.

By fitting the Arrhenius equation (Equation 26) to the rate constant values in Table 5, one finds that the estimated value of the rate constant at a reference temperature of 23°C (73°F) equals 0.47 day^{-1}. The ages in Table 5 can be converted to equivalent ages at 23°C (73°F). Plotting strength vs. equivalent age to compare strength values at equal maturities, one obtains the results in Figure 13B. This plot clearly shows that the concrete does not have a unique strength-maturity relationship. Thus, the "maturity rule" will result in erroneous strength estimates, especially at later ages. At this point, the reader might conclude that the maturity method is not very useful. However, this is not so.

The hyperbolic strength gain equation, Equation 20, can be converted to a relative strength gain equation:

$$\frac{S}{S_\infty} = \frac{k_T(t - t_0)}{1 + k_T(t - t_0)} \tag{39}$$

Introducing the age conversion factor

$$\alpha = \frac{k_T}{k_r} \tag{40}$$

the relative strength gain can be expressed in terms of equivalent age:

$$\frac{S}{S_\infty} = \frac{k_r\alpha(t - t_0)}{1 + k_r\alpha(t - t_0)} \tag{41}$$

Equation 41 states that there should be a unique relative strength vs. equivalent age curve having an initial slope equal to k_r.* To test the validity of this assertion, the strength values in Table 5 were converted to relative strength by dividing them by the corresponding limiting strength. Figure 13C is a plot of the relative strength values vs. equivalent age. The data points are indeed grouped around a single curve, whose equation is

* This has been called the "rate constant model" by Tank.[59]

TABLE 5
Cylinder Strength vs. Age for Constant Temperature Curing

12°C			21°C			32°C		
Age (days)	Strength (psi)	Strength (MPa)	Age (days)	Strength (psi)	Strength (MPa)	Age (days)	Strength (psi)	Strength (MPa)
0.92	705	4.86	0.51	510	3.52	0.37	980	6.76
1.83	1935	13.34	1.02	1680	11.59	0.79	1925	13.28
5.45	3710	25.59	3.01	3075	21.21	1.90	3040	20.97
12.07	4720	32.55	7.00	4210	29.03	5.00	4045	27.90
25.97	5390	37.17	13.20	4990	34.41	10.00	4530	31.24
46.02	6075	41.90	28.00	5490	37.86	20.76	4770	32.90
S_∞ =	6470	44.62		6060	41.79		5080	35.03
k =	0.265 day^{-1}			0.361 day^{-1}			0.815 day^{-1}	

From Carino, N. J., *ASTM J. Cem. Concr. Aggr.*, 6, 61, 1984. With permission.

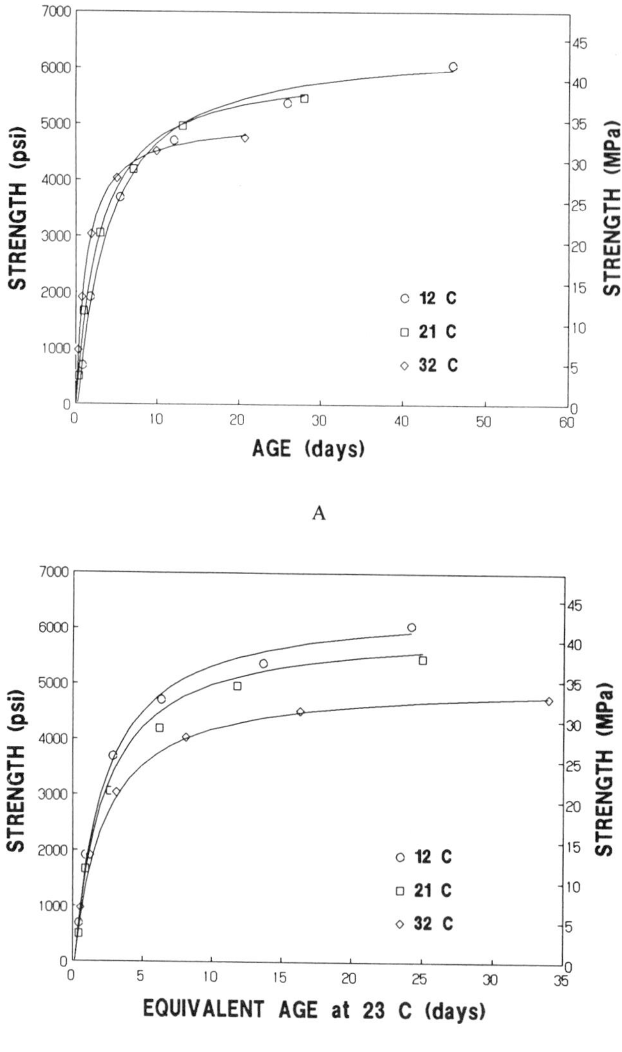

FIGURE 13. (A) Compressive strength vs. age for concrete cylinders cured at three temperatures (Type I cement, W/C = 0.55). (B) Compressive strength vs. equivalent age. (C) Relative compressive strength vs. equivalent age.

$$\frac{S}{S_\infty} = \frac{0.47\ (t_e - 0.16)}{1 + 0.47\ (t_e - 0.16)} \tag{42}$$

where the offset equivalent age of 0.16 days was obtained by trial and error to achieve the best fit of the hyperbola.

The above discussion demonstrates that, while the early-age temperature affects the long-term strength, there is no significant effect of curing temperature on the relationship between equivalent age and relative strength. This leads to the following modified maturity rule:

> Samples of a given concrete mixture which have the same equivalent age and which have had a sufficient supply of moisture for hydration will have developed equal fractions of their limiting strength irrespective of their actual temperature histories.

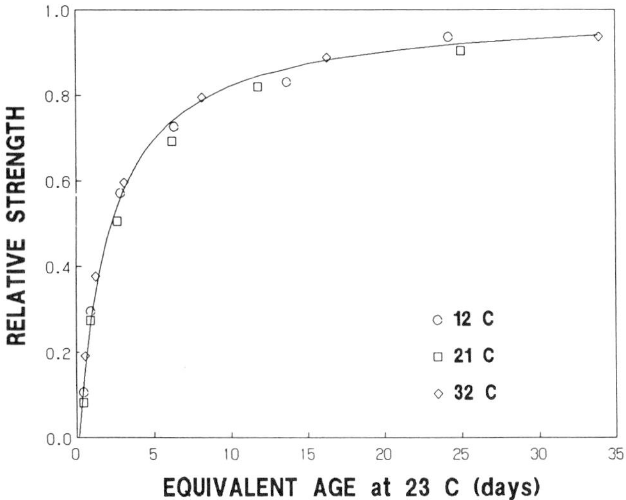

FIGURE 13C.

The significance of this modified maturity rule is that if one measures only the temperature of concrete while it is curing, only the relative strength gain can be estimated. Additional information is needed to estimate absolute strength values. This subject is discussed further in the next section dealing with applications of the maturity method.

SUMMARY

This section has discussed the theoretical basis of the maturity method, taking into account recent research on the subject. It was shown that under isothermal conditions, the strength gain of concrete can be described by a hyperbolic curve defined by three parameters: (1) the age when strength development is assumed to begin, t_0; (2) a rate constant, k_T, which is related to the initial slope of the curve; and (3) the limiting strength, S_∞. These parameters can be estimated from strength tests by transforming strength data and using linear regression analyses. It has been further shown that the parameters are temperature dependent.

To describe strength gain under variable temperature conditions, a maturity function is needed to account for the effect of time and temperature. It has been shown that the product of the rate constant and age is the general form of the maturity function. Thus the key element in arriving at a valid maturity function is describing the relationship between the rate constant and the curing temperature.

The equivalent age approach is a convenient method for accommodating a variety of proposed maturity functions. Curing time intervals at a given temperature are converted to equivalent intervals at a reference temperature. The age conversion factor for each time interval is simply the value of the rate constant at the actual temperature divided by the value at the reference temperature.

If the rate constant is assumed to be a linear function of temperature, the resulting maturity function is identical to the Nurse-Saul function. The key parameter in this approach is the datum temperature, T_0, and it has been shown that the traditional value of $-10°C$ is not necessarily the best value to use.

Tests have shown that over a wide temperature range, the rate constant is a nonlinear function of temperature. The Arrhenius equation has been found to describe accurately the relationship between the rate constant and curing temperature. In this case, the key parameter is the activation energy, which defines the temperature sensitivity of the rate constant. The value of activation energy depends on the cement chemistry, the cement fineness, and the type and quantity of cement replacements and admixtures. It is not clear whether water-cement ratio has a consistent effect on activation energy. For concretes made with ordinary portland cement and without admixtures, it appears that the activation energy is between 40 and 45 kJ/mol.

For greater accuracy in estimated strength, the activation energy for the particular concrete should be determined experimentally. However, the actual concrete mixture does not have to be tested. The required information can be obtained by monitoring the heat of hydration of cement paste, which includes the intended admixtures and cement replacements that may be used. Alternatively, the information can be obtained from the compressive strength development of mortar cubes. The mortar should have the same water-cement ratio as the concrete, and the ratio of cement to sand should be the same as the ratio of cement to coarse aggregate in the concrete.

Finally, it was shown that the limiting strength of a concrete mixture is affected by the early-age temperature history. Thus there is not a unique strength-equivalent age relationship for a given concrete. However, there is a unique relative strength vs. equivalent age relationship. This relationship can be developed from strength gain data obtained at a reference curing temperature. A modified maturity rule is proposed which suggests that relative strength can be reliably estimated from the measured temperature history. In order to estimate the absolute strength level, additional information about the concrete is required.

APPLICATION OF MATURITY METHOD

As with the other methods for estimating in-place strength of concrete which are discussed in this book, the maturity method has various applications in concrete construction.[68] It may be used to estimate in-place strength to assure that critical construction operations, such as form removal or application of post-tensioning, can be performed safely. It may be used to decide when a sufficient amount of curing has occurred and the concrete can be exposed to ambient conditions without endangering its long-term performance. In addition, it may serve as a tool in planning construction activities.

The method is also useful in laboratory work involving test specimens of different sizes. Due to heat of hydration, specimens with lower surface to volume ratios experience higher early-age temperature rise than specimens with higher ratios. The maturity method can be used to ensure that differently sized specimens are tested at the same maturity. A good example is a testing program conducted to establish the correlation relationship between an in-place test method and cylinder strength. Often the in-place tests are performed on large specimens which experience higher early-age temperatures than the companion cylinder specimens. Failure to perform companion tests at the same maturity will result in inaccurate correlation relationships.

BASIC PRINCIPLE

Proper curing procedures[69] must be used to apply the maturity method for estimating strength development. It is essential that there is an adequate supply of moisture for hydration. If concrete dries out, strength gain ceases but the computed maturity value continues to increase with time. In such a case, strength estimates based upon the maturity method are meaningless.

The basic principle in applying the maturity method is illustrated in Figure 14. Two phases are involved: (1) laboratory testing and (2) field measurement of the in-place temperature history. The laboratory testing must be performed prior to estimating in-place strength. Two types of laboratory tests may be required: one is used to establish the temperature sensitivity of the rate constant for the particular materials in the concrete. This information, which can be obtained from mortar tests, is needed to develop the correct

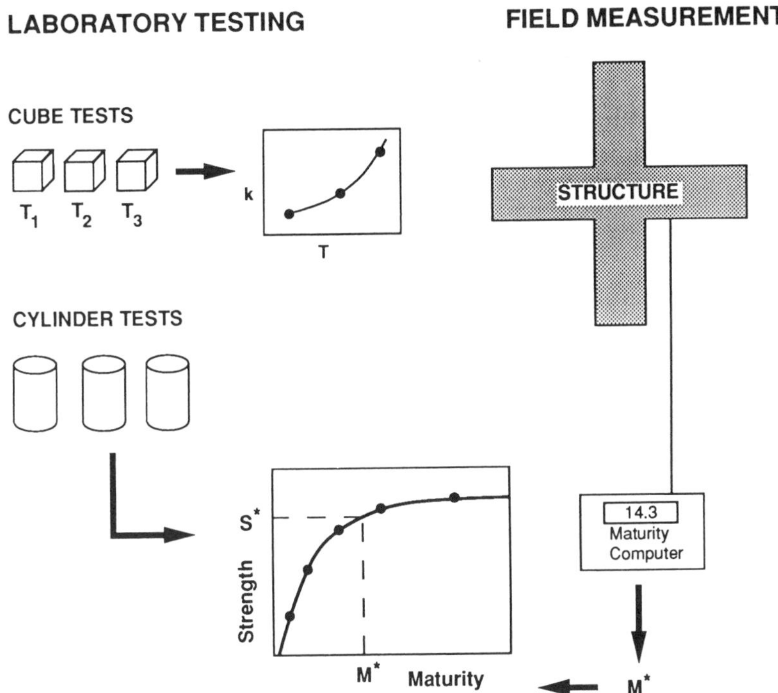

FIGURE 14. Procedures for using maturity method involve laboratory testing and field measurements.

maturity function for the concrete. This second type of testing establishes the strength-maturity relationship for the concrete. The second type of testing is mandatory, while the need for the first type depends on the desired accuracy of the estimated strength. If maximum accuracy is desired, the activation energy value should be determined experimentally. If this is not the case, typical values of activation energy may be used.

In the field, the temperature history of the structure must be monitored to evaluate the in-place maturity. The in-place maturity is used with the strength-maturity relationship obtained from laboratory testing to estimate the in-place strength. Some of the available instruments for measuring in-place maturity are discussed in the next part of this section. The contractor, testing agency, and design engineeer should hold a pre-construction meeting to select appropriate locations for temperature sensors. Consideration should be given to those portions of the structure which are critical because of unfavorable combinations of exposure and required strength. For example, in flat-plate, multi-story construction, the slab-column joints are often the most highly stressed regions. During cold weather, corners of structures experience the most severe exposure.

MATURITY INSTRUMENTS

The temperature history of the structure is the basic information needed to evaluate the in-place maturity (expressed as the temperature-time factor or equivalent age). Therefore, a device is needed to record temperature as a function of time. Analog strip chart recorders or digital dataloggers connected to thermocouples embedded in the concrete are suitable. The measured thermal history is converted to a maturity value using maturity functions such as Equation 1 or Equation 6. The calculations can be automated using a personal computer with spreadsheet software.[70] However, this automation still requires manual entry of temperature values at regular time intervals.

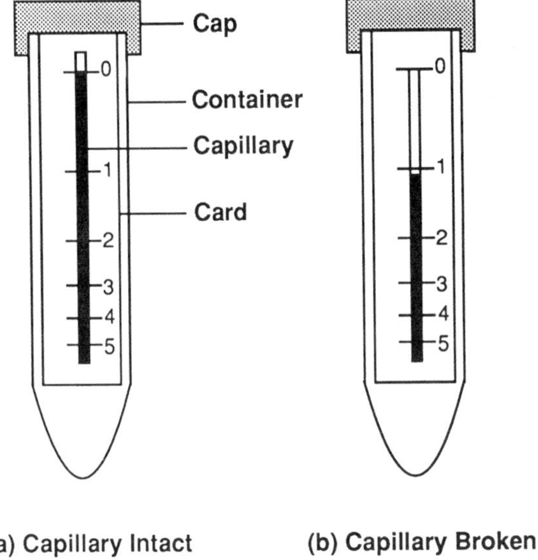

FIGURE 15. A disposable maturity meter which uses the evaporation of a liquid from a capillary tube as an indicator of maturity.

A more convenient approach is to use commercial "maturity meters." These are instruments which monitor the temperature history and automatically perform the maturity calculations. Concrete temperatures are monitored with reusable probes or with expendable thermocouple wires. The earliest models were single channel instruments based on the Nurse-Saul function with a datum temperature of -10°C (14°F). Later, multi-channel models were introduced whereby each channel could be independently activated as the corresponding sensor was embedded in the concrete.

In 1977, Freiesleben Hansen and Pederson[28] introduced a single-channel maturity computer based on the Arrhenius equation. This instrument computes the equivalent age according to Equation 6 and uses the relationship for activation energy given in Equations 6a and 6b.

Recently, a four-channel maturity computer has been developed which uses thermocouple wires as sensors, and permits the use of either the Nurse-Saul function or the Arrhenius equation. In addition, the user can specify the value of the datum temperature or the activation energy.

Hansen[71] developed a disposable "mini maturity meter" as an alternative to high cost maturity computers (multi-channel models cost about $1300 U.S. in 1988). The active component of this device is a glass capillary containing a fluid, as shown in Figure 15a. The instrument is based on the principle that the effect of temperature on the rate constant for evaporation of the fluid from the capillary tube is governed by the Arrhenius equation. Thus the evaporation of the fluid and strength development of concrete are influenced by temperature in a similar manner. By choosing a fluid with an activation energy similar to the concrete, the amount of evaporation at a given time is indicative of strength development in the concrete. The meter contains a fluid which is reported to have an activation energy of 40 kJ/mol. The capillary tube of the mini-maturity meter is attached to a card which is marked in units of equivalent age at 20°C (68°F). The device was developed primarily for use during construction, and hence the scale is limited to an equivalent age of 5 days at 20°C (68°F). The card is attached to the removable cap of the plastic container. The meter is activated by removing the cap and breaking the capillary at the "0 days" mark (Figure 15b). The cap is replaced on the plastic container and the container is inserted into the fresh

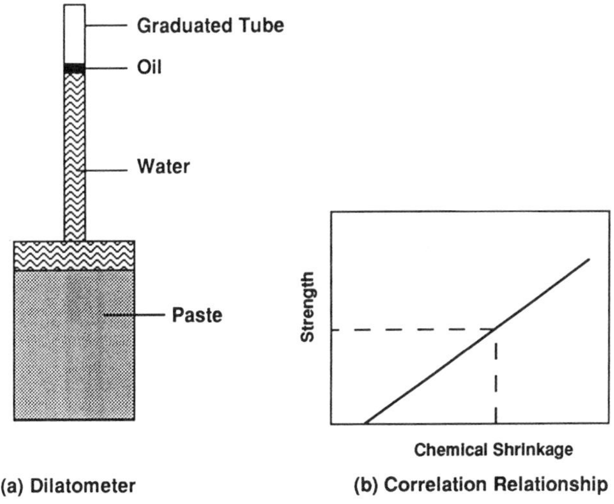

FIGURE 16. (a) Schematic of a maturity meter based on measuring chemical shrinkage; (b) schematic of correlation relationship for estimating in-place strength.

concrete. When it is desired to read the meter, the cap is removed and the position of the fluid is noted in units of equivalent days at 20°C (68°F). In 1988, the disposable meters were priced at about $10 U.S.

Another novel idea for a ''maturity meter'' is based on measuring the chemical shrinkage of cement paste as it hydrates.[72] In this approach, cement paste at the same water-cement ratio as the concrete is placed in a vessel and covered with water. The vessel, known as a dilatometer, contains a small diameter tube to monitor the decrease in water level as the paste hydrates and undergoes chemical shrinkage (Figure 16a). A layer of low volatility fluid such as oil is used to prevent evaporation of water from the tube. The vessel would be placed in the fresh concrete so that the paste would experience the same temperature history as the concrete. The fall in water level with time would indicate the chemical shrinkage. Geiker[60] has shown that there is nearly a linear relationship between chemical shrinkage and strength, and that this relationship is independent of the curing temperature. In addition, the ultimate value of chemical shrinkage was found to be affected by the initial curing temperature in a similar manner as ultimate strength of mortar specimens. Thus a meter based on chemical shrinkage would automatically account for the effects of early-age temperature on limiting strength. This would be a significant advancement over maturity instruments based solely on temperature history, because it would allow estimation of absolute strength rather than relative strength. To use such a device, a strength vs. chemical shrinkage correlation curve would be developed for the particular concrete mixture. Such a relationship is shown schematically in Figure 16b. The dilatometer would be filled with a sample of the field concrete, and measurement of in-place chemical shrinkage would be used to estimate the in-place strength. Presently, a workable field instrument and test procedure based on chemical shrinkage has not been developed.

In summary, a variety of commercial devices are available which can automatically compute the in-place maturity. The user should keep in mind that the maturity calculations are based on specific values of datum temperature or activation energy. Hence, they will only correctly account for temperature effects if these values are applicable to the materials being used. For instruments based on the Nurse-Saul function it is possible to correct the displayed temperature-time factors for a different value of the datum temperature using the following equation:

$$M_c = M_d (T_0 - T_{0d})\, t \tag{43}$$

where,

M_c = corrected temperature-time factor,
M_d = displayed temperature-time factor,
T_0 = the desired datum temperature,
T_{0d} = the datum temperature used by the instrument, and
t = the elapsed time from when instrument was turned on.

The readings of maturity computers based on the Arrhenius equation cannot be corrected for a different value of the activation energy. The user of such an instrument should refer to Figure 11 for an understanding of the effect of activation energy on the age factor used to compute equivalent age.

MATURITY METHOD COMBINED WITH OTHER METHODS

There are several factors that can lead to errors in the estimated in-place strength based upon the maturity method:

1. Errors in batching which reduce the potential strength of the concrete
2. High early-age temperatures which reduce the limiting strength of the concrete
3. Improper curing procedures which cause concrete to dry below a critical level and cause hydration to cease
4. Use of activation energy or datum temperature values which are not representative of the concrete mixture

Because of these limitations, it is not prudent to rely solely on measurements of in-place maturity to verify the attainment of a required level of strength before performing a critical construction operation. Therefore, as is discussed later, the standard practice requires that maturity testing be supplemented with other tests. One approach is to use the maturity method along with other in-place tests of the concrete. For example, the maturity method has been used along with pullout tests.[73,74] As will be discussed, by combining the maturity method with other tests, the amount of required testing may be reduced without compromising safety.

One of the considerations in using in-place tests (such as pullout, probe penetration, or break-off) is knowing when these tests should be performed. If they are performed before the concrete reaches the required strength, they have to be repeated after additional curing. Premature testing is possible if a period of unusually cold weather occurs during the time between placement and testing. On the other hand, if the tests are performed after the concrete is well above the required strength level, unnecessary delays in construction activities may have occurred.

The proper time to perform the other in-place tests can be determined by measuring the in-place maturity. From the pre-established strength-maturity relationship, the user determines the maturity value corresponding to the required strength. The in-place maturity is monitored. When the necessary maturity value is attained, the in-place tests are performed and the concrete strength is estimated from the correlation curve for that test. If the resulting estimated strength equals or exceeds the required strength, the construction activity can proceed. If the estimated strength is significantly less than the required value, the engineer may consider the following questions before deciding on the course of action:

1. Were the sites of the in-place tests close enough to the locations of the temperature

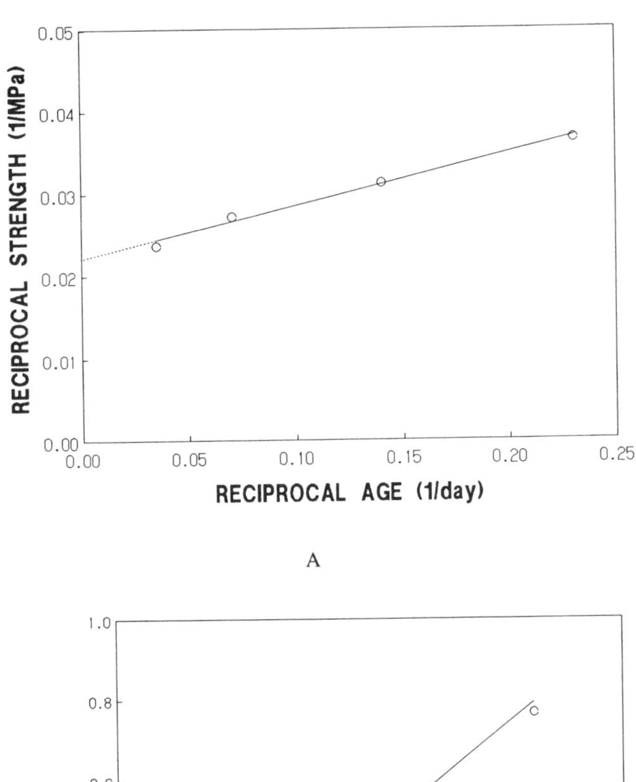

A

B

FIGURE 17. (A) Plot of reciprocal strength vs. reciprocal age to evaluate the limiting strength. (B) Plot to evaluate the rate constant and the age when strength development begins. (C) Relative strength vs. equivalent age (left scale is based on limiting strength and right scale is based on 28-day strength).

sensors, so that the maturity values were indicative of the maturity of the concrete that was tested?

2. Were proper curing procedures used to ensure an adequate supply of moisture?
3. Is the value of the activation energy or datum temperature used to compute maturity a reasonably accurate value for the materials being used?

If the answer to any of these questions is ''no'', engineering judgment is required to decide whether the construction activity may proceed after additional curing, or whether additional curing and re-testing is needed before beginning the activity.

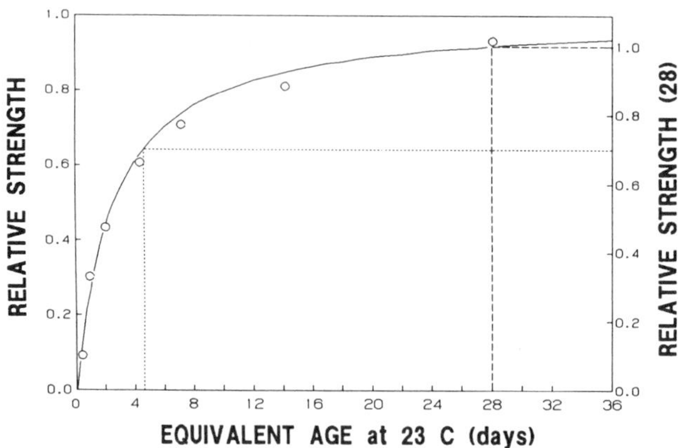

FIGURE 17C.

If the answer to all of the above questions is "yes", it is reasonable to conclude that the low estimate of the in-place strength resulted because the concrete tested does not have the same potential strength as the concrete used to develop the strength-maturity relationship. In this case, the engineer must decide whether to require additional curing before beginning the scheduled activity, or whether there is sufficient concern to question the quality of the concrete.

ILLUSTRATIVE EXAMPLES

Strength-Maturity Relationship

The data reported in Table 1 are used to construct a strength-maturity relationship to illustrate how the maturity method could be applied on a project. Assume that the strength-maturity relationship obeys the hyperbolic equation presented in the previous section. The two-step procedure associated with Equations 28 and 29 is used to obtain the values of the parameters for the relative strength vs. equivalent age relationship. Assume that the temperature of the test specimens was 23°C (73°F) throughout the testing period, so that the ages in Table 1 are equivalent ages at 23°C.

First, the limiting strength is estimated by considering the data for tests beyond 4 days. Figure 17A shows the reciprocal of strength plotted against the reciprocal of age. From linear regression analysis, the intercept is 0.0222, which equals the reciprocal of the limiting strength. Therefore, the limiting strength is S_∞ = 1/0.0222 MPa = 45.0 MPa (6530 psi). Next, estimate the rate constant and the age when strength development begins by considering the first three test results in Table 1. In accordance with Equation 29, the quantity $S/(S_\infty - S)$ is plotted against age as shown in Figure 17B. From linear regression analysis, the slope of the line is found to be 0.40 day^{-1}, which is the value of the rate constant at 23°C (73°F). The intercept of the line with the age axis is 0.08 days. Thus, the hyperbolic strength-equivalent age relationship for this concrete mixture is

$$S(\text{MPa}) = 45.0 \frac{0.4\ (t_e - 0.08)}{1 + 0.4\ (t_e - 0.08)} \tag{44}$$

This relationship would correctly predict in-place strength development if the early-age temperature in the field were close to 23°C (73°F). However, as was discussed in the previous section, the following relative strength vs. equivalent age curve would be applicable for other early-age temperatures:

$$RS_\infty = \frac{S}{S_\infty} = \frac{0.4\,(t_e - 0.08)}{1 + 0.4\,(t_e - 0.08)} \quad (45)$$

where

RS_∞ = fraction of limiting strength.

Figure 17C shows Equation 45 along with the relative strength values obtained by dividing the strengths in Table 1 by the limiting strength of 45 MPa (6530 psi). The fit is good.

A more useful relationship would be to express relative strength in terms of the 28-day strength. If 28 days is substituted for the equivalent age in Equation 45, we obtain $S_{28}/S_\infty = 0.92$. Thus the 28-day strength is 92% of the limiting strength, or conversely the limiting strength is 1/0.92 = 1.09 times the 28-day strength. Thus Equation 45 can be rewritten as follows:

$$RS_{28} = \frac{S}{S_{28}} = \frac{(1.09)\,0.4\,(t_e - 0.08)}{1 + 0.4\,(t_e - 0.08)} \quad (46)$$

where

S_{28} = the 28-day strength, and
RS_{28} = fraction of the 28-day strength.

To obtain the graph of the relationship between RS_{28} and equivalent age, the left-hand scale on Figure 17C is multiplied by 1.09. This new scale is shown on the right of Figure 17C.

Planning construction — Next, the application of the maturity method for planning construction activities is discussed. First, Equation 46 is solved for equivalent age in terms of a desired strength ratio. Using general terms, one obtains the following:

$$t_e = \frac{RS_{28}}{k_r(\phi - RS_{28})} + t_0 \quad (47)$$

where

ϕ = $\frac{S_\infty}{S_{28}}$ (= 1.09 in this example),
k_r = the rate constant as the reference temperature (= 0.40 days^{-1} in this example), and
t_0 = the equivalent age at start of strength development (= 0.08 days in this example).

Good engineering practice[75] recommends that, unless approved by the engineer/architect, supporting forms and shores should not be removed from horizontal members (beams and slabs) until at least 70% of the design strength has been attained. Assuming that the strength gain of concrete to be used for construction is described by Equation 45 and Figure 17C, the equivalent age at 23°C (73°F) needed to reach 70% of the 28-day design strength would be

$$t_e = \frac{0.70}{0.40(1.09 - 0.70)} + 0.08 \approx 4.6 \text{ days} \tag{48}$$

The equivalent age of 4.6 days would be the actual age only if the curing temperature were 23°C (73°F).

Now consider the time it would take to reach an equivalent age of 4.6 days for different constant curing temperature conditions. The maturity function based on the Arrhenius equation is used and it is assumed that the concrete has an activation energy of 40 kJ/mol. The age conversion factor would be

$$\alpha = e^{\frac{-40000}{8.314}\left[\frac{1}{273 + T} - \frac{1}{296}\right]} \tag{49}$$

The following age factor values are obtained for different concrete temperatures. The number of days needed to reach an equivalent age of 4.6 days at 23°C (73°F) is obtained by dividing the required equivalent age by the age factors.

Temperature °C (°F)	Age factor	Days to reach t_e = 4.6 days at 23°C (73°F)
5 (41)	0.35	13.1
15 (59)	0.64	7.2
23 (73)	1.00	4.6
35 (95)	1.88	2.4
45 (113)	3.08	1.5

With this information, the contractor can analyze the economics of different construction schedules. For example, if the job were to be performed during cold weather, and if a 3-day cycle was desired for the re-use of forms, protection would have to be provided to ensure that the concrete temperature was maintained above about 30°C (86°F). The contractor could then decide whether it would be more economical to provide the needed protection or use a longer cycle time.

Evaluation of in-place tests — Finally, we discuss the use of the maturity method for rational interpretation of the results of other in-place tests used to verify the attainment of required strength during construction. The procedure is as follows:

1. Based on the required strength, the 28-day design strength, and the relative strength-maturity relationship, compute the maturity (time-temperature factor or equivalent age) at which the required strength is expected to be achieved.
2. Perform the in-place tests and estimate the in-place concrete strength using the appropriate correlation relationship for that test method.
3. If the estimated strength based on the in-place tests equals or exceeds the required strength, perform the planned construction activity.
4. If the estimated strength is less than the required strength, use the maturity method to analyze the in-place results and to help decide the appropriate action to take.

To illustrate the procedure, assume that concrete with a 28-day design strength of 40 MPa (5800 psi) is used in a post-tensioned structure, and the concrete must have a strength of 25 MPa (3600 psi) before post-tensioning can be applied. The equivalent age to achieve the required strength is computed using Equation 46. Thus we have the following:

Concrete properties:

k_r	=	0.4 day^{-1}	— Rate constant at reference temperature of 23° C (73°F)
t_0	=	0.08 days	— Equivalent age at start of strength gain
S_∞/S_{28}	=	1.09	— Ratio of limiting strength to 28-day strength
S_{28}	=	40 MPa	— Given 28-day strength

Construction requirements:

S_{req}	=	25 MPa	— Strength required for post-tensioning
RS_{28}	=	25/40 = 0.63	— Required strength ratio

Equivalent age to attain required strength:

t_e	=		3.4 days at 23°C (73°F) (from Equation 47)

Suppose that in-place tests are performed when the in-place equivalent age is 3 days, and based on these tests the estimated compressive strength of the concrete is 23 MPa (3300 psi). Since the estimated strength is less than the required strength and the tests were performed at less than the required maturity, the maturity method can be used to analyze the results.

First, Equation 46 is used to compute the relative strength that should have developed at an equivalent age of 3 days. Using the estimated in-place strength, the 28-day strength of the in-place concrete is estimated. Finally, the equivalent age needed to achieve the required strength is calculated. Thus we have the following:

Evaluation of in-place testing results: Case 1

t_e	=	3 days	— Equivalent age when in-place test performed
RS_{28}	=	0.59	— Theoretical strength ratio at test age (Equation 46)
S_e	=	23 MPa	— Estimated strength from in-place test result
S_{28}	=	23/0.59	= 39 MPa — Estimated 28-day strength of in-place concrete
RS_{28}	=	25/39	= 0.64 — Needed strength ratio to achieve S_{req}
t_e	=	3.6 days	— Equivalent age to achieve S_{req}

In this case, the in-place strength was lower than the required strength primarily because the tests were performed at an equivalent age of 3 days rather the computed required age of 3.4 days. The calculations indicate that the in-place concrete has the necessary strength potential and the required strength will be attained after additional curing. Thus the post-tensioning can proceed according to plans.

Now, consider the case where the estimated strength from the in-place tests is only 20 MPa (3200 psi). The evaluation of the results is as follows:

Evaluation of in-place testing results: Case 2

t_e	=	3 days	— Equivalent age when in-place test performed
RS_{28}	=	0.59	— Theoretical strength ratio at test age
S_e	=	20 MPa	— Estimated strength from in-place test result
S_{28}	=	20/0.59	= 34 MPa — Estimated 28-day strength of in-place concrete
RS_{28}	=	25/34	= 0.74 — Needed strength ratio to achieve S_{req}
t_e	=	5.3 days	— Equivalent age to achieve S_{req}

In this case, it appears that the in-place concrete does not have the expected potential strength because the estimated 28-day strength of the in-place concrete is only 34 MPa (4900 psi) rather than the design value of 40 MPa (5800 psi). At this point the engineer has to take a closer look at the three questions discussed at the end of the previous sub-section. If the engineer is assured that sufficient moisture was provided and that the measured in-place maturity was representative of the maturity at the locations of the in-place tests, there is good reason to suspect a problem in the quality of the concrete. If the lower than expected design strength can be tolerated, it is only necessary to delay the post-tensioning until the equivalent age of 5.3 days has been attained. If the lower potential strength cannot be tolerated, additional investigation may be needed, and the questionable concrete may have to be removed.

The mathematical manipulations involved in the above analyses can be added to computer programs developed for the analysis of in-place test results. Spreadsheet software is a very convenient tool for analyzing in-place test data[70,76] and the above procedures can be incorporated easily into such spreadsheets.

SUMMARY

This section has discussed the essential elements for application of the maturity method. Three elements are required: (1) the maturity function for the concrete materials; (2) a strength-maturity relationship for the concrete mixture to be used in construction; and (3) measurement of the in-place thermal history. Selection of the proper maturity function is necessary to obtain the most accurate results, and it requires knowledge of the effects of curing temperature on the rate constant for strength development. A testing procedure using mortar cubes can be used to obtain this information.

The strength-maturity relationship is obtained from strength development data of concrete specimens. The temperature history of the specimens is recorded to evaluate the maturity at the corresponding test ages. Regression analysis of the data can be used to construct a mathematical description for the relative strength-maturity relationship.

Various instruments are available for monitoring the in-place temperature and computing maturity values. In selecting a device, the user should be aware of the maturity function used by the equipment. State-of-the-art instruments permit the user to select the parameters for the maturity function.

Because of its inherent limitations, the maturity method should not be used alone to decide whether critical construction activities can be performed. Examples have been given to illustrate how the maturity method can be used with other in-place tests to verify the attainment of the required in-place strength.

The next section summarizes the current ASTM Standard Practice governing the use of the maturity method.

STANDARD PRACTICE

Procedures for using the maturity method have been standardized in ASTM C 1074.[11] The standard practice permits the user to express maturity either as a temperature-time factor using the Nurse-Saul function, Equation 1, or as an equivalent age using the Arrhenius equation, Equation 6. For the Nurse-Saul function, it is recommended that the datum temperature be taken as 0°C (32°F) if ASTM Type I cement is used without admixtures and the expected curing temperature is within 0 and 40°C (32 and 104°F). For the Arrhenius equation, an activation energy of 41.5 kJ/mol is recommended. For other conditions or when maximum accuracy is desired, the best value of the datum temperature or activation energy should be determined experimentally.

The ASTM standard provides procedures for developing the strength-maturity relationship and for estimating the in-place strength. In addition, a procedure is provided for obtaining the datum temperature or activation energy if that is desired.

DATUM TEMPERATURE OR ACTIVATION ENERGY

The procedure for determining the datum temperature or activation energy follows the approach discussed in the previous section on the theoretical basis of the maturity method. Basically, mortar cubes made with the materials to be used in construction are cured at three temperatures. Two of the curing temperatures should be the minimum and maximum curing temperatures expected for the in-place concrete, and the third temperature should be midway between the extremes. The cubes are tested for compressive strength at regular time intervals.

The standard also requires the determination of the final setting times at the three curing temperatures using the penetration resistance method.[78] The final setting times are used to represent the ages when strength development is assumed to begin. The strength-age data for the mortar cubes are then analyzed to obtain the rate constant as a function of temperature. The reciprocals of the cube strengths are plotted as a function of reciprocals of the ages beyond the times of final setting. This approach is similar to using Equation 27 to represent the strength-age data. For each temperature, the least-squares best-fit straight line is determined, and the rate constant is obtained by dividing the intercept by the slope of line.

To obtain the datum temperature, the rate constants are plotted as a function of temperature and the best-fit straight line is determined. The intercept of the line with the temperature axis gives the datum temperature. To obtain the activation energy, the natural logarithms of the rate constants are plotted against the reciprocals of the absolute curing temperatures. The negative of the slope of the straight line equals the activation energy divided by the gas content (referred to as Q in the standard).

STRENGTH-MATURITY RELATIONSHIP

To develop the strength-maturity relationship, cylindrical concrete specimens are prepared using the mixture proportions and constituents of the concrete to be used in construction. These specimens are prepared according to the usual procedures for making and curing test specimens in the laboratory.

After the cylinders are molded, temperature sensors are embedded at the centers of at least two cylinders. The sensors are connected to instruments which automatically compute maturity or to temperature recording devices such as data-loggers or strip-chart recorders.

The specimens are cured in a water bath or in a moist curing room. At ages of 1, 3, 7, 14 and 28 days, compression tests are performed on at least three specimens. At the time of testing, the average maturity value for the instrumented specimens is recorded. If maturity instruments are used, the average of the displayed values is recorded. If temperature recorders are used, the maturity is evaluated according to Equation 1 or Equation 6. A time interval of one-half hour or less should be used for the first 48 h, and longer time intervals are permitted for the remainder of the curing period.

A plot is made of the average compressive strength as a function of the average maturity value. A best-fit smooth curve is drawn through the data, or regression analysis may be used to determine the best-fit curve for one of the strength-maturity relationships discussed in the previous section. The resulting curve would be used to estimate the in-place strength of that concrete mixture.

The ASTM standard assumes that the initial temperature of the concrete in the field is approximately the same as the laboratory temperature when the cylinders were prepared. If the actual early-age temperatures are significantly greater than the laboratory temperatures, the limiting in-place strength is reduced. Thus the in-place strength may be over estimated by the strength-maturity relationship.

ESTIMATING IN-PLACE STRENGTH

The procedure for estimating the in-place strength requires measuring the in-place maturity. As soon as is practicable after concrete placement, temperature sensors are placed in the fresh concrete. As previously mentioned, the sensors should be installed at locations in the structure which are critical in terms of exposure conditions and structural requirements. The importance of this step cannot be overemphasized when the strength estimates are being used for timing the start of critical construction operations.

The sensors are connected to maturity instruments or temperature recording devices which are activated as soon as is practicable after concrete placement. When a strength estimate is desired, the maturity value from the maturity instrument is read or maturity is evaluated from the temperature record. Using the maturity values and the previously established strength-maturity relationship, compressive strengths at the locations of the sensors are estimated.

Because the temperature history is the only measurement made in the field, there is no assurance that the in-place concrete has the correct mixture proportions. Therefore, the ASTM standard requires verification of the potential strength of the in-place concrete before performing critical operations, such as formwork removal or post-tensioning. Failure to do this can lead to drastic consequences in the event of such undetected batching errors as using excessive amounts of cement replacements or retarding admixtures. Alternative methods for verification of concrete strength include: (1) other in-place tests which measure an actual strength property of the in-place concrete; (2) early-age compressive strength tests of standard-cured specimens molded from samples of the concrete in the structure; or (3) compressive strength tests on specimens molded from samples of the concrete in the structure and subjected to accelerated curing.

SUMMARY

This chapter has (1) reviewed the historical development of the maturity method; (2) provided a theoretical basis for the method; and (3) discussed how the method may be applied during construction.

The maturity method was initially proposed as a means to estimate strength development of concrete during accelerated curing, such as steam or electric curing. The idea was subsequently extended to ordinary curing conditions. The early work was empirical in nature. Recent work has attempted to establish a theoretical basis for the method and to explain the inherent approximations and limitations of the method.

A variety of so-called maturity functions have been proposed to account for the effect of time and temperature on strength development. It has been shown that the product of the rate constant and age is the general form of an appropriate maturity function. The rate constant is related to the rate of strength development during the acceleratory period immediately following setting. Thus the key element of a suitable maturity function is having the correct representation of the effect of temperature on the rate constant.

If the rate constant is assumed to be a linear function of temperature, the resulting maturity function is the traditional Nurse-Saul function. However, test data show that, over a wide temperature range, the rate constant is not a linear function of temperature. Therefore, the Nurse-Saul function is inherently approximate and will either overestimate or underestimate the effects of temperature on strength gain. It has been shown that if a straight-line approximation is used, the best-fit value of the datum temperature (temperature at which the rate constant equals 0) is not necessarily −10°C (14°F) as has been traditionally used in the Nurse-Saul function. Rather, the datum temperature depends on the temperature sensitivity of the rate constant and the temperature range over which the linear approximation is used.

A nonlinear function, such as the Arrhenius equation, can better represent the effect of temperature on strength development over wide temperature ranges. For the Arrhenius equation, the activation energy is the parameter which defines the temperature sensitivity of the rate constant. Recent work indicates that the activation energy is dependent on the cementitious components of the concrete and may also depend on water-cement ratio.

The equivalent age approach is the most flexible technique to represent maturity. In this case, the age factor is used to convert a curing time interval at any temperature to an equivalent time interval at a reference temperature. The age factor is simply the ratio of the value of the rate constant at any temperature to its value at the reference temperature.

An ASTM standard exists for application of the maturity method. The standard provides a procedure, based on testing mortar cubes, for obtaining the values of activation energy (or datum temperature) for the particular concrete mixture. This testing is required for maximum accuracy in estimating strength gain. The standard also provides a procedure for establishing the strength-maturity relationship of the concrete mixture. This relationship is obtained experimentally by testing cylindrical specimens at various values of maturity. It must be emphasized that to accurately estimate the in-place strength, one needs the correct maturity function. In addition, the early-age temperature of the in-place concrete must be similar to that of the specimens used to develop the strength-maturity relationshiop. The traditional maturity method cannot account for the effects of early-age temperature on limiting strength.

Because of the dependence of the limiting strength on the early-age curing temperature, a unique strength-maturity relationship does not exist for a given concrete mixture. However, it appears that there is a unique relative strength vs. maturity relationship. Thus the only reliable information that can be obtained from measuring in-place maturity is relative strength gain. It is for this reason, and because the maturity method cannot detect batching errors, that the maturity method must be supplemented with other tests before performing critical construction operations.

The maturity method can be used along with other in-place tests. This approach provides the needed assurance of the in-place strength, and it permits rational interpretation of low estimated strengths based on in-place tests. A potential application is to use the maturity method along with other in-place methods to determine whether the in-place concrete meets contract strength requirements.

The maturity method is amenable to computer application because the maturity function and strength-maturity relationship can be represented by simple mathematical expressions. The Nordic countries have pioneered in the use of the maturity method in computer programs designed to simulate the expected outcome of alternative construction schemes and concrete mixtures. For example, one can simulate the effects of cement content and amount of insulation on the in-place temperature history of a structural component exposed to different ambient conditions. The maturity method can then be used to estimate in-place strength based on the computed thermal history. Predictions based on this approach have been in good agreement with actual strength measurements.[75]

In summary, the maturity method provides a simple procedure to account for the effects of temperature and time on strength development. In combination with other in-place tests, the maturity method is expected to play an important role in advanced concrete technology.

REFERENCES

1. **Saul, A. G. A.,** Principles underlying the steam curing of concrete at atmospheric pressure, *Mag. Concr. Res.*, 2 (6), 127, 1951.
2. **Bergstrom, S. G.,** Curing temperature, age and strength of concrete, *Mag. Concr. Res.*, 5(14), 61, 1953.
3. **Plowman, J. M.,** Maturity and the strength of concrete, *Mag. Concr. Res.*, 8(22), 13, 1956.
4. **McIntosh, J. D.,** The effects of low-temperature curing on the compressive strength of concrete, Proc. RILEM Symp. on Winter Concreting (Copenhagen 1956), Danish Institute for Building Research Copenhagen, Session BII.
5. **Klieger, P.,** Effects of mixing and curing temperatures on concrete strength, *J. Am. Concr. Inst.*, 54(12), 1063, 1958.
6. **Alexander, K. M. and Taplin, J. H.,** Concrete strength, cement hydration and the maturity rule, *Austr. J. Appl. Sci.*, 13, 277, 1962.

7. Cold weather concreting (ACI 306R-88), Rep. by Committee 306, *J. Am. Concr. Inst.*, Vol. 85, No. 4, July/August 1988, 280.
8. RILEM Recommendations for concreting in cold weather, Kukko, H. and Koskinen, I., Eds., Technical Research Centre, Research Notes 827, Espoo, January, 1988.
9. **Freiesleben Hansen, P. and Pederson, E. J.,** Vinterstobning af beton, Anvisning 125, Statens, Byggeforskningsinstitut, Copenhagen, 1982, 96 (in Danish).
10. In-place methods for determination of strength of concrete (ACI 228R-88), Rep. by Committee 228, *J. Am. Concr. Inst.*, 85(5), 446, 1988.
11. Standard practice for estimating concrete strength by the maturity method, ASTM C 1074-87, Annual Book of ASTM Standards, Vol. 04.02 Concrete and Aggregates, 1988.
12. Malhotra, V. M., Maturity concept and the estimation of concrete strength, Information Circular IC 277, Department of Energy Mines Resources (Canada), November 1971.
13. **McIntosh, J. D.,** Electrical curing of concrete, *Mag. Concr. Res.*, 1(1), 21, 1949.
14. **Nurse, R. W.,** Steam curing of concrete, *Mag. Concr. Res.*, 1(2), 79, 1949.
15. **Rastrup, E.,** Heat of hydration in concrete, *Mag. Concr. Res.*, 6(17), 79, 1954.
16. **Wastlund, G.,** Hardening of concrete as influenced by temperature, General Report of Session BII, Proceedings, RILEM Symposium Winter Concreting (Copenhagen 1956), Danish Institute for Building Research Copenhagen,
17. **Verbeck, G. J. and Helmuth, R. H.,** Structure and physical properties of cement paste, Proc. 5th Int. Symp. on the Chemistry of Cement, Part III, 1-32 (Tokyo 1968).
18. **Swenson, E. G.,** Estimation of strength gain of concrete, *Eng. J.*, September, 1967.
19. **Hudson, S. B. and Steele, G. W.,** Prediction of potential strength of concrete from early tests, *Highw. Res. Rec.*, 370, 25, 1971.
20. **Hudson, S. B. and Steele, G. W.,** Developments in the prediction of potential strength of concrete from results of early tests, *Transp. Res. Rec.*, 558, 1, 1975.
21. Standard test method for developing early age compression test values and projecting later age strengths, ASTM C 918-80, Annual Book of ASTM Standards, Vol. 04.02 Concrete and Aggregates, 1987.
22. **Weaver, J. and Sadgrove, B. M.,** Striking times of formwork-tables of curing periods to achieve given strengths, Construction Industry Research and Information Association, Rep. 36, (London, October 1971).
23. **Sadgrove, B. M.,** Prediction of strength development in concrete structures, *Transp. Res. Rec.*, 558, 19, 1975.
24. **Bickley, J. A.,** Practical application of the maturity concept to determine in-situ strength of concrete, *Transp. Res. Rec.*, 558, 45, 1975.
25. **Mukherjee, P. K.,** Practical application of maturity concept to determine in-situ strength of concrete, *Transp. Res. Rec.*, 558, 87, 1975.
26. **Nisbet, E. G. and Maitland, S. T.,** Mass concrete sections and the maturity concept, *Can. J. Civil Eng.*, Vol. 3, 47, 1976.
27. **Copeland, L. E., Kantro, D. L., and Verbeck, G.,** Chemistry of hydration of portland cement. Part III-Energetics of the hydration of portland cement, Proc. 4th Int. Symp. on Chemistry of Cement, NBS Monograph 43, Washington, 453, 1962.
28. **Freiesleben Hansen, P. and Pedersen, E. J.,** Maturity computer for controlled curing and hardening of concrete, *Nordisk Betong*, 1, 19, 1977.
29. **Byfors, J.,** Plain concrete at early ages, Swedish Cement and Concrete Research Institute, Rep. 3:80, 1980.
30. **Naik, T. R.,** Maturity functions concrete cured during winter conditions, *Temperature Effects on Concrete*, Naik, T. R., Ed., ASTM STP 858, 107, 1985.
31. **Lew, H. S., and Reichard, T. W.,** Mechanical properties of concrete at early ages, *J. Am. Concr. Inst.*, 75(10), 533, 1978.
32. **Carino, N. J., Lew, H. S., and Volz, C. K.,** Early age temperature effects on concrete strength prediction by the maturity method, *J. Am. Concr. Inst.*, 80(2), 93, 1982.
33. **Carino, N. J.,** Temperature effects on strength-maturity relation of mortar, NBSIR 81-2244, U.S. Natl. Bur. of Stand., March 1981.
34. **Carino, N. J. and Lew, H. S.,** Temperature effects on strength-maturity relations of mortar, *J. Am. Concr. Inst.*, 80(3), 177, 1983.
35. **Carino, N. J.,** Maturity functions for concrete, Proceedings, RILEM International Conference on Concrete at Early Ages (Paris, 1982), Ecole Nationale des Ponts et Chausses, Paris, Vol. I, 123.
36. **Carino, N. J.,** The maturity method: theory and application, *ASTM J. Cement, Concr. Aggregates*, 6(2), Winter 61, 1984.
37. **Nykanen, A.,** Hardening of concrete at different temperatures, especially below the freezing point, Proceedings, RILEM Symposium on Winter Concreting (Copenhagen 1956), Danish Institute for Building Research Copenhagen, Session BII.
38. Discussion of Reference 3, *Mag. Concr. Res.*, 8(24), 169, 1956.

39. **Bernhardt, C. J.,** Hardening of concrete at different temperatures, Proceedings, RILEM Symposium on Winter Concreting (Copenhagen, 1956), Danish Institute for Building Research Copenhagen, Session BII.
40. **Goral, M. L.,** Empirical time-strength relations of concrete, *J. Am. Conc. Inst.*, 53(2), 215, 1956.
41. ACI Committee 209, Prediction of creep, shrinkage, and temperature effects in Concrete Structures, American Concrete Institute, Detroit, MI, SP-27, 51, 1971.
42. **Chin, F. K.,** Relation between strength and maturity of concrete, *J. Am. Concr. Inst.*, 68(3), 196, 1971.
43. **Chin, F. K.,** Strength tests at early ages and at high setting temperatures, *Transp. Res. Rec.*, 558, 69, 1975.
44. **Carino, N. J.,** Closure to discussion of Reference 32, *J. Am. Concr. Inst.*, 81(1), 98, 1984.
45. **Lew, H. S. and Reichard, T. W.,** Prediction of strength of concrete from maturity, in *Accelerated Strength Testing*, SP-56, American Concrete Institute, Detroit, MI, 229, 1978
46. **Freiesleben Hansen, P. and Pederson, E. J.,** Curing of concrete structures, CEB Information Bulletin 166, May 1985.
47. **Knudsen, T.,** On particle size distribution in cement hydration, Proc. 7th Int. Congr. on the Chemistry of Cement (Paris, 1980), Editions Septima, Paris, 1980, Vol. II, 170.
48. **Knudsen, T.,** Modelling hydration of portland cement: the effects of particle size distribution, Proc. Engineering Foundation Conf. on Characterization and Performance Prediction of Cement and Concrete, July 1982, Henniker, NH, 125.
49. **Knudsen, T.,** The dispersion model for hydration of portland cement. I. General concepts, *Cement Concr. Res.*, 14, 622, 1984.
50. **Bezjak, A. and Jelenic, I.,** On the determination of rate constants for hydration processes in cement pastes, *Cement Concr. Res.*, 10(4), 553, 1980.
51. **Copeland, L.E. and Kantro, D. L.,** Chemistry of hydration of portland cement at ordinary temperature, in *The Chemistry of Cements*, Vol. I, Taylor, H. F. W. Ed., Academic Press, London, 1964, chap. 8.
52. **Regourd, M.,** Structure and behavior of slag portland cement hydrates, Proc. 7th Int. Congr. on the Chemistry of Cement (Paris, 1980), Editions Septima, Paris, 1980, Vol. I.
53. **Filliben, J. J.,** DATAPLOT — Introduction and overview, U.S. Natl. Bur. of Stand., Spec. Publ. (SP) 667, June 1984. (Program available from NTIS.)
54. **Alexander, K. M., Taplin, J. H., and Wardlaw, J.,** Correlation of strength and hydration with composition of portland cement, Proc. 5th Int. Symp. on the Chemistry of Cement (Tokyo, 1968), Cement Association of Japan, Tokyo, 1969, Vol. III, 152.
55. **Seki, S., Kasahara, K., Kuriyama, T., and Kawasumi, M.,** Effects of hydration of cement on compressive strength, modulus of elasticity and creep of concrete, Proc. 5th Int. Symp. on the Chemistry of Cement (Tokyo, 1968), Cement Association of Japan, Tokyo, 1969, Vol. III, 175.
56. **Gauthier, E. and Regourd, M.,** The hardening of cement in function of temperature, Proc. RILEM Int. Conf. on Concrete at Early Ages (Paris, 1982), Ecole Nationale des Ponts et Chausses, Paris, Vol. I, 145.
57. **Regourd, M., Mortureux, B., Gauthier, E., Hornain, H., and Volant, J.,** Characterization and thermal activation of slag cements, 7th Int. Congr. on the Chemistry of Cement (Paris, 1980), Editions Septima, Paris, 1980, Vol. III.
58. **Bresson, J.,** Prediction of strength of concrete products, Proc. RILEM Int. Conf. on Concrete at Early Ages (Paris, 1982), Ecole Nationale des Ponts et Chausses, Paris, Vol. I, 111.
59. **Tank, R. C.,** The Rate Constant Model for Strength Development of Concrete, Ph.D. dissertation, Polytechnic University, Brooklyn, NY, June 1988.
60. **Geiker, M.,** Studies of Portland Cement Hydration by Measurements of Chemical Shrinkage and Systematic Evaluation of Hydration Curves by Means of the Dispersion Model, Ph.D. dissertation, Technical University of Denmark, 1983.
61. **Geiker, M. and Knudsen, T.,** Chemical shrinkage of portland cement pastes, *Cement Concr. Res.*, 12 (5), 603, 1982.
62. **Roy, D. M. and Idorn, G. M.,** Hydration, structure, and properties of blast furnace slag cements, mortars and concrete, *J. Am. Concr. Inst.*, 79 (6), 444, 1982.
63. **Barnes, B. D., Orndorff, R. L., and Roten, J. E.,** Low initial temperature improves the strength of concrete cylinders, *J. Am. Concr. Inst.*, 74(12), 612, 1977.
64. **Dodson, C. J. and Rojagoplan, K. S.,** Field tests verify temperature effects on concrete strength, *Concr. Int.*, 1(12), 26, 1979.
65. **Tashiro, C. and Tanaka, H.,** The effect of the lowering of initial curing temperature on the strength of steam cured mortar, *Cement Concr. Res.*, 7(5), 545, 1977.
66. **Al-Rawi, R. S.,** Effects of cement composition and w/c ratio on strength of accelerated cured concrete, *Cement Concr. Res.*, 7(3), 313, 1977.
67. **Butt, Y. M., Kolbasov, V. M., and Timashev, V. V.,** High temperature curing of concrete under atmospheric pressure, Proc. 5th Int. Symp. on the Chemistry of Cement (Tokyo 1968), Cement Association of Japan, Tokyo, 1969, Vol. III, 437.

68. **Naik, T. R.,** Concrete strength prediction by the maturity method, *ASCE Eng. Mech. Div. J.*, 106, No. EM3, 465, 1980.
69. Standard practice for curing concrete, ACI 308-81(Rev. 1986), American Concrete Institute, Detroit MI.
70. **Dilly, R. L., Beizai, V., and Vogt, W. L.,** Integration of time-temperature curing histories with PC spreadsheet software, *J. Am. Concr. Inst.*, 85(5), 375, 1988.
71. **Hansen, A. J.,** COMA-Meter — The mini maturity meter, Nordisk Betong, September 1981.
72. **Knudsen, T. and Geiker, M.,** Chemical shrinkage as an indicator of the stage of hardening, Proc. RILEM Int. Conf. on Concrete at Early Ages (Paris, 1982), Ecole Nationale des Ponts et Chausses, Paris, Vol. I, 163.
73. **Peterson, C. G. and Hansen, A. J.,** Timing of loading determined by pull-out and maturity tests, RILEM Int. Conf. on Concrete at Early Ages (Paris, 1982), Ecole Nationale des Ponts et Chausses, Paris, Vol. I, 173.
74. **Parsons, T. J. and Naik, T. R.,** Early age concrete strength determination by pullout testing and maturity, *In Situ/Nondestructive Testing of Concrete*, Malhotra, V. M., Ed., ACI SP-82, American Concrete Institute, Detroit MI, 1984, 177.
75. Guide for concrete formwork, Rep. of ACI Committee 347, *J. Am. Concr. Inst.*, 85(5), 530, 1988.
76. **Carino, N. J. and Stone, W. C.,** Analysis of in-place test data with spreadsheet software, *Computer Use for Statistical Analysis of Concrete Tests Data*, Eds., Balaguru P. and Ramakrishnan, V. ACI SP-101, American Concrete Institute, Detroit, 1987.
77. **Maage, M. and Helland, S.,** Cold weather concrete curing planned and controlled by microcomputer, *Concr. Int.*, 10(10), 34, 1988.
78. Standard test method for time of setting of concrete mixtures by penetration resistance, ASTM C 403-88, Annual Book of ASTM Standards, Vol. 04.02 Concrete and Aggregates, 1988.

Chapter 6

RESONANT FREQUENCY METHODS*

V. M. Malhotra and V. Sivasundaram

ABSTRACT

This chapter presents a review of resonant frequency method of testing concrete nondestructively. The background and the theoretical basis of this method are briefly discussed, and the test method standardized by ASTM is detailed. A proposed new method of resonant frequency testing is also described. The factors affecting resonant frequency and dynamic modulus of elasticity, study of durability of concrete by means of resonant frequency testing, and the correlations between dynamic modulus of elasticity and (1) static modulus of elasticity, (2) strength properties of concrete are presented. A brief discussion of the damping properties of concrete is also given. Further, specialized applications of this test method and the usefulness and limitations are discussed. A list of 44 references is included.

INTRODUCTION

An important dynamic property of any elastic system is the natural frequency of vibration. For a vibrating beam of given dimensions, the natural frequency of vibration is mainly related to the dynamic modulus of elasticity and density. Hence, the dynamic modulus of elasticity of a material can be determined from the measurement of the natural frequency of vibration of prismatic bars and the mathematical relationships existing between the two. These relationships were derived for solid media considered to be homogeneous, isotropic, and perfectly elastic, but they may be applied to heterogeneous systems, such as concrete, when the dimensions of the specimens are large in relation to the size of the constituents of the material.

For flexural vibrations of a long, thin rod, the following equation or its equivalent may be found in any complete textbook on sound:[1]

$$N = \frac{m^2 k}{2\Pi L^2} \sqrt{\frac{E}{d}} \tag{1}$$

and solving for E

$$E = \frac{4\Pi^2 L^4 N^2 d}{m^4 k^2} \tag{2}$$

where

E	=	dynamic modulus of elasticity
d	=	density of the material
L	=	length of the specimen
N	=	fundamental flexural frequency
k	=	radius of gyration of the section about an axis perpendicular to the plane of bending ($k = t/\sqrt{12}$ for rectangular cross section where t = thickness)
m	=	a constant (4.73 for the fundamental mode of vibration)

The dynamic modulus of elasticity can also be computed from the fundamental longitudinal frequency of vibration of a specimen, according to the following equation:[2]

$$E = 4L^2dN^2 \tag{3}$$

Equations 1 and 3 were obtained by solving the respective differential equations for the motion of a bar vibrating: (1) in flexure in the free-free mode, and (2) in the longitudinal mode.

Thus, the resonant frequency of vibration of a concrete specimen or structure directly relates to its dynamic modulus of elasticity and, hence, its mechanical integrity. The method of determining the dynamic elastic moduli of solid bodies from their resonant frequencies has been in use for the past 45 years. However, until up to the last few years, resonant frequency methods had been used almost exclusively in laboratory studies. In these studies, natural frequencies of vibration are determined on concrete prisms and cylinders in order to calculate the dynamic moduli of elasticity and rigidity, the Poisson's ratio, and for monitoring the degradation of concrete during durability tests.

RESONANT FREQUENCY METHOD

This method was first developed by Powers[3] in the U.S. in 1938. He determined the resonant frequency by matching the musical tone created by concrete specimens, usually 51 × 51 × 241-mm prisms, when tapped by a hammer, with the tone created by one of a set of orchestra bells calibrated according to frequency. The error likely to occur in matching the frequency of the concrete specimens to the calibrated bells was of the order of 3%. The shortcomings of this approach, such as the subjective nature of the test, are obvious. But this method laid the groundwork for the subsequent development of more sophisticated methods.

In 1939 Hornibrook[4] refined the method by using electronic equipment to measure resonant frequency. Other early investigations on the development of this method included those by Thomson[5] in 1940, by Obert and Duvall[2] in 1941, and by Stanton[6] in 1944. In all the tests that followed the work of Hornibrook, the specimens were excited by a vibrating force. Resonance was indicated by the attainment of vibrations having maximum amplitude as the driving frequency was changed. The resonant frequency was read accurately from the graduated scale of the variable driving audio oscillator. The equipment is usually known as a sonometer.

TEST EQUIPMENT

The testing apparatus required by ASTM C 215-85, entitled "Standard Test Method for Fundamental Transverse, Longitudinal, and Torsional Frequencies of Concrete Specimens",[7] is shown schematically in Figure 1. Equipment meeting the ASTM requirements has been designed by various commercial organizations. One of the commercially available sonometers is shown in Figure 2. The resonant frequency test equipment presently used in monitoring the long-term deterioration of massive concrete blocks (dimension 305 × 305 × 915-mm) exposed to sea water in Treat Island, Maine is shown in Figure 3.

The testing apparatus consists primarily of two sections, one generates mechanical vibrations and the other senses these vibrations.[8]

VIBRATION GENERATING SECTION

The principal part of this section is an electronic audio-frequency oscillator, which generates electrical audio-frequency voltages. The oscillator output is amplified to a level suitable for producing mechanical vibrations. The relatively undistorted power output of the amplifier is fed to the driver unit for conversion into mechanical vibrations.

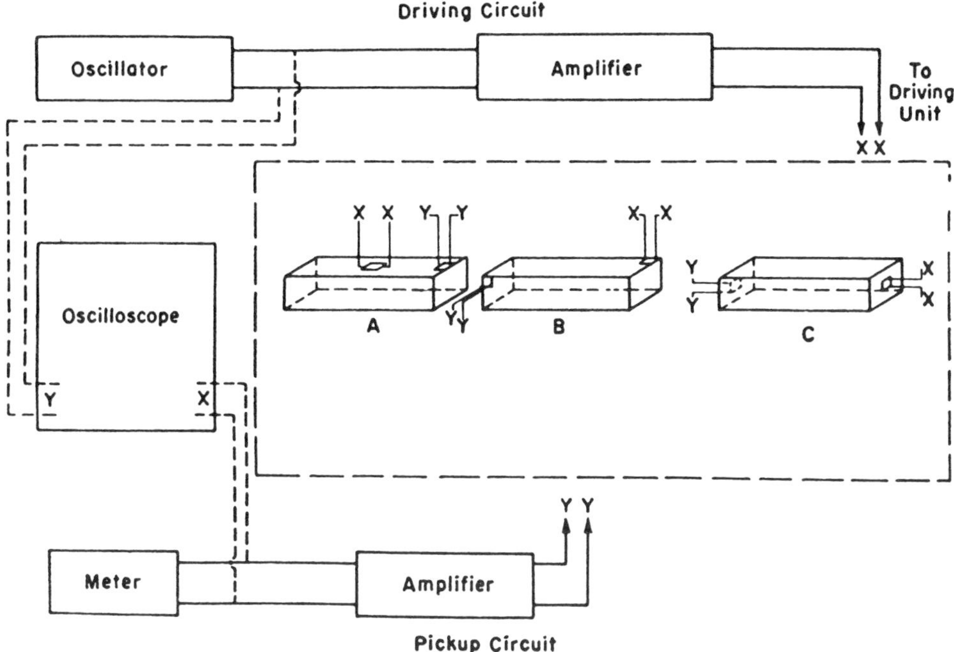

FIGURE 1. Schematic diagram of a typical apparatus showing driver and pick-up positions for the three types of vibration. (A) Transverse resonance. (B) Torsional resonance. (C) Longitudinal resonance. (Adapted from ASTM C 215-85.)

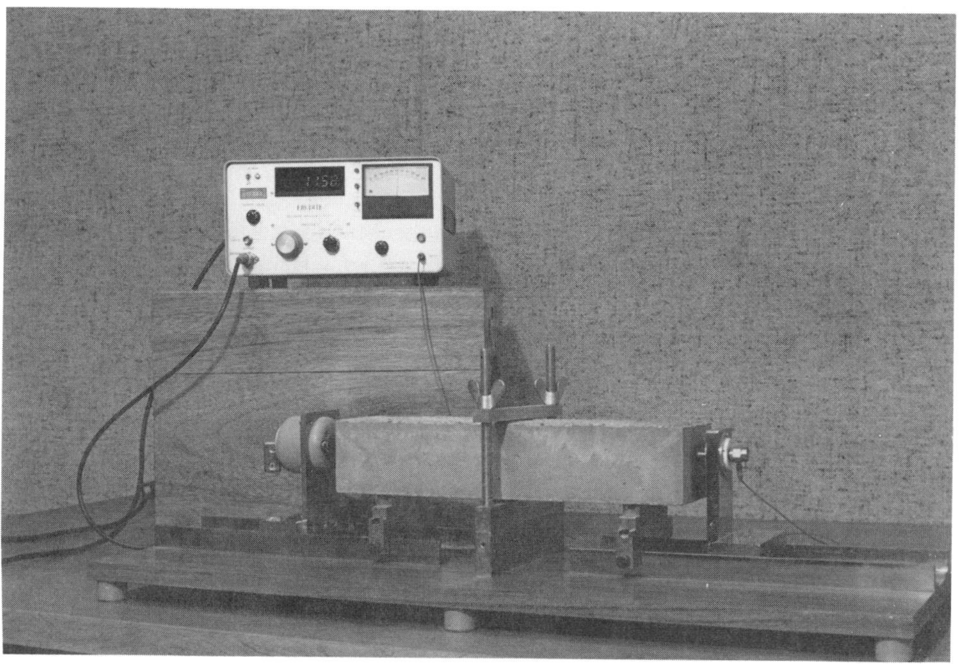

FIGURE 2. Longitudinal resonance testing of a 76 × 102 × 406-mm concrete prism by a sonometer.

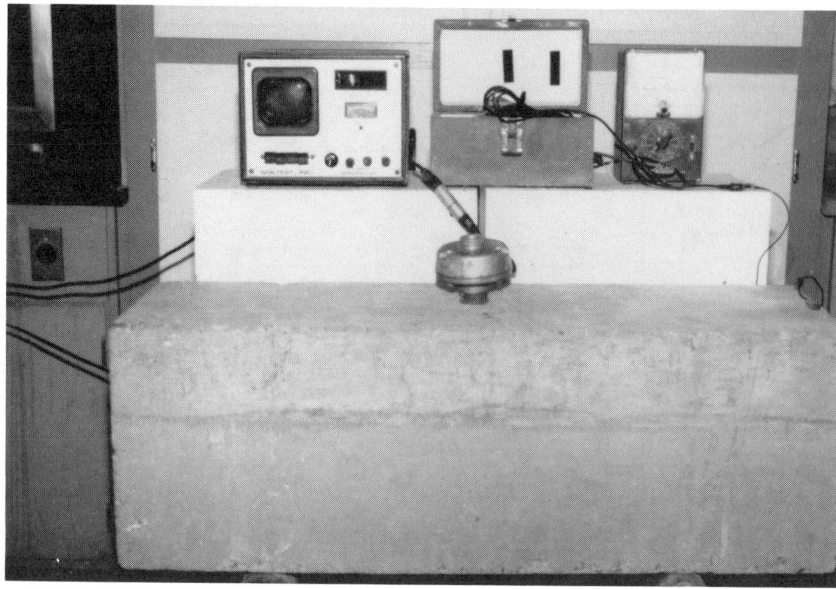

FIGURE 3. The resonant frequency test equipment used in monitoring the deterioration of 305 × 305 × 915-mm concrete blocks exposed to sea water in Treat Island, Maine.

VIBRATION SENSING SECTION

The mechanical vibrations are sensed by a piezoelectric transducer. The transducer is contained in a separate unit and converts mechanical vibrations to electrical AC voltage of the same frequencies. These voltages are amplified for the operation of a panel-mounted meter which indicates the amplitude of the transducer output. As the frequency of the driver oscillator is varied, maximum deflection of the meter needle indicates when resonance is attained. Visible indications that the specimens are vibrating at their fundamental modes can be obtained easily through the use of an auxiliary cathode-ray oscilloscope, and its use is generally recommended.

OPERATION OF THE SONOMETER

Some skill and experience are needed to determine the fundamental resonant frequency using a meter-type indicator because several resonant frequencies may be obtained corresponding to different modes of vibration. Specimens having either very small or very large ratios of length to maximum transverse direction are frequently difficult to excite in the fundamental mode of transverse vibration. It has been suggested that the best results are obtained when this ratio is between three and five.

The supports for the specimen under test should be of a material having a fundamental frequency outside the frequency range being investigated and should permit the specimen to vibrate without significant restriction. Ideally, the specimens should be held at the nodal points, but a sheet of soft sponge rubber is quite satisfactory and is preferred if the specimens are being used for freezing and thawing studies.

The fundamental transverse vibration of a specimen has two nodal points, at distances from each end of 0.224 times the length. The vibration amplitude is maximum at the ends, about three fifths of the maximum at the center, and zero at the nodal points. Therefore, movement of the pickup along the length of the specimen and observation of the meter reading will show whether the specimen is vibrating at its fundamental frequency.

For fundamental longitudinal and torsional vibrations, there is a point of zero vibration (node) at the midpoint of the specimen and the maximum amplitude is at the ends.

TABLE 1
Approximate Ranges of Resonant Frequencies of Concrete Prism and Cylinder Specimens

Size of specimens (mm)	Approximate range of resonant frequency, Hz	
	Transverse	Longitudinal
152 × 152 × 710-mm prism	550—1150	1800—3200
102 × 102 × 510-mm prism	900—1600	2500—4500
152 × 305-mm cylinder	2500—4500	4000—7500

From Jones, R., *Non-Destructive Testing of Concrete*, Cambridge University Press, London, 1962. With permission.

Sometimes in resonance testing of concrete specimens, two resonant frequencies may appear which are close together. Kesler and Higuchi[12] believed this to be caused by a nonsymmetrical shape of the specimen that causes interference due to vibration of the specimen in some direction other than that intended. Proper choice of specimen size and shape should practically eliminate this problem; for example, in a specimen of rectangular cross section the above problem can be eliminated by vibrating the specimen in the direction parallel to the short side.

In performing resonant frequency tests, it is helpful to have an estimate of the expected fundamental frequency. Table 1 shows the approximate ranges of fundamental longitudinal and flexural resonant frequencies of standard concrete specimens given by Jones.[13]

CALCULATION OF DYNAMIC MODULI OF ELASTICITY AND RIGIDITY AND POISSON'S RATIO

The dynamic moduli of elasticity and rigidity and the Poisson's ratio of the concrete can be calculated by equations given in ASTM C 215-85. These are modifications of theoretical equations applicable to specimens that are very long in relation to their cross section, and were developed and verified by Pickett[9] and Spinner and Tefft.[10] The corrections to the theoretical equations in all cases involve Poisson's ratio and are considerably greater for transverse resonant frequency than for longitudinal resonant frequency. For example, a standard 102 × 102 × 510-mm prism requires a correction factor of about 27% at the fundamental transverse resonance, as compared with less than 0.5% at the fundamental longitudinal resonance.[13,15] The longitudinal and flexural modes of vibration give nearly the same value for the dynamic modulus of elasticity. The dynamic modulus of elasticity may range from 14.0 GPa, for low quality concretes at early ages, to 48.0 GPa for good quality concrete at later ages.[11] The dynamic moduls of rigidity is about 40% of the modulus of elasticity.[14]

OTHER METHODS OF RESONANT FREQUENCY TESTING

A new method for determining fundamental frequencies has been proposed by Gaidis and Rosenberg[16] as an alternative to the ASTM C 215 method. In this method, the concrete specimen is struck with a small hammer. The impact causes the specimen to vibrate at its natural frequencies. The amplitude and frequency of the resonant vibrations are obtained using a spectrum analyzer that determines the component frequencies via the fast Fourier transform. The amplitude of the specimen response vs. frequency is displayed on the screen of a cathode ray tube, and the frequencies of major peaks can be read directly.

In operation, the pick-up accelerometer is fastened to the end of the specimen with microcrystalline wax, and the specimen is struck lightly with a hammer. The output of the

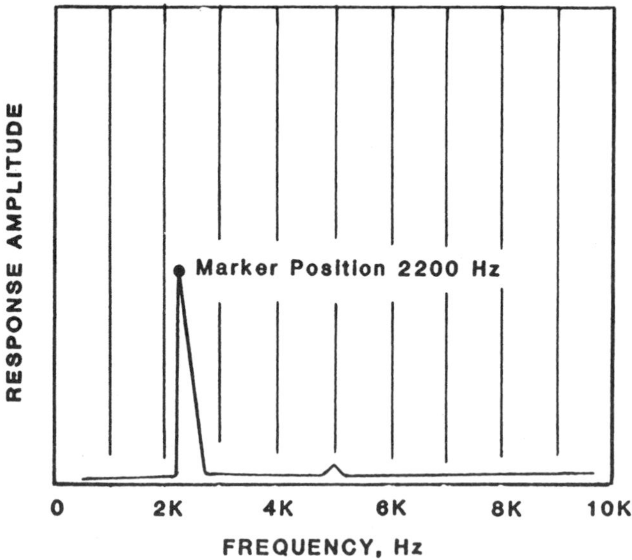

FIGURE 4. The amplitude of specimen response vs. frequency, displayed on a cathode-ray tube. (Adapted from Reference 16.)

TABLE 2
Fundamental Transverse Frequencies of Concrete Determined Using ASTM C 215 and the Impact-Resonance Method

Prism size, mm	ASTM C 215 method frequency, Hz	Spectrum analyzer frequency, Hz
76.2 × 102 × 406	1762	1760
76.2 × 102 × 406	1725	1720
76.2 × 102 × 406	728	620
76.2 × 102 × 406	1936	1860
76.2 × 102 × 406	1912	1920
76.2 × 102 × 406	1243	980
25.4 × 25.4 × 279	1148	1160
25.4 × 25.4 × 279	1077	1060
25.4 × 25.4 × 279	977	980
25.4 × 25.4 × 279(γ)	501	400

γ damaged by freezing and thawing cycling.

From Gaidis, J. M. and Rosenburg, M., *Cem. Concr. and Aggr.*, 8, 117, 1986. With permission.

accelerometer is recorded by the waveform analyzer and recorded signal is processed to obtain the frequency response. On the resulting amplitude vs. frequency curve, a dot marker may be moved to coincide with the peak, and the frequency value of the peak is displayed on the screen (Figure 4). Some typical results obtained with this method and those obtained with ASTM C 215 are shown in Table 2. The advantages of this method over the forced-resonance procedure in ASTM C 215 are the greater speed of testing, and capability of testing specimens having a wide range of dimensions. However, the initial high cost of equipment appears to be a disadvantage. This impact resonance procedure is being considered by ASTM as an alternative to the existing procedure.

FACTORS AFFECTING RESONANT FREQUENCY AND DYNAMIC MODULUS OF ELASTICITY

A number of factors affect the resonant frequency measurements, the dynamic modulus of elasticity, or both. Some of these are discussed below.

INFLUENCE OF MIX PROPORTIONS AND PROPERTIES OF AGGREGATES

The dynamic modulus of elasticity of concrete is affected by the elastic moduli of its constituent materials and their relative proportions. According to Jones,[13] for a given composition of cement paste, that is, the same water-cement ratio, the elastic modulus of hardened concrete increases with an increase in the percentage of total aggregate. It has also been reported that an increase in the amount of mixing water or in the volume of entrapped air reduces the dynamic modulus of elasticity.[13]

In a recent CANMET study* of concretes incorporating high volumes of low-calcium fly ashes, the values of static and dynamic elastic moduli were found to be relatively high in relation to the strength. The higher values are believed to be caused by the unhydrated fly ash particles acting as fine filler in the concrete.

SPECIMEN-SIZE EFFECT

Obert and Duvall[2] demonstrated that for a given concrete, the value of the dynamic modulus of elasticity varies depending on the size of specimen used in the measurements. The larger specimens, because of their dimensions and weight, have lower resonant frequencies. Kesler and Higuchi[12] found that longer beams resonating at lower frequencies gave higher elastic moduli than did proportionately smaller beams. On the other hand, Jones[13] found little change in the dynamic modulus for different specimens having a frequency range of 70 to 10,000 Hz. Thornton and Alexander[17] pointed out that, if other parameters remain unchanged, the resonant frequency of the fundamental flexural mode will increase with the increase in thickness or with the decrease in length of the specimen.

INFLUENCE OF CURING CONDITIONS

Obert and Duvall[2] have shown that although the dynamic modulus of elasticity depends on the moisture content, the change in the elastic modulus with drying is rather small after about 3 or 4 days of air drying. Further, it has been shown that a large decrease in the dynamic modulus of elasticity occurs over the first 48 h of oven drying but the subsequent change is small. Oven drying, even at as low a temperature as 34°C, causes an irreversible reduction of the elastic modulus. A possible explanation is that shrinkage results in microcracking of paste with subsequent reduction in its stiffness and thus affecting the value of the dynamic modulus of elasticity.

Kesler and Higuchi[18] in their studies have concluded:

1. For the same curing conditions, the dynamic modulus of elasticity increases as the strength increases.
2. If the concrete is kept moist, the modulus of elasticity increases with age, and if the concrete is allowed to dry, the modulus of elasticity decreases with age.

However, studies at CANMET* on large concrete blocks incorporating large amounts of supplementary cementing materials have indicated that longer periods of air-drying do not cause any detrimental effects on the static and dynamic moduli of elasticity. These findings have been confirmed by pulse-velocity measurements as well. The strength gain in

* Unpublished CANMET data.

these concretes over longer periods of time due to the pozzolanic reaction of the supplementary materials appears to cause the increase in elastic moduli.

Jones[13] has reported significant differences in the values of elastic moduli determined from flexural and longitudinal resonance tests on beam specimens moist-cured for 25 days and subsequently air-cured for 150 days. The above difference is believed to have been caused by the loss of moisture resulting in gradients for moisture content, elastic modulus, and density in each dimension of the beam. These gradients would affect the flexural and longitudinal modes of vibration in different ways.

The effects of curing conditions on the resonant frequency and dynamic modulus of elasticity are rather critical. Unless special curing conditions are required, water-curing is to be preferred and the specimen should be in a water-saturated or saturated-surface-dry condition at the time of test. This will help in achieving more reproducible results.

RESONANT FREQUENCY AND DURABILITY OF CONCRETE

The determination of flexural resonance has been employed to advantage in studying the effects of successive accelerated freezing and thawing cycles and aggressive environments on concrete specimens. The advantages of resonance methods in this regard are

1. The repeated tests can be carried out on the same specimens over a very long period, and the number of test specimens to be cast is therefore greatly reduced.
2. The results obtained with flexural resonance methods on the same specimen are more reproducible than those obtained with destructive tests on a group of supposedly duplicate specimens.

DETERIORATION OF CONCRETE IN FREEZING AND THAWING CYCLING

Extensive studies of changes in dynamic modulus of elasticity with the deterioration of concrete subjected to repeated cycles of freezing and thawing have been reported by Hornibrook,[4] Thomson,[5] Long and Kurtz,[19] Axon et al.,[20] and Malhotra and Zoldners.[21-23] Results of one such study are shown in Figure 5.

The ASTM Method C 666-84, entitled "Standard Test Method for Resistance of Concrete to Rapid Freezing and Thawing", specifies resonant frequency methods for studying the deterioration of concrete specimens subjected to repeated cycles of freezing and thawing. The standard requires the calculation of the relative dynamic modulus of elasticity and durability factor.

Wright and Gregory[24] suggested the use of the resonant frequency values rather than the values of dynamic elastic modulus as a criterion for evaluating the results of freezing and thawing tests. This is because the calculations for the dynamic elastic modulus are based on the assumption that concrete is isotropic and homogeneous material, which it is not. The calculations in ASTM C 666 use the square of the resonant frequency values to evaluate the test results. Changes in the shape factor, due to dimensional changes during the test, are ignored.

CORROSION OF CONCRETE IN AGGRESSIVE MEDIA

A number of studies have been reported in which resonance methods have been used to determine the damage sustained by concrete in aggressive media such as acidic or alkaline environments.

Studies dealing with the determination of the dynamic elastic modulus of concrete specimens made from five different types of cements and subjected to the action of ammonium nitrate solution have been reported by Chefdeville.[25] In his investigation, the percentage reduction in the dynamic elastic modulus of elasticity of the concrete specimens was calculated by subjecting the specimens to resonant frequency testing at predetermined intervals.

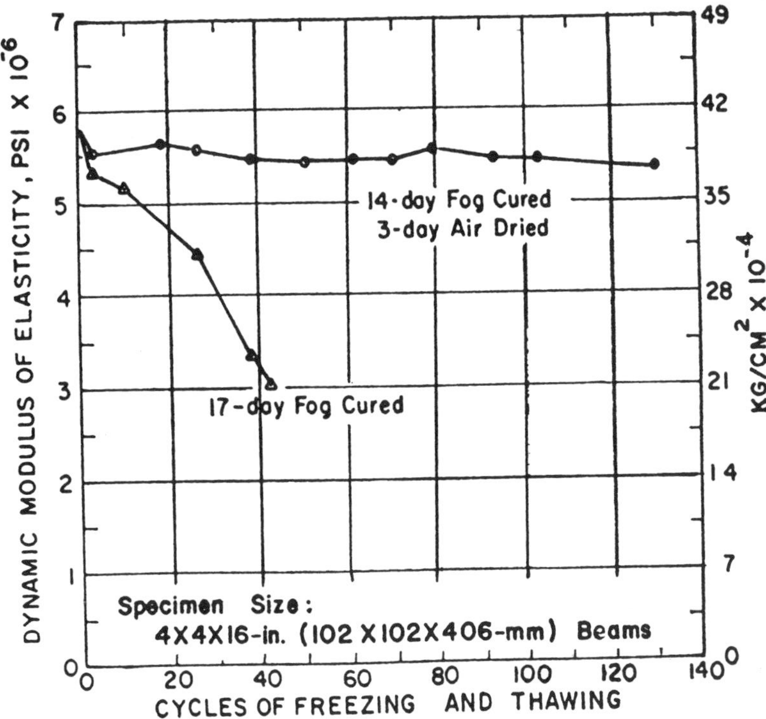

FIGURE 5. Effect of cycles of freezing and thawing on dynamic modulus of elasticity. (Adapted from Reference 19.)

It was found that the reduction in the dynamic elastic modulus is somewhat similar to that obtained in the freezing and thawing studies. The corrosive effects of dilute sulfuric acid and acetic acids on concrete prisms have also been reported by Stuterheim et al.[12-b]

In a recent study, the corrosive effect of ammonium nitrate-based fertilizers on blended natural pozzolan-portland cement concretes was investigated by testing for dynamic modulus of elasticity and density for 8 weeks.[26] It was observed that the corrosive effect increased with nitrification of the alkaline medium, and the concretes made with pozzolan-substituted cements were more damaged than the portland cement concretes.

REPRODUCIBILITY OF TEST RESULTS

Limited data are available on the reproducibility of the dynamic modulus of elasticity based on resonance tests. Jones[13] has published data indicating that for standard-size prisms and cylinders the reproducibility of dynamic modulus of elasticity is superior to that obtained in static tests (Table 3). According to Jones, the greater variability of the static elastic modulus results is due to greater errors introduced in the testing procedure rather than to greater variability between specimens. On the other hand, each of the measurements needed to determine the dynamic elastic modulus by resonance methods, that is, the resonant frequency, length, and density, can be measured accurately.

ASTM C 215 gives the following precision statement for fundamental transverse frequency only, determined on concrete prisms as originally cast.

Single-Operator Precision — Criteria for judging the acceptability of measurements of fundamental transverse frequency obtained by a single operator in a single laboratory on concrete specimens made from the same materials and subjected to the same conditions are given in below. These limits apply over the range of fundamental transverse frequency from 1400 to 3300 Hz.

TABLE 3
Comparison of Reproducibility of the Standard Methods of Measuring Static and Dynamic Young's Modulus of Elasticity

Size of specimen (mm)	Young's modulus of elasticity	No. of specimens	Standard error of 3 results (MPa)
152 × 305-mm cylinder	Static	3	604
76 × 76 × 305-mm prisms	Static	3	1000
152 × 152 × 710-mm prisms	Dynamic	3	270
102 × 102 × 510-mm prisms	Dynamic	3	350
76 × 76 × 305-mm prisms	Dynamic	3	370

From Jones, R., *Non-Destructive Testing of Concrete,* Cambridge University Press, London, 1962. With permission.

Test Results for Single Operator in a Single Laboratory

	Coefficient of variation, %[A]	Acceptable range of two results, % of average[A]
Within-batch single specimen	1.0	2.8
Within-batch, average of 3 specimens[B]	0.6	1.7
Between-batch, average of 3 specimens per batch	1.0	2.8

[A] These numbers represent, respectively, the 1S% and D2S% limits as described in Practice C 670.

[B] Calculated as described in Practice C 670.

Note 1 — The coefficients of variation for fundamental transverse frequency have been found to be relatively constant over the range of frequencies given for a range of specimen sizes and age or condition of the concrete, within limits.

The different specimen sizes represented by the data include the following (the first dimension is the direction of vibration):

76 by 102 by 406 mm (3 by 4 by 16 in.)
102 by 76 by 406 mm (4 by 3 by 16 in.)
89 by 114 by 406 mm ($3^1/_2$ by $4^1/_2$ by 16 in.)
76 by 76 by 286 mm (3 by 3 by $11^1/_4$ in.)
102 by 89 by 406 mm (4 by $3^1/_2$ by 16 in.)
76 by 76 by 413 mm (3 by 3 by $16^1/_4$ in.)

The multilaboratory coefficient of variation for averages of three specimens from a single batch of concrete has been found to be 3.9% for fundamental transverse frequencies over the range from 1400 to 3300 Hz (Note 2). Therefore, two averages of three specimens from the same batch tested in different laboratories should not differ by more than 11.0% of their common average (see Note 6).

Note 2 — These numbers represent, respectively, the 1S and D2S limits as described in Practice C 670, where 1S is the estimate of the standard deviation characteristic of the total statistical population and D2S is the difference between two individual test results that would be equaled or exceeded in the long run in only 1 case in 20 in the normal and correct operation of the method."

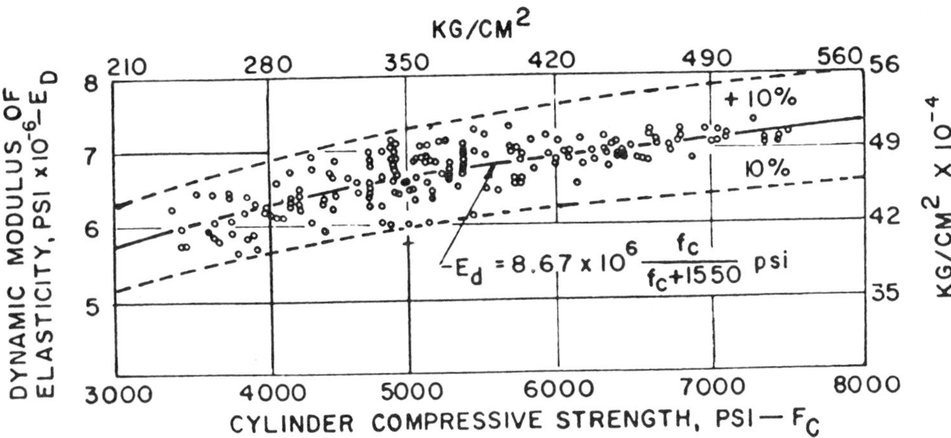

FIGURE 6. Relationship between dynamic modulus of elasticity and compressive strength of concrete. (Adapted from Reference 31.)

CORRELATION BETWEEN DYNAMIC MODULUS OF ELASTICITY AND STRENGTH PROPERTIES OF CONCRETE

Several investigators[27-34] have attempted to establish empirical relationships between the dynamic modulus of elasticity and strength of concrete. Some of these correlations appear to hold for the particular type of concrete investigated, but it is doubtful that any generalized relationships can be given. Therefore, if the flexural and compressive strengths of concrete are to be estimated from the dynamic modulus of elasticity, it is essential first to establish an experimental relationship between these strengths and the dynamic modulus of elasticity.

Jones[13] stated that no general relationship exists between the dynamic elastic modulus of concrete and its flexural or compressive strength. Nevertheless, he concluded that limited correlations are obtained when the changes in the dynamic elastic modulus and strength are produced by changes in the age of the concrete, the degree of compaction, the water-cement ratio, or by deterioration.

Notwithstanding the above limitations, relationships between the strength parameters and the dynamic modulus of elasticity have been reported by various researchers.[31,34] Two such relationships are illustrated in Figures 6 and 7.

COMPARISON OF MODULI OF ELASTICITY DETERMINED FROM LONGITUDINAL AND TRANSVERSE FREQUENCIES

In routine calculation of dynamic elastic modulus, only the transverse frequencies are determined because it is generally easier to determine the transverse resonant frequency. However, Batchelder and Lewis[35] have shown that excellent correlation exists between the elastic moduli calculated from the transverse and longitudinal frequencies (Figure 8).

In his studies, Jones[13,15] found that for wet concretes there was no appreciable difference in the dynamic modulus of elasticity determined from the transverse and longitudinal modes of vibration. However, when the beams were allowed to dry, the dynamic elastic modulus calculated from the transverse vibrations was lower than that calculated from longitudinal vibrations. This was attributed to the moisture gradients within the concrete beams.

COMPARISON OF DYNAMIC AND STATIC MODULI OF ELASTICITY

A considerable amount of work has been carried out by various researchers to establish

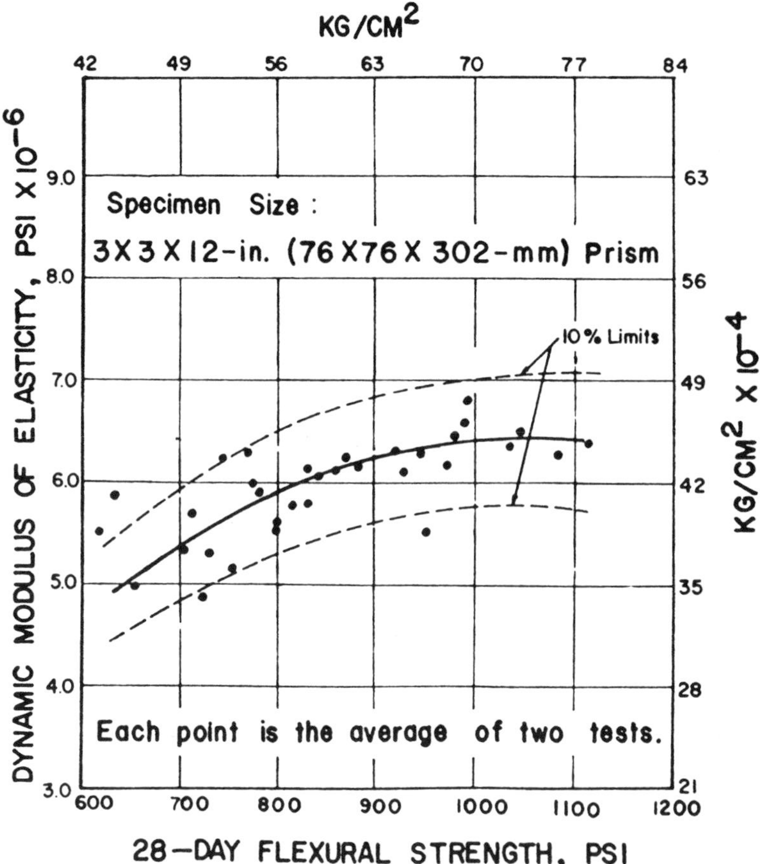

FIGURE 7. Relationship between dynamic modulus of elasticity and 28-day flexural strength of concrete. (Adapted from Reference 34.)

the relationship between the dynamic elastic modulus and static elastic modulus obtained from conventional stress-strain tests conducted at low rates of loading as in ASTM C 469. It should be noted that resonance tests subject the concrete to very low strains compared to static load tests. The following observations may be made from the investigations by Powers,[3] Stanton,[6] Sharma and Gupta,[31] Whitehurst,[36] Klieger,[37] and Philleo:[38]

1. The dynamic modulus of elasticity is generally somewhat higher than the static elastic modulus; the difference depends upon the degree of precautions taken during the conduct of the experiments and the applications of the correction factors allowed for in the equations for the computations of the dynamic elastic modulus.
2. As the age of the specimen increases, the ratio of static elastic modulus to dynamic modulus also increases and more nearly approaches 1.0.*
3. For higher static moduli of elasticity, the values for both dynamic and static moduli of elasticity show close agreement.

* On the basis of tests of 2-year-old specimens reported by Witte and Price,[6] the static elastic modulus in compression was equivalent to 89% of the dynamic modulus, while the static elastic modulus in flexure was equal to 88% of the dynamic modulus. When the tests were repeated after the specimens were 3 years old, these values were found to be 96 and 87%, respectively.

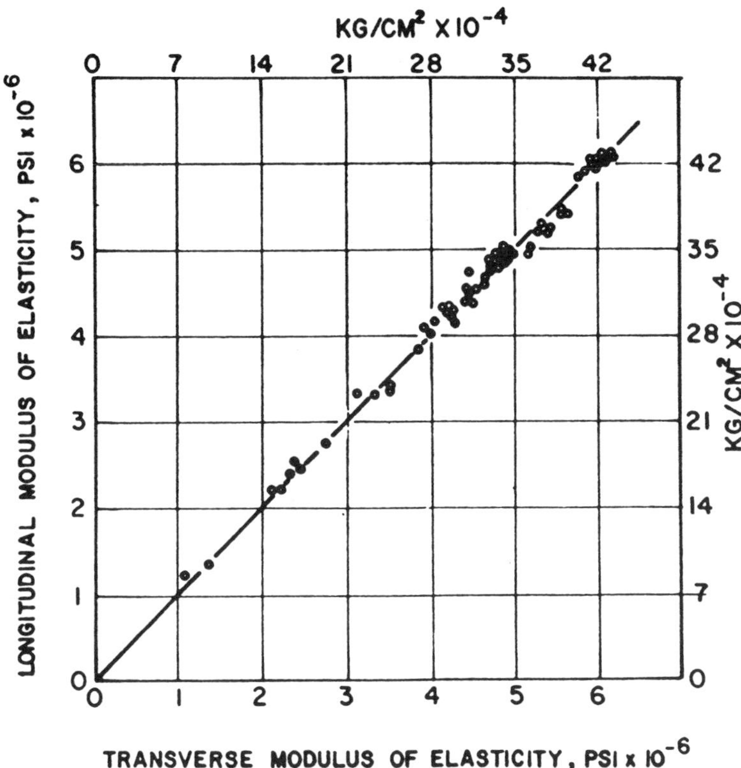

FIGURE 8. Comparison of dynamic moduli of elasticity determined from longitudinal and transverse frequencies. (Adapted from Reference 35.)

Figure 9 shows the relationship between static and dynamic elastic moduli for high-strength concrete as developed by Sharma and Gupta.[31]

Hansen[39] reported an investigation in which concretes with paste contents of 32 and 50%, each having water-cement ratios of 0.40 and 0.60, were investigated together with a concrete with 20% replacement of cement by silica fume. This study showed that the static elastic modulus determined according to ASTM C 469 could be predicted from the dynamic elastic modulus computed using the resonance technique. It was also shown that the entire static elastic modulus vs. strength relationship for each concrete could be predicted from early age measurements of dynamic elastic modulus and compressive strength. Figure 10 shows the excellent agreement of the results of this study to that obtained by Kesler and Higuchi.[18]

Figure 11 shows the elastic modulus-compressive strength curves obtained from early-age results (less than 3 days) as well as those obtained over the entire age span investigated. The predicted modulus of elasticity values are within 5% and 1% for the two systems with and without silica fume, respectively, when compared with those calculated from regression curves, based on data up to 28 days.

SPECIALIZED APPLICATIONS OF RESONANCE TESTS

Gatfield[40] has reported a relatively simple and inexpensive technique for fatigue tests of plain concrete using the resonant frequency method. In his method, significant vibratory strains up to 120×10^{-6} are induced on concrete specimens in flexure. The local failures caused by the loading are determined by measuring the changes in flexural resonant fre-

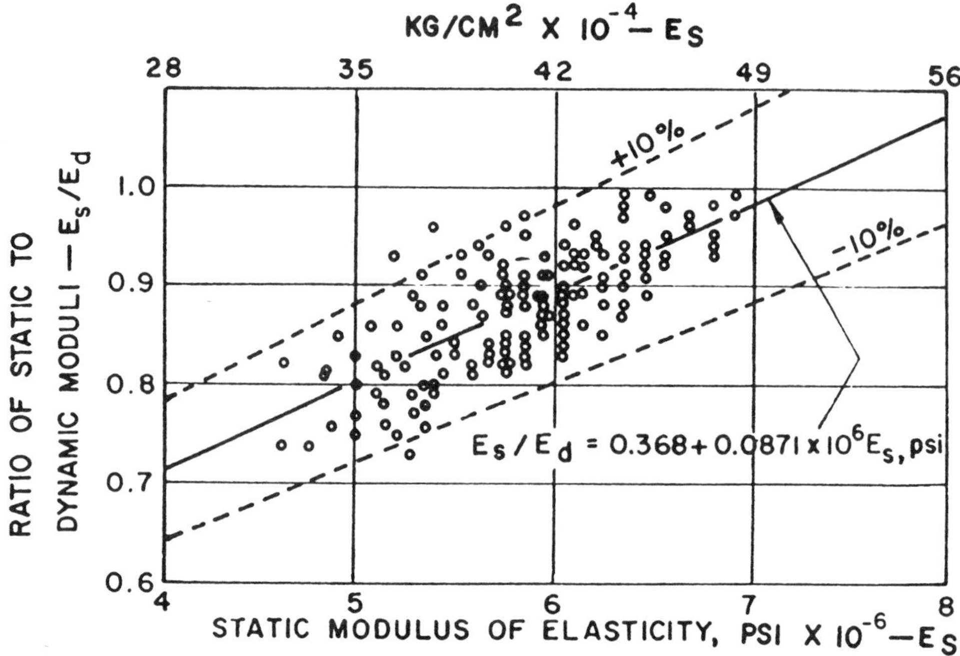

FIGURE 9. Relationship between static modulus of elasticity and ratio of static to dynamic moduli. (Adapted from Reference 31.)

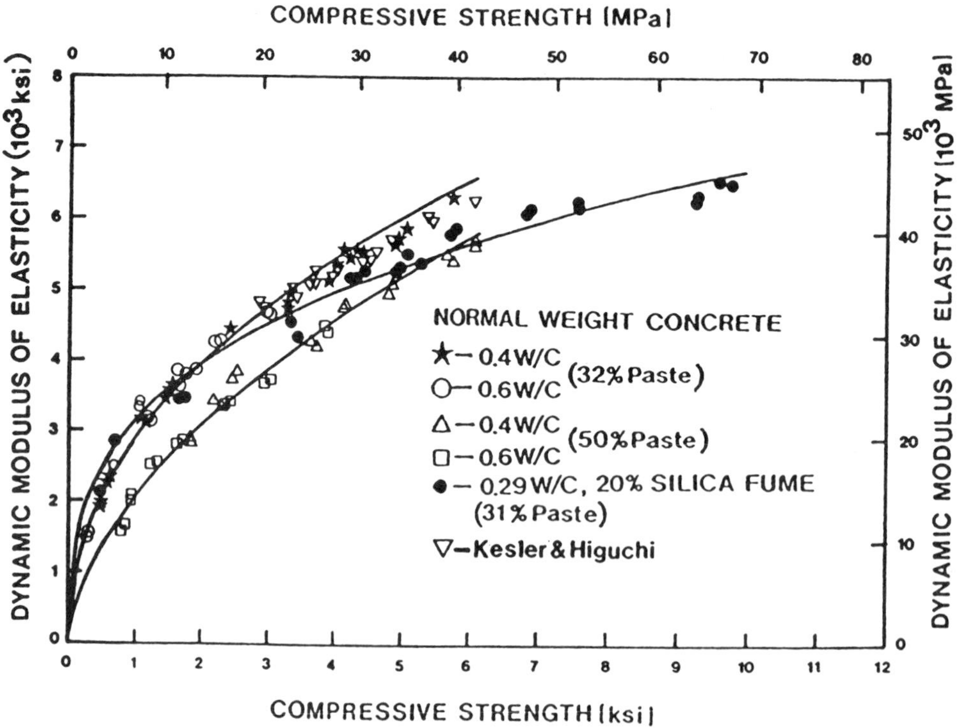

FIGURE 10. Dynamic modulus of elasticity for concretes containing river gravel. (Adapted from Reference 39.)

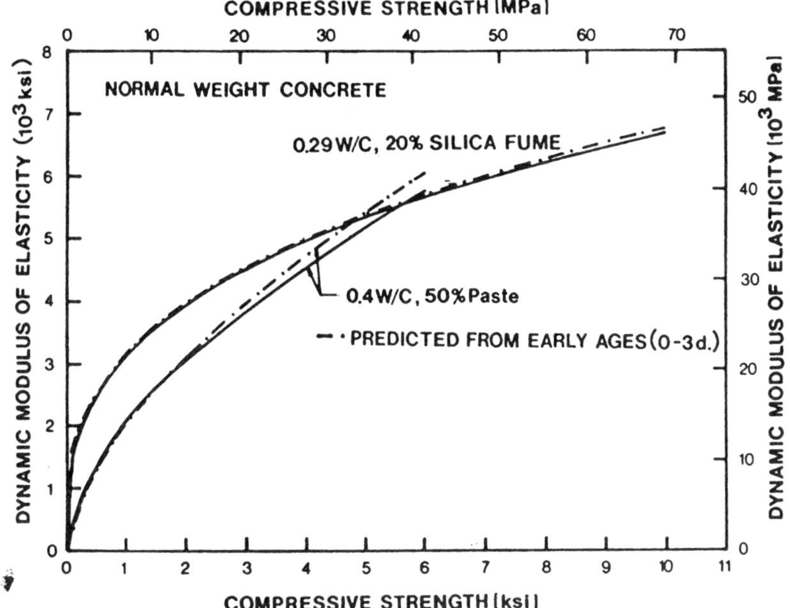

FIGURE 11. A comparison between dynamic elastic modulus vs. compressive strength curves predicted from early ages, and curing times up to about 28 days. (Adapted from Reference 39.)

quency. Also, studies on creep and relaxation of concrete using the resonant frequency techniques have been reported by Chang and Kesler.[41,42]

The variations in the dynamic elastic modulus measured before and after the test specimens were subjected to loading-unloading cycles have been published by Daxelhoffer.[12-k] It was concluded that the stressing of concrete even for a short time affected the dynamic elastic modulus, which showed a decrease of 3 to 3.5% at 28 days.

Thornton and Alexander[17] observed in a study that the dynamic elastic modulus of large concrete structures, and hence, their mechanical integrity, can be assessed by measuring the fundamental resonant frequency of vibration. In this investigation, mathematical and physical modeling tests of large concrete piers were performed in order to establish the range of frequencies expected when actual tests were performed. Following the laboratory modeling, resonant frequency measurements were performed on the actual piers by setting the structures in motion by impact loading, and the frequency content of the vibrations was determined by waveform analyzers. The results indicated that the piers were structurally sound, and it was shown that by exciting a structure with an impact force and analyzing the output, reliable data of overall structural integrity could be obtained.

DAMPING PROPERTIES OF CONCRETE

Damping is the property of a material causing free vibrations in a specimen to decrease in amplitude as a function of time. In concrete technology it is short-term decay that is of interest rather than long-term effects. Several investigators, particularly Thomson,[5] Obert and Duvall,[2] Kesler and Higuchi,[18] and Shrivastava and Sen,[30] Swamy and Rigby[14] have shown that certain properties of concrete can be related to its damping ability.

There are several methods of determining the damping characteristics of a material, but two common methods used for concrete are

1. The determination of logarithmic decrement, δ, which is the natural logarithm of the ratio of any two successive amplitudes in the free vibration of the specimen
2. Calculation of the damping constant Q from the resonance curve of the test specimen

LOGARITHMIC DECREMENT

The logarithmic decrement is the natural logarithm of the ratio between the amplitudes of successive oscillations in the damped sine wave produced by the decay of free vibrations of a specimen, and it is given by the following equation:

$$\delta = \ln \frac{h_1}{h_2} \tag{4}$$

where

δ	=	logarithmic decrement
h_1 and h_2	=	amplitudes of two successive vibrations after the driving force has been removed from the specimen

The amplitudes h_1 and h_2 can be obtained by using an oscilloscope to record the decay of vibrations at resonance after the driving oscillator is turned off. The image on a cathode-ray oscilloscope can be photographed, and the amplitudes h_1 and h_2 can easily be measured off the developed film. If a digital oscilloscope is used, the amplitudes can be measured directly from the display using the cursor controls.

DAMPING CONSTANT

The damping constant Q is given by the equation

$$Q = \frac{f_0}{f_2 - f_1} \tag{5}$$

where

Q	=	damping constant
f_o	=	resonant frequency of vibrations
f_1, f_2	=	frequencies on either side of resonance at which the amplitude is 0.707 times the amplitude at resonance

The values of f_1 and f_2 can be determined if an output meter is employed for resonance indication. After locating the fundamental resonance, the oscillator is de-tuned on each side of the resonance frequency until the output meter reads 0.707 times the reading at resonance. The frequencies at which this occurs are the frequencies f_1 and f_2.[13] In general, hardened concrete has Q values between 50 and 200; the higher the value of Q the less is the decrease in successive amplitudes of vibration.

Figure 12 is a resonance curve to illustrate how the damping constant is determined. In this case the value of f_0 is 1990 Hz and the values of f_2 and f_1, at which the amplitude was 0.707 times the amplitude at resonance are 1996 and 1984 Hz, respectively. This gives a Q value of 1990/12 = 166.

The relationship between the damping constant and the logarithmic decrement is as follows:

$$Q = \frac{\pi}{\delta} \tag{6}$$

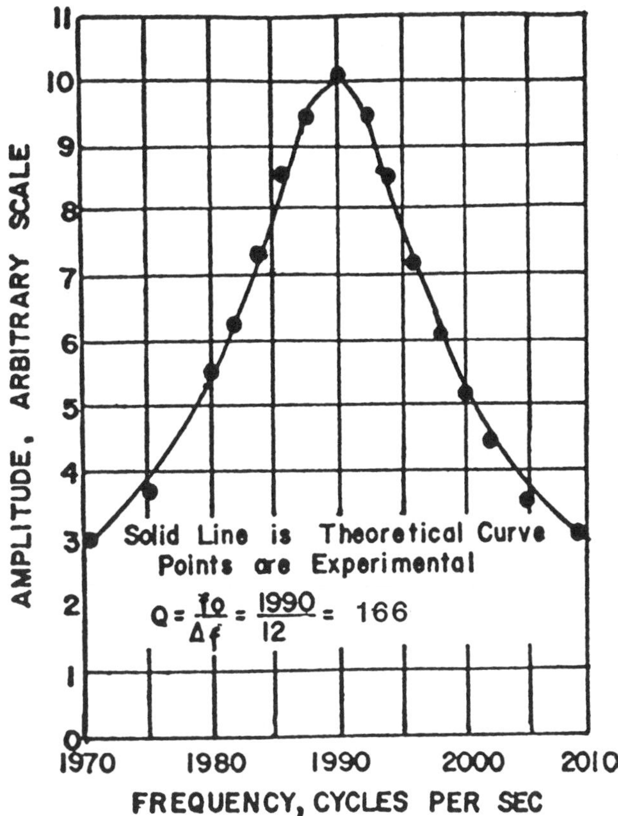

FIGURE 12. Representative resonance curve. (Adapted from Reference 2.)

Substituting the value of δ from Equation 4:

$$Q = \frac{\pi}{\ln \frac{h_1}{h_2}} \tag{7}$$

Table 4 gives values of the damping constant for rock specimens obtained using the resonance curve method and the decay procedure. According to Obert and Duvall,[2] of the above two methods, the resonance curve approach is the more accurate and the less difficult to measure.

The damping in concrete is a complex phenomenon. According to Swamy and Rigby,[14] most of the damping in concrete occurs in the matrix, with some in the interfacial boundaries, and less in the aggregate. The presence of air voids in dry specimens contributes little to the damping, whereas moisture in the matrix is a major contributor to the damping ability of concrete.

In measuring the parameters associated with the damping properties of concrete, considerable care should be taken to use supports which exert low restraint on the concrete specimens. Otherwise, substantial energy losses would be introduced, resulting in erroneous test data.

The study of the vibrational characteristics of structures, such as frequency, damping, and mode shape is called modal analysis.[17] The relationship between the modal properties

TABLE 4
Damping Constant as Determined From the Width of Resonance Curve and From the Logarithmic Decay of Free Vibration

Material	Damping constant Q	
	Resonance	Decay
Granite No. 50	456	444
Granite No. 51	435	417
Granite No. 47	297	286
Granite No. 48	280	256
Granite No. 49	250	270
Sandstone No. 39	83	86
Sandstone No. 41	59	54
Sandstone No. 42	61	61

From Obert, L. and Duvall, W. I., in *Proc. ASTM*, 41, 1053, 1941. With permission.

and the factors influencing them is complex. Generally factors such as geometry, elastic modulus, and boundary conditions significantly affect the modal properties.

Thornton and Alexander[43] observed that the resonant frequencies and the damping functions can be calculated by the input of parameters of geometry, restraint, and dynamic elastic modulus of the total structure into a finite element program, calibrated from measurements made at the completion of construction, when the structure is known to be sound. Then, these values can be compared with actual data measured from the structure at a later time to detect any anomalies.

In one field investigation, impact-resonance measurements were made on the walls of both the prototype and the model of a concrete building.[43] Measurements of mode shapes, resonant frequencies, and damping factors were made on the prototype before soil was moved against the outside walls and afterward. When the wall was covered by soil, the resonant frequency of the fundamental mode was found to increase by 30% together with an increase in damping. Later, the prototype wall was subjected to blast loading and sustained minor structural damage. Again, the impact-resonance measurements were repeated after removing the soil away from the wall. Compared with the initial measurements (without soil), the resonant frequency was found to be decreased slightly, and the damping was found to be decreased significantly.

Thornton and Alexander[43] have reported another investigation in which tests made on a dam showed that the damping was about 3 or 4% of critical damping. When damping is less than critical, the motion is oscillatory, and it is non-oscillatory otherwise. When cylinders made from the dam concrete were tested, supported at the nodes with narrow supports, the damping of the specimens was found to be only 0.37% of critical damping. Although size may have been a factor, the boundary conditions were considered more critical in influencing the damping. The damping was found to be (1) 0.55%, (2) 0.86%, and (3) 1.5% of critical damping, when the cylinders were tested (1) lying on the soil, (2) embedded 50 mm deep in the soil, and (3) embedded 75 mm deep in the soil, respectively. In all the above, damping was determined from measurements of the fundamental resonant frequency of the flexural mode, and the frequency was not found to change significantly. The above indicates the ability of the boundary conditions to influence damping.

STANDARDIZATION OF RESONANT FREQUENCY METHODS

The ASTM Test Method C 215 "Standard Test Method for Fundamental Transverse, Longitudinal, and Torsional Frequencies of Concrete Specimens" was published in 1947 and since then has been revised six times.[7] The last revision to this Standard was in 1985, and the standard is to be further revised by the ASTM C 215 subcommittee in the near future. The impact-resonance procedure will probably be added as an alternative to the forced resonance procedure currently used.

The significance and use statement of the test method as given in ASTM Method C 215 is as follows:

"3.1 This test method is intended primarily for detecting significant changes in the dynamic modulus of elasticity of laboratory or field test specimens that are undergoing exposure or weathering or other types of potentially deteriorating influences.

3.2 This test method may be used to assess the uniformity of field concrete, but it should not be considered as an index of compressive or flexural strength nor as an adequate test for establishing the compliance of the modulus of elasticity of field concrete with that assumed in design.

3.3 The conditions of manufacture, the moisture content, and other characteristics of the test specimens materially influence the results obtained.

3.2 Different computed values for the dynamic modulus of elasticity may result from widely different resonant frequencies of specimens of different sizes and shapes of the same concrete. Therefore, comparison of results from specimens of different sizes or shapes should be made with caution."

LIMITATIONS AND USEFULNESS OF RESONANT FREQUENCY METHODS

Although the basic equipment and testing procedures associated with the resonant frequency techniques have been standardized in various countries, and commercial testing equipment is easily available, the usefulness of the tests is seriously limited because:

1. Generally, these tests are carried out on small-sized specimens in a laboratory rather than on structural members in the field because resonant frequencing is affected considerably by boundary conditions and the properties of concrete. The size of specimens in these tests is usually 152 × 305-mm cylinders or 76 × 76 × 305-mm prisms.
2. The equations for the calculation of dynamic elastic modulus involve "shape factor" corrections. This necessarily limits the shape of the specimens to cylinders or prisms. Any deviation from the standard shapes can render the application of shape factor corrections rather complex.

Notwithstanding the above limitations, the resonance tests provide an excellent means for studying the deterioration of concrete specimens subjected to repeated cycles of freezing and thawing and to deterioration due to aggressive media. The use of resonance tests in the determination of damage by fire and the deterioration due to alkali-aggregate reaction have also been reported by Chefdeville[25] and Swamy and Al-Asali.[44]

The resonant frequency test results are often used to calculate the dynamic modulus of elasticity of concrete but the values obtained are somewhat higher than those obtained with standard static tests carried out at lower rates of loading. The use of dynamic modulus of elasticity in design calculations is not recommended.

Various investigators have published correlations between the strength of concrete and its dynamic modulus of elasticity. The indiscriminate use of such correlations to predict compressive and/or flexural strength of concrete is discouraged unless similar relationships have been established in the laboratory for the particular concrete under investigation.

REFERENCES

1. **Rayleigh, J. W.,** *Theory of Sound,* 2nd ed., Dover Press, New York, 1945.
2. **Obert, L. and Duvall, W. I.,** Discussion of dynamic methods of testing concrete with suggestions for standardization, *Proc. ASTM,* 41, 1053, 1941.
3. **Powers, T. C.,** Measuring Young's modulus of elasticity by means of sonic vibrations, *Proc. ASTM,* 38, Part II, 460, 1938.
4. **Hornibrook, F. B.,** Application of sonic method to freezing and thawing studies of concrete, ASTM Bull. No. 101, December 1939, 5.
5. **Thomson, W. T.,** Measuring changes in physical properties of concrete by the dynamic method, *Proc. ASTM,* 40, 1113, 1940. Also, discussion by T. F. Willis and M. E. de Reus, pp. 1123-1129.
6. **Stanton, T. E.,** Tests comparing the modulus of elasticity of portland cement concrete as determined by the dynamic (sonic) and compression (secant at 1000 psi) methods, ASTM Bull. No. 131, Dec. 1944, 17. Also, discussion by L. P. Witte and W. H. Price, pp. 20-22.
7. Standard Test Method for Fundamental Transverse, Longitudinal, and Torsional Frequencies of Concrete Specimens, (ASTM C 215-85), 1987 Annual Book of ASTM Standards, Section 4, American Society for Testing and Materials, Philadelphia, 154.
8. **Malhotra, V. M.,** Testing of hardened concrete: nondestructive methods, Monogr. No. 9, American Concrete Institute, Detroit, 1976, 52.
9. **Pickett, G.,** Equations for computing elastic constants from flexural and torsional resonant frequencies of vibration of prisms and cylinders, *Proc. ASTM,* 45, 846, 1945.
10. **Spinner, S. and Teftt, W. E.,** A method for determining mechanical resonance frequencies and for calculating elastic moduli from these frequencies, *Proc. ASTM,* 61, 1221, 1961.
11. **Orchard, D. F.,** *Concrete Technology,* Vol. 2, Practice, John Wiley & Sons, New York, 1962, 181.
12. **Kesler, C. E. and Higuchi, Y.,** Problems in the sonic testing of plain concrete, Proc. Int. Symp. on Nondestructive Testing of Materials and Structures, Vol. 1, RILEM, Paris, 1954. 45. Other contributions dealing with sonic tests and published in this symposium are listed below:

Vol. 1

(a) **Cabarat, R.,** Measurement of elastic constants by an acoustical procedure, 9.
(b) **Stutterheim, N., Lochner, J. P. A., and Burger, J. F.,** A method for determining the dynamic Young's modulus of concrete specimens developed for corrosion studies, 18.
(c) **Takabayashi, T.,** Comparison of dynamic Young's modulus and static modulus for concrete, 34.
(d) **Arredi, F.,** Nondestructive tests on concrete specimens performed in the hydraulic structures laboratory of the engineering faculty of Rome, 55.
(e) **Takano, S.,** Determination of concrete strength by a nondestructive method, 61.
(f) **Higuchi, Y.,** Studies presented by the author, 69.
(g) **Fujita, K. I.,** Nondestructive method for concrete, 71.
(h) **Ban, S.,** Activity of the committee on nondestructive concrete inspection of the ASTM, 74.
(i) **Kilian, G.,** Evolution of the mechanical and elastic properties of concretes as a function of age, the proportion of binder and the nature of the aggregates, 75.
(j) Report by Lazard, 80.
(k) **Daxelhofer, J. P.,** Concrete anisotropy brought out by the measurement of the dynamic modulus, 89.
(l) **Daxelhofer, J. P.,** Remarks on the use of sonic methods for investigating the liability to frost damage of concretes at the materials testing laboratory of the Lausanne Polytechnical School, 98.
(m) **Daxelhofer, J. P.,** Note on the static and dynamic moduli of a concrete, 106.
(n) **Daxelhofer, J. P.,** Note on the variation of the dynamic modulus in function of the water content of a lightweight concrete, 108.

(o) **Elvery, R. H.,** Symposium on the non-destructive testing of concrete, 111.
(p) **Beauzee, C.,** Errors of measurement in the determination of the modulus of elasticity by the sonic method, 120.
(q) **Jones, R.,** The testing of concrete by an ultrasonic pulse technique, 137.
(r) **Chefdeville, J.,** The qualitative control, 166.
(s) **Andersen, J.,** Apparatus for determination of sound velocity in concrete and execution of the measurements, 179.
(t) **Nerenst, P.,** Speed of propagation in concrete determined by a condensing chronograph, 184.
(u) **Nerenst, P.,** Wave velocity as influenced by curing conditions and age, 200.
(v) **Andersen, J.,** The use of sound-velocity measurements for practical tests of concrete, 205.

Vol. 2
(w) **Kameda, Y., Awaya, K., and Yokoyama, I.,** The nondestructive testing of concrete, 209.
(x) **Voellmy, A.,** Vibration testing of concrete in structures in Switzerland, 216.
(y) **Magnel, G. and Huyghe, G.,** Determination of the strength of a concrete by a nondestructive process, 219.
(z1) **Okushima, M. and Kosaka, Y.,** Four reports, 248.
(z2) **Borges, F.,** Some uses of ultrasounds at the laboratorio Nacional de Engenharia Civil, 252.
(z3) **Mamillan, M.,** The use of sonic methods for the study of freestones, 259.
(z4) **Dawance, G.,** Application of the vibration test to the study of rocks and rock masses, 275.
(z5) Nondestructive testing of bituminous concrete. Relationship between the speed of sound propagation and the temperature of the concrete, 277.
(z6) **Moles, A.,** A geoseismic apparatus for investigating the compactness of soils, 282.

13. **Jones, R.,** *Non-Destructive Testing of Concrete,* Cambridge University Press, London, 1962.
14. **Swamy, N. and Rigby, G.,** Dynamic properties of hardened paste, mortar, and concrete, Materials and Structures/Research and Testing (Paris), 4 (19), 13, 1971.
15. **Jones, R.,** The effect of frequency on the dynamic modulus and damping coefficient of concrete, *Mag. Concr. Res. (London),* 9 (26), 69, 1957.
16. **Gaidis, J. M. and Rosenburg, M.,** New test for determining fundamental frequencies of concrete, *Cement Concr. Aggregates,* CCAGDP, 8 (2), 117, 1986.
17. **Thornton, H. and Alexander, A.,** Development of impact and resonant vibration signature for inspection of concrete structures, ACI Spec. Publ. SP 100, American Concrete Institute, 1987, 667.
18. **Kesler, C. E. and Higuchi, Y.,** Determination of compressive strength of concrete by using its sonic properties, *Proc. ASTM,* 53, 1044, 1953.
19. **Long, B. G. and Kurtz, H. J.,** Effect of curing methods on the durability of concrete as measured by changes in the dynamic modulus of elasticity, *Proc. ASTM,* 43, 1051, 1943.
20. **Axon, E. O., Willis, T. F., and Reagel, F. V.,** Effect of air-entrapping portland cement on the resistance to freezing and thawing of concrete containing inferior coarse aggregate, *Proc. ASTM,* 43, 981, 1943, Also, discussion by C. E. Wuerpel, 995.
21. **Malhotra, V. M. and Zoldners, N. G.,** Durability Studies of Concrete for Manicouagan-2 Project, Mines Branch Investigation Rep. IR 64-69, Department of Energy, Mines and Resources, Ottawa, July 1964.
22. **Malhotra, V. M. and Zoldners, N. G.,** Durability of Non-Air-Entrained Concrete Made with Type I and Modified Type II Cements, Mines Branch Investigation Rep. IR 65-86, Department of Energy, Mines and Resource, Ottawa, September 1965.
23. **Malhotra, V. M. and Zoldners, N. G.,** Durability of Non-Air-Entrained and Air-Entrained Concretes Made with Type I and Modified Type II Cements, Mines Branch Investigation Rep. IR 67-29, Department of Energy, Mines and Resources, Ottawa, February 1967.
24. **Wright, P. J. F. and Gregory, J. M.,** An investigation into methods of carrying out accelerated freezing and thawing tests on concrete, *Mag. Concr. Res. (London),* 6 (19), 39, 1955.
25. **Chefdeville, J.,** Application of the Method toward Estimating the Quality of Concrete, RILEM Bull. (Paris), No. 15, August 1953, Special Issue-Vibrating Testing of Concrete, 2nd part, 61.
26. **Akman, M. S. and Yildirim, M.,** Loss of durability of concrete made from portland cement blended with natural pozzolans due to ammonium nitrate, *Durability of Building Mater. (Amsterdam),* 4 (4), 357, 1987.
27. **Long, B. G., Kurtz, H. J., and Sandenaw, T. A.,** An instrument and a technique for field determination of the modulus of elasticity and flexural strength of concrete (pavements), *ACI J. Proc.,* 41 (3), 217, 1945.
28. **Sweet, H. S.,** Research on concrete durability as affected by coarse aggregate, *Proc. ASTM,* 48, 988, 1948.
29. **L' Hermite, R.,** The strength of concrete and its measurement, *Ann. L'Institut Technique Bâtiment Travaux Publics (Paris),* 12, 3, 1950.
30. **Shrivastava, J. P. and Sen, B.,** Factors affecting resonant frequency and compressive strength of concrete, *Indian Concr. J. (Bombay),* 37, (1), 27, 1963. and 37 (3), 105, 1963.
31. **Sharma, M. R. and Gupta, B. L.,** Sonic modulus as related to strength and static modulus of high strength concrete, *Indian Concr. J. (Bombay),* 34 (4), 139, 1960.

32. **Kaplan, M. F.,** Effects of incomplete consolidation on compressive and flexural strengths, ultrasonic pulse velocity, and dynamic modulus of elasticity of concrete, *ACI J. Proc.*, 56 (9), 853, 1960.
33. **Kaplan, M. F.,** Ultrasonic pulse velocity, dynamic modulus of elasticity, Poisson's ratio and the strength of concrete made with thirteen different coarse aggregates, RILEM Bull. (Paris), New Series No. 1, March 1959, 58.
34. **Malhotra, V. M. and Berwanger, C.,** Correlations of Age and Strength with Values Obtained by Dynamic Tests on Concrete, Mines Branch Investigation Rep. IR 70-40, Department of Energy, Mines and Resources, Ottawa, June 1970.
35. **Batchelder, G. M. and Lewis, D. W.,** Comparison of dynamic methods of testing concretes subjected to freezing and thawing, *Proc. ASTM,* 53, 1053, 1953.
36. **Whitehurst, E. A.,** Evaluation of concrete properties from sonic tests, ACI Monogr. No. 2, American Concrete Institute/Iowa State University Press, Detroit, 1966.
37. **Klieger, P.,** Long-term study of cement performance in concrete. Chapter 10 — Progress Report on Strength and Elastic Properties of Concrete, *ACI J. Proc.*, 54 (6), 481, 1957.
38. **Philleo, R. E.,** Comparison of results of three methods for determining Young's modulus of elasticity of concrete, *ACI J. Proc.*, 51 (5), 461, 1955.
39. **Hansen, W.,** Static and dynamic modulus of concrete as affected by mix composition and compressive strength, ACI Spec. Publ. SP 95, American Concrete Institute 1986, 115.
40. **Gatfield, E. N.,** A Method of Studying the Effect of Vibratory Stress, Including Fatigue, on Concrete in Flexure, paper presented to the RILEM Technical Committee on Nondestructive Testing of Concrete, Varna, Bulgaria, September 3—6, 1968.
41. **Chang, T. S. and Kesler, C. E.,** Correlation of sonic properties of concrete with creep and relaxation, *Proc. ASTM,* 56, 1257, 1956.
42. **Chang, T. S. and Kesler, C. E.,** Prediction of creep behavior in concrete from sonic properties, Proc. Highw. Res.. Board, 35, 436, 1956.
43. **Thornton, H. and Alexander, A.,** Development of Nondestructive Testing Systems for In Situ Evaluation of Concrete Structures, Tech. Rep. REMR-CS-10, Waterways Experiment Station, Corps of Engineers, Vicksburg, MI, December 1987.
44. **Swamy, R. N. and Al-Asali, M. M.,** Engineering properties of concrete affected by alkali-silica reaction, *ACI Mater. J.*, 85(5), 367, 1988.

Chapter 7

THE ULTRASONIC PULSE VELOCITY METHOD

Tarun R. Naik and V. M. Malhotra

ABSTRACT

The ultrasonic pulse velocity method has been used successfully to evaluate the quality of concrete for over 50 years. This method can be used for detecting internal structure changes in mortar and concrete such as deterioration due to aggressive chemical environment, cracking, and changes due to freezing and thawing. By using the pulse velocity method it is also possible to obtain the dynamic modulus of elasticity, Poisson's ratio, thickness of concrete slabs, and estimating the strength of concrete test specimens as well as in-place concrete.

The pulse velocity method is a truly nondestructive method, as the technique involves the use of sonic waves resulting in no damage to the concrete element being tested. The same test sample can be tested again and again, which is very useful for testing concrete undergoing internal structural changes over a long period of time.

HISTORICAL BACKGROUND

Concrete technologists have been interested in determining properties of concrete by nondestructive tests for many decades. Many test methods have been proposed for laboratory test specimens using vibrational methods beginning in the 1930s. Powers,[1] Obert,[2] Hornibrook,[3] and Thomson[4] were the first to conduct extensive research using vibrational techniques such as the resonant frequency method.

World War II accelerated the research regarding nondestructive testing using sonic methods. The development of the pulse velocity method began in Canada and England at about the same time. In Canada, Leslie and Cheesman[5] developed an instrument called the soniscope. While in England, Jones[6] developed an instrument called the ultrasonic tester. In principle, both the soniscope and the ultrasonic tester were quite similar, with only differences in minor details. Since the 1960s, pulse velocity methods have moved out of laboratories and to construction sites. Malhotra[9] has compiled an extensive list of papers published on this subject.

THEORY OF WAVE PROPAGATION

Three basic types of stress waves are created when a solid medium is disturbed by a vibratory load. These waves are called longitudinal or compressional waves, transverse or shear waves, and Rayleigh or surface waves. These stress waves travel through an elastic medium in a similar fashion as sound waves travel through air. Compression waves travel the fastest, followed by shear waves and surface waves. The velocity of these waves depends upon the elastic properties of the medium. Therefore, if the mass of the medium and the velocity of the wave propagation is known, the elastic properties of the medium can be estimated. A pulse originating at a point inside, or through, an infinite, homogeneous, isotropic, elastic medium creates two types of waves — compression and shear waves. If a pulse is created at the surface of this medium, then surface waves are also created, in addition to the compression and shear waves.

In the ultrasonic pulse velocity test method an ultrasonic pulse is created at a point on the test object, and the time of its travel from that point to another is measured. Knowing the distance between these two points, the velocity of the pulse can be determined. Portable pulse velocity equipment available today for concrete testing measures the time of arrival of the first wave. This is the compression wave. The compression wave velocity for an infinite, homogeneous, isotropic, elastic medium can be shown to be[10]

$$V = (KE/D)^{1/2} \tag{1}$$

where:

V	=	Compression wave velocity
K	=	$(1-u)g/(1+u)(1-2u)$
E	=	Modulus of elasticity
D	=	Unit weight
g	=	Acceleration due to gravity
u	=	Poisson's ratio

The K value varies within a very narrow band. For example, for u = 0.15, 0.20, and 0.25, K = 34.00, 35.78, and 38.74 ft/S^2, respectively. Therefore, error in determining the Poisson's ratio within 0.05 leads to an error in the calculated the velocity of about 6%. Conversely, if the velocity and the density can be measured reasonably accurately, say within 1%, then an error of 0.05 in estimating the Poisson's ratio will lead to less than 7% total error in the computed dynamic modulus of elasticity.

PULSE VELOCITY TEST INSTRUMENT

The test instrument consists of a means of producing and introducing a pulse into the concrete (pulse generator and transmitter), and a means of accurately measuring the time taken by the pulse to travel through the concrete. The equipment may also be connected to a cathode ray (CR) oscilloscope. A schematic circuit diagram is shown in Figure 1. A complete description is provided in ASTM Test Method C 597.[11]

Portable ultrasonic testing units have become available worldwide. One of the units available in the U.S. is called the V-Meter.[12]* This equipment is portable and simple to operate. The V-Meter carrying case is 7 × 4.5 × 6.5 in. (180 × 110 × 160-mm) and it weighs 7 lbs (3.2 kg), including a rechargeable battery. The operating ambient temperature range for the test equipment is 30° to 100°F (0° to 40°C).

The pulse travel time is displayed in three numerical digits, which can be varied for three different ranges (1) 0.1 to 99.9 microseconds (MS) with an accuracy of 0.1 MS; (2) 1 to 999 MS with the accuracy of 1 MS; and (3) 10 to 9990 MS with the accuracy of 10 MS. The instrument comes equipped with a 1 to 4 MS variable control unit to zero adjust the instrument each time it is used. This control is used in conjunction with a standard reference bar provided with the instrument. An accurate pulse travel time through the bar is stated on the bar.

The V-Meter comes with two transducers, one each for transmitting and receiving the ultrasonic pulse. They are 1.97 in. (50 mm) in diameter and 1.65 in. (42 mm) long, and have a typical resonant frequency of 54,000 cycles per second ** (cps) for concrete testing. These consist of lead zirconate titanate ceramic (PZT-4) piezoelectric elements housed in stainless steel cases. Similar transducers having different resonance frequencies are also

* This is the same unit which is manufactured in England and is known as PUNDIT.

** or Hertz.

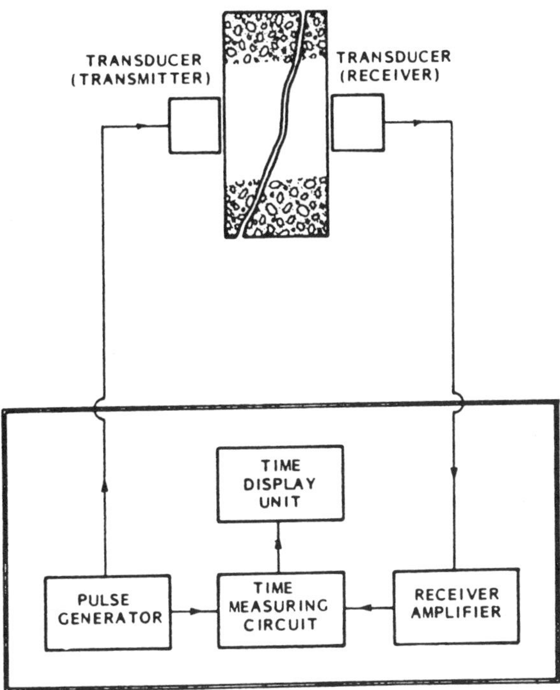

FIGURE 1. Schematic diagram of pulse velocity test circuit. (Adapted from Reference 10.)

available for special applications, e.g., a high frequency, 150,000 cps transducer is used for small-size specimens, relatively short path lengths, or high-strength concrete; while a low frequency, 24,000 cps transducer is used for larger specimens and relatively longer path lengths, or larger size aggregates.

The instrument comes with an internal, rechargeable, nickel-cadmium battery, which can provide power for about 6 h of continuous operation. A constant current charger is built into the instrument to allow the battery to be recharged from an A.C. power supply outlet. The V-Meter can also be operated directly from the A.C. power supply. For other details about the instrument see Figure 2.

THE PULSE VELOCITY METHOD

The basic idea upon which the pulse velocity method is established is that if the velocity of a pulse of longitudinal waves through a medium can be determined, and if the density and the Poisson's ratio of the medium is known, then the dynamic modulus of elasticity of the medium can be computed from Equation 1. Furthermore, knowing the modulus of elasticity, other mechanical properties can be estimated from empirical correlation with the dynamic modulus of elasticity.

The transmitting transducer of the pulse velocity instrument transmits a pulse to one face of the concrete and the receiving transducer, at a distance L, receives the pulse through the concrete. The pulse velocity instrument display indicates the time, T, it takes for the pulse to travel through the concrete. The longitudinal vibration pulse velocity, therefore, is

$$V = L/T \tag{2}$$

The longitudinal vibration pulse transmitted to the concrete undergoes many reflections

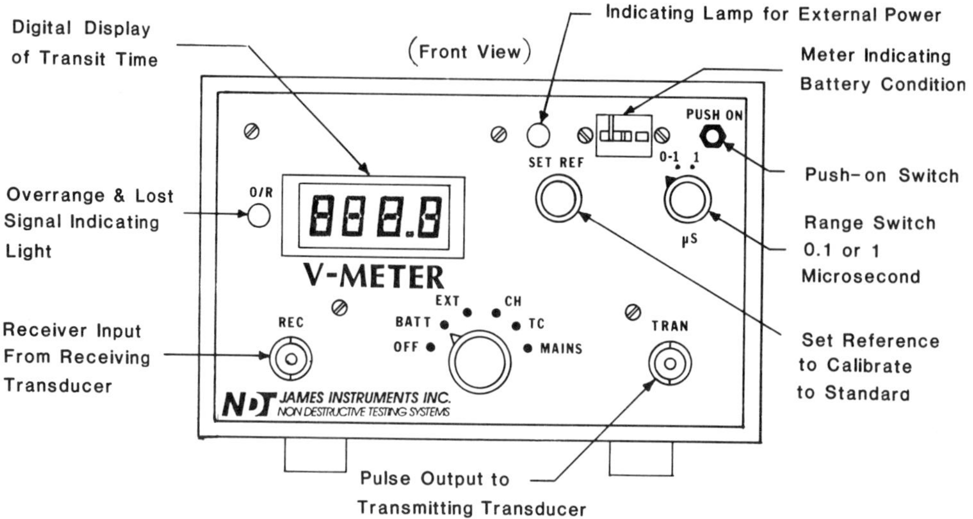

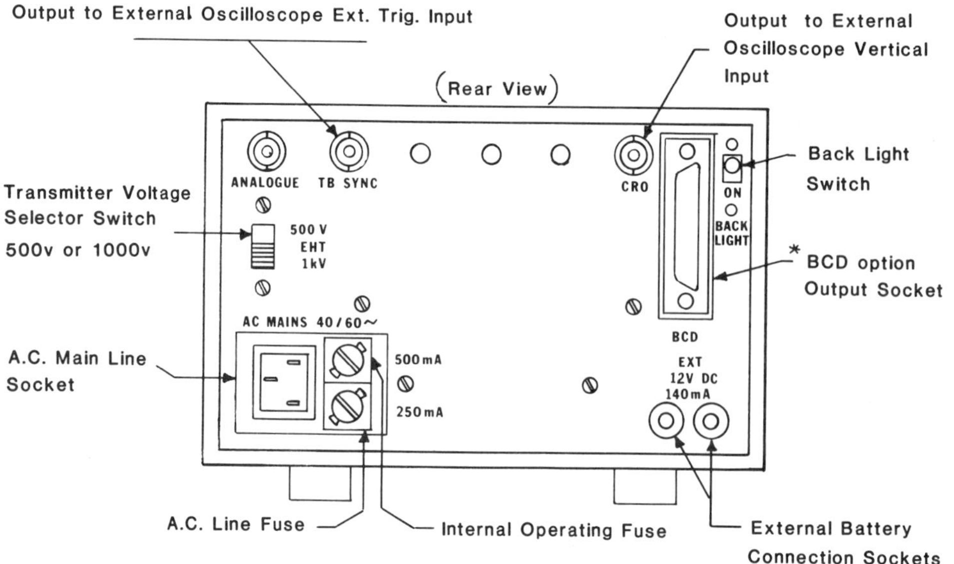

FIGURE 2. Pulse velocity instrument, V-Meter. (Adapted from Reference 12.)

at various aggregate-mortar boundaries. By the time the pulse reaches the receiving transducer it gets transformed into a complex wave form, which contains compression waves and shear waves. Of course, compression waves travelling the fastest arrive first at the receiver.

In order to transmit or receive the pulse, the transducers must be in full contact with the test medium, otherwise the air pocket between the transducer and the test medium will introduce an error in the indicated transit time. This error is introduced because only a negligible amount of pulse can be transmitted through the air. Many couplants available in the market can be used to assure a good contact; petroleum jelly has proven to be one of the superior couplants. Other couplants are grease, liquid soap, and kaolin-glycerol paste. The couplant should be as thin as possible. Repeated readings at a particular location should be taken until a minimum value of transit time is obtained. If the concrete surface is very rough then it may have to be ground smooth, or smooth surface may have to be established with the use of plaster of Paris or suitable quick setting cement paste or quick setting epoxy mortar. Time for setting of the paste should be allowed before proceeding with the pulse velocity test. An exponential receiver probe with a tip diameter of only 0.24 in. (6 mm) may also be used to receive the pulse in very rough surfaces, e.g., locations where the surface mortar is scaled off due to fire or weathering action. It must be emphasized, however, that this probe is good only for receiving the signal. A smooth surface is still required for the transmitting transducer.

The pulse velocity for ordinary concrete is typically of the order of 12000 ft/s (3660 m/s). Therefore, for about a 12-in. (300-mm) path length, the travel time is approximately 80 microseconds. It is obvious that the instrument must be very accurate to measure such a small transit time. The path length should also be carefully measured. The pulse velocity method is a vibration technique. Any sources creating even the slightest vibration in the element under test, e.g., jack hammers, should be eliminated during the time of the test. Many other factors also affect the pulse velocity. They are discussed in detail in the next section.

There are three possible ways in which the transducers may be arranged,[12] as shown in Figure 3 A—C. These are (a) direct transmission; (b) semidirect transmission; and (c) indirect or surface transmission.

The direct transmission method, Figure 3A, is the most desirable and the most satisfactory arrangement because maximum energy of the pulse is transmitted and received with this arrangement.

The semidirect transmission method, Figure 3B, can also be used quite satisfactorily. However, care should be exercised that these transducers are not too far apart, otherwise the transmitted pulse might attenuate and a pulse might not be received. This method is useful in avoiding concentrations of reinforcements.

The indirect or surface transmission method, Figure 3C, is least satisfactory because the amplitude of the received signal may only be about 3% or less than that received by the direct transmission method. This method is also more prone to errors. A disadvantage of this method is that a special procedure may be necessary for determining the pulse velocity, (Figure 4). First the location of the transmitting transducer is fixed and the receiver location is changed in fixed increments along a line, and a series of readings are taken. The direct distance between the two transducers is plotted on the X-axis and the corresponding pulse transit time is plotted on the Y-axis, Figure 4. The slope of this plot is the surface pulse velocity along the line.

Another disadvantage of the surface method is that the pulse propagates in the concrete layer near the surface. This concrete is sometimes of slightly different composition than the concrete in the lower layer. For example, the concrete near the surface of a slab has higher amounts of fine materials than the concrete in the lower portion of the slab. This disadvantage, however, can be turned into a very significant use for detecting and estimating the thickness of a layer of different quality material. A layer of different quality concrete may occur due to improper construction practices (e.g., poor vibration and finishing, cold joints due to delay, incorrect placement, etc.), damage due to weathering action (e.g., frost, sulfate attack, corrosion of reinforcement, and other embedded items, etc.), damage by fire and earthquake, etc. The layer thickness can be estimated by following a similar procedure as described for the determination of the surface pulse velocity, Figure 4. When the two transducers are closer together, the pulse travels through the upper layer of concrete, and as the transducers are moved further apart, the pulse travels through the lower layer. The pulse velocity through the upper layer (V1) and the lower layer (V2) will be indicated on the plot by the different slopes of the two straight lines fitted to the data, Figure 4. The distance X in Figure 4 at which the change in these slopes occur is measured and the thickness of the upper layer, t, is calculated from the following equation:[10]

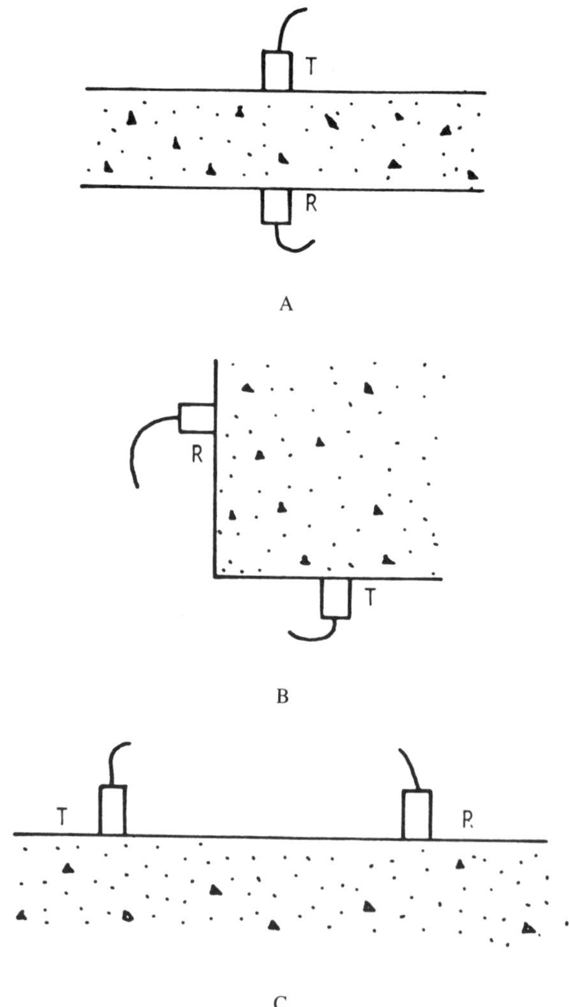

FIGURE 3. Methods of pulse velocity measurements. (A) Direct method. (B) Indirect method. (C) Surface method.

$$t = (x/2)\,[(V2 - V1)/V2 + V1]^{1/2} \tag{3}$$

This method is particularly suitable when the upper layer (the poor quality layer) is distinct and of reasonably uniform thickness.

FACTORS AFFECTING THE PULSE VELOCITY

It is relatively easy to conduct a pulse velocity test. It is, however, important that the test be conducted such that the pulse velocity readings are reproducible and that they are affected only by the properties of the concrete under test rather than by other factors. Factors influencing the pulse velocity readings are discussed below.

The factors affecting the pulse velocity can be divided into two categories: (1) factors affecting the concrete property that affect the pulse velocity; and (2) factors affecting the pulse velocity measurement regardless of the properties of concrete.

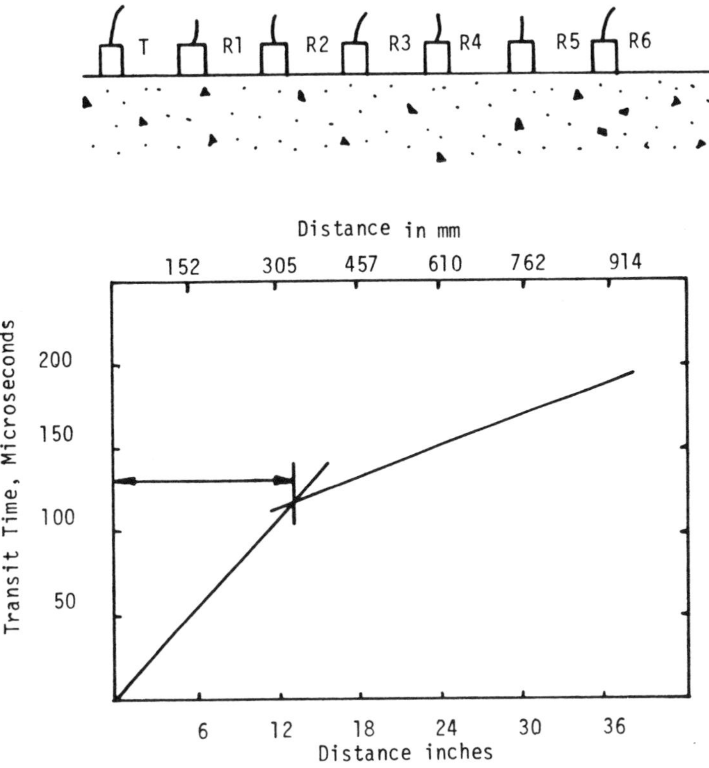

FIGURE 4. Use of surface method to determine depth of deterioration.

FACTORS AFFECTING CONCRETE PROPERTIES

Aggregate Size, Grading, Type and Content

Many investigators found that the pulse velocity vs. compressive strength relationship varies with aggregate characteristics.[12-19] Jones[19] reported that for the same concrete mix (1:1.5:3) and at the same compressive strength level, concrete with rounded gravel had the least pulse velocity, crushed limestone resulted in the highest pulse velocity, while crushed granite gave a high velocity that was between these two. On the other hand, type of aggregate had no significant effect upon the relationship between the pulse velocity and the modulus of rupture. Additional test results by Jones,[13] Kaplan,[20] and Bullock and Whitehurst[15] indicate that at the same strength level the concrete having the higher aggregate content gave a higher pulse velocity. The effects of varying the proportion of coarse aggregate in a concrete mix on the pulse velocity vs. compressive strength relationship are shown in Figure 5.[13] The figure shows that for a given value of pulse velocity, the higher the aggregate-cement ratio, the lower the compressive strength.

Cement Type

Jones[19] has reported that the type of cement did not have a significant effect upon the pulse velocity. The rate of hydration, however, is different for different cements and it will influence the pulse velocity. As degree of hydration increases the strength will increase and the pulse velocity will also increase. Facaoaru[14] found that the use of rapid-hardening cements leads to an increase in the strength corresponding to a given pulse velocity level.

Water-to-Cement Ratio (w/c)

Kaplan[20] has studied the effect of w/c on the pulse velocity. He has shown that as the

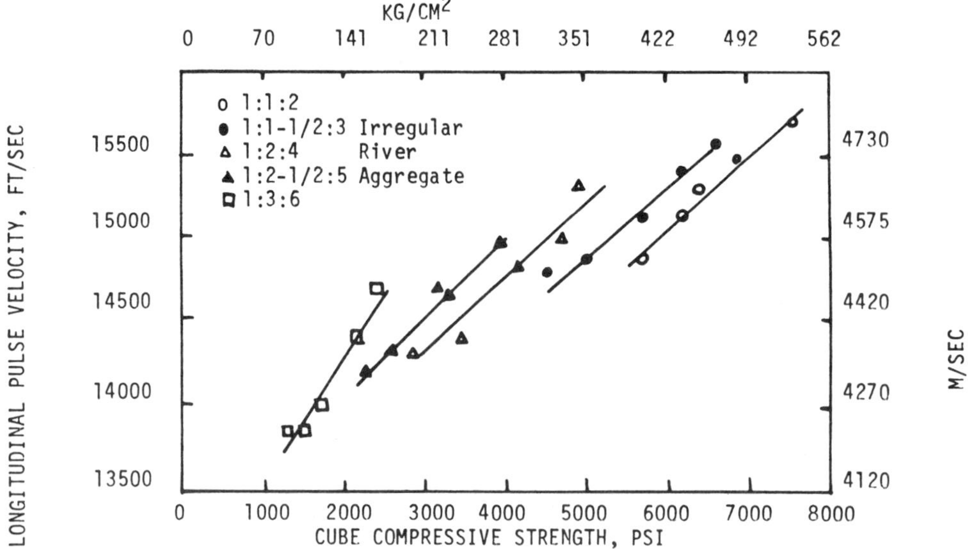

FIGURE 5. Effect of aggregate-cement ratio on the relationship between pulse velocity and compressive strength. (Adapted from Reference 13.)

w/c increases, the density, compressive and flexural strengths, and the corresponding pulse velocity decreases.

Admixtures

Air entrainment does not appear to influence the relationship between the pulse velocity and the compressive strength of concrete.[19] Other admixtures will influence the pulse velocity in approximately the same manner as they would influence the rate of hydration. For example, the addition of calcium chloride will accelerate the setting time of concrete and will increase the early-age strength and the pulse velocity.

Degree of Compaction

Hand rodded or inadequately vibrated concrete is less dense than well compacted concrete. Pulse velocity, as can be seen from Equation 1, is related to the density. Therefore, a decrease in density due to inadequate compaction, honeycomb, etc. will result in a proportionate decrease in the pulse velocity.

Curing Conditions and Age of Concrete

Effect of curing conditions and age of concrete on the pulse velocity is similar to their effect upon the strength development of concrete. Kaplan[21] found that the pulse velocity for laboratory-cured specimens were higher than for the site-cured specimens. He also found that pulse velocity in columns cast from the same concrete were lower than in the site and laboratory specimens. Jones[19] has reported the relationship between the pulse velocity and age. He showed that velocity increases very rapidly initially and at later ages it flattens out, similar to the strength vs. age curve for a particular type of concrete. He further concluded that because pulse velocity and strength are similarly affected by age, the relationship between pulse velocity and strength is independent of age.

FACTORS AFFECTING THE PULSE VELOCITY MEASUREMENT REGARDLESS OF PROPERTIES OF CONCRETE

Acoustical Contact

Influence of improper acoustical contact was discussed in the Pulse Velocity Method

TABLE 1
Corrections for Pulse Velocity Due to Temperature Changes

Concrete temperature, C	Correction, % Air dried concrete	Correction, % Water saturated concrete
60	+5	+4
40	+2	+1.7
20	0	0
0	−0.5	−1
under −4	−1.5	−7.5

From RILEM Recommendation NDT 1, Paris, France, December 1972. With permission.

section. If sufficient care is not exercised in obtaining a good contact, a wrong pulse velocity reading may result.

Temperature of Concrete

Temperature variations between 40° to 90°F (5° and 30°C) have been found to have an insignificant effect upon the pulse velocity.[22] For other temperatures the corrections in Table 1 are recommended.[23]

Moisture Condition of Concrete

Table 1 also gives corrections for different moisture conditions of concrete. It is apparent that the effect of moisture condition on the pulse velocity is small. It can be observed that the pulse velocity for saturated concrete is higher than for the air dry concrete. Moisture generally had less influence on the velocity in high-strength concrete than on low-strength concrete.[22]

Path Length

Theoretically the length of the path traveled by the wave and the frequency of the wave (which is the same as the frequency of the transducer) should not affect the propagation time; and, therefore, it should not affect the pulse velocity. However, in practice smaller path lengths tend to give a slightly higher pulse velocity. RILEM[23] has recommended the following minimum path lengths:

1. 4 in. (100 mm) for concrete having maximum aggregate size of 1.3 in. (30 mm)
2. 6 in. (150 mm) for concrete having maximum aggregate size of 1.75 in. (45 mm)

Size and Shape of a Specimen

In most cases the pulse velocity is not dependent upon the size and the shape of a specimen. However, Equation 1 is valid only for a medium having an infinite extent. This requirement is easily satisfied for a finite dimension test specimen by requiring that the least lateral dimension of the specimen be greater than the wavelength of the pulse. The wavelength W is given by W = V/f, where V = pulse velocity, and f = frequency of vibration. For a concrete having a pulse velocity of about 12,000 ft/s (3660 m/s) and frequency of the transducers of 54,000 cps, the wavelength is about 2.7 in. (68 mm). Therefore, a specimen made from this concrete, when tested with a transducer having a frequency of 54,000 cps, should have a minimum lateral dimension of 2.7 in. (68 mm). If the least lateral dimension of this specimen is less than 2.7 in. (68 mm), then Equation 1 will not give an accurate

TABLE 2
Frequency of Transducer vs. Least Lateral Dimension

Minimum frequency of transducer, cps	Least lateral dimension (or maximum size aggregate)		Range of path length	
	in.	mm	in.	mm
60,000	3	70	4—28	100—700
40,000	6	150	8—60	200—1500
20,000	12	300	>60	>1500

From RILEM Recommendation NDTI, Paris, France, December 1972. With permission.

answer. For such situations, it would be advisable to use a higher frequency transducer, thus reducing the wavelength and the corresponding least lateral dimension requirement. The maximum size aggregate should also be smaller than the wavelength, otherwise the wave energy will attenuate to the point that no clear signal may be received at the receiving transducer. RILEM[23] recommendations about least lateral dimension, and corresponding maximum size aggregate are given in Table 2.

Level of Stress

Pulse velocity is generally not affected by the level of stress in the element under test. However, when the concrete is subjected to a very high level of stress, say 65% of the ultimate stress or greater, microcracks develop within the concrete, which will reduce the pulse velocity considerably.

Presence of Reinforcing Steel

One of the most significant factors which influences the pulse velocity of concrete is the presence of reinforcement. The pulse velocity in steel is $1^1/_2$ to 2 times the pulse velocity in plain concrete. Therefore, pulse velocity readings in the vicinity of reinforcing is usually higher than that in the plain concrete. Whenever possible, test readings should be taken such that the reinforcement is avoided from the pulse length. If reinforcements cross the pulse path, correction factors should be used. The correction factors that are used are those currently recommended by RILEM[23] and British Standards.[24] Appendix 1 provides the information about the correction factors proposed from RILEM.[23] Chung[25,26] has demonstrated the importance of including bar diameters as a basic parameter in the correction factors. However, the RILEM recommendations involve only two basic parameters which are the pulse velocity in the surrounding concrete and the path lengths within the steel and concrete. A recent paper by Bungey[27] also provides correction factors that include bar diameters. It should be emphasized, however, that in heavily reinforced sections, it may not be possible to obtain satisfactory readings.

STANDARDIZATION OF THE PULSE VELOCITY METHOD

ASTM Committee C-9 initiated the development of a standard for pulse velocity in the late 1960s. A tentative standard was issued in 1968. A standard test method was issued in 1971 and no significant changes have been made in the standard since then.[11] The Significance and Use statement of the test method, as given in the ASTM C 597-83, "Standard Test Method for Pulse Velocity Through Concrete",[11] is as follows:

This test method may be used to advantage to assess the uniformity and relative quality of concrete, to indicate the presence of voids and cracks, to estimate the depth of cracks, to indicate changes in the properties of concrete, and in the survey of structures, to estimate the severity of deterioration or cracking.

The results obtained by the use of this test method should not be considered as a means of measuring strength nor as an adequate test for establishing compliance of the modulus of elasticity of field concrete with that assumed in design.

The procedure is applicable in both field and laboratory testing regardless of size or shape of the specimen within the limitations of available pulse-generating sources.

APPLICATIONS

The pulse velocity method has been used successfully in laboratory as well as in the field.[28-48] Furthermore, it can be used for quality control and quality assurance, as well as for the analysis of deterioration.

DETERMINATION OF DYNAMIC MODULUS OF ELASTICITY AND POISSON'S RATIO

One of the most direct uses, and theoretically the most correct use of the pulse velocity method, is in determining dynamic modulus of elasticity and Poisson's ratio of concrete. Using Equation 1, Leslie and Cheesman,[5] Whitehurst,[7] Philleo,[28] Goodell,[29] and Swamy[30] have published detailed test results. The dynamic Poisson's ratio of concrete can be assumed to be between 0.2 to 0.3 for most concretes. This assumed value will lead to an error of about 10%, or less, for the value of the dynamic modulus of elasticity.

ESTIMATION OF STRENGTH OF CONCRETE

The pulse velocity method provides a convenient means of estimating the strength of both *in situ* and precast concrete. The strength can be estimated from the pulse velocity by a pre-established graphical correlation between the two parameters, for example, as shown in Figure 6. The relationship between strength and pulse velocity is not unique, but is affected by many factors, e.g., aggregate size, type and content, cement type and content, water-to-cement ratio, moisture content, etc. The effect of such factors has been studied by many researchers.[10,16,18,20,21] They have clearly pointed out that no attempts should be made to estimate compressive strength of concrete from pulse velocity values unless similar correlations have been previously established for the type of concrete under investigation. RILEM[23] and British Standard[24] have provided recommended practices to develop the pre-established relationship between pulse velocity and compressive strength, which can be later used for estimating the *in situ* strength based upon the pulse velocity.

ESTABLISHING HOMOGENEITY OF CONCRETE

The pulse velocity method is very suitable for the study of homogeneity of concrete, and therefore for relative assessment of quality of concrete. Heterogeneities in a concrete member will cause variations in the pulse velocity. *In situ* concrete strength varies in a structure because of the variations in materials, supply and in mixing, and due to inadequate or variable compaction. The pulse velocity method is extremely effective in establishing comparative data and for qualitative evaluation of concrete. For obtaining these qualitative data, a system of measuring points, i.e., a grid pattern, may be established. Depending upon the quantity of the concrete to be evaluated, the size of the structure, the variability expected, and the accuracy required, a grid of 6-in. (300-mm) spacing, or greater, should be established. Generally about 3-ft. (1-m) spacing is adequate. Other applications of this qualitative comparison of *in situ* or test specimen concrete are (1) to check the density of concrete in order to evaluate the effectiveness of consolidation, and (2) for locating areas of honeycombed concrete.

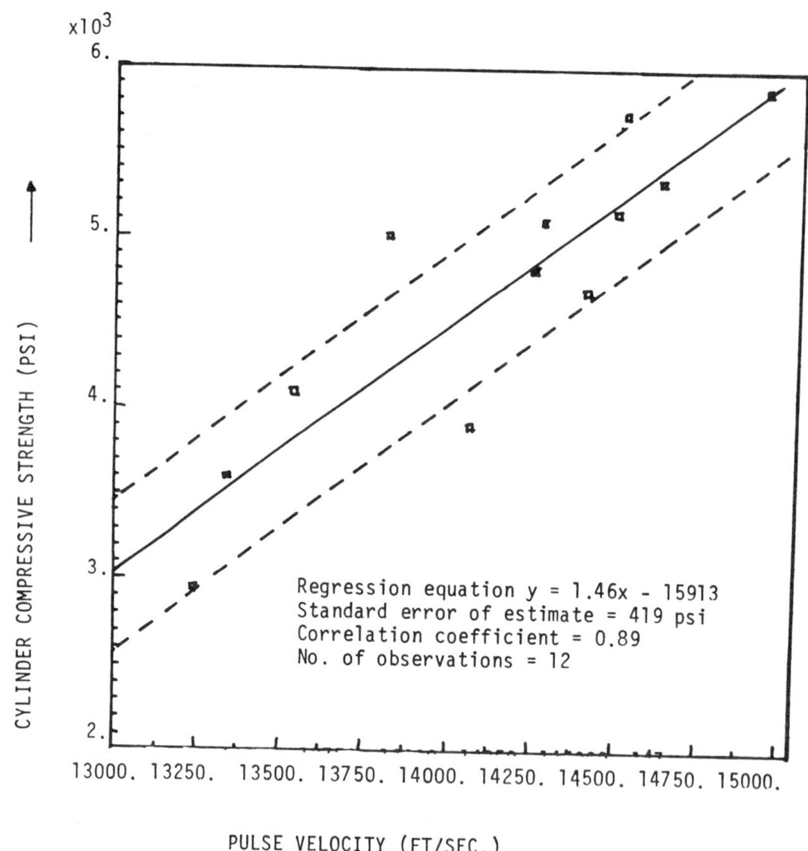

FIGURE 6. Velocity strength relationship for estimation of strength of concrete.

Many researchers[10,19,32-34,48] have reported results of carefully conducted surveys for determining the homogeneity of concrete in various types of structures. Of many such surveys carried out on existing structures, one that deserves mention is that reported in 1953 by Parker[32] of the Hydro-Electric Power Commission of Ontario, Canada. It was made on a dam built in 1914. A total of 50,000 readings were taken, most of them 1-ft (6.3-m) spacings. The pulse velocities measured on the structure ranged from below 5000 to over 17,000 ft/s (1525 to over 5185 m/s) and these values were used, with success, to determine the areas of advance deterioration. Recently, Naik[48] has reported a similar investigation on a dam built in 1906. Figure 7 shows field testing of mass concrete with the soniscope.[9] Some thousands of pulse velocity measurements have been made on 29 concrete dams during the period 1948—1965. McHenry and Oleson[35] cite ten of these case histories in which velocity measurements have been a valuable supplement to other observations in settling questions regarding repair or maintenance of dams.

STUDIES ON THE HYDRATION OF CEMENT

The pulse velocity method has the advantage that it is truly nondestructive. Therefore, he changes in the internal structure of concrete can be monitored on the same test specimen.

Whitehurst,[7] Chefdeville,[36] Van de Winden and Brant,[37] and others have published nformation on successful application of the pulse velocity method for monitoring the hard-

FIGURE 7. Field testing of mass concrete with the soniscope. (Left) soniscope transmitter in rack an upstream face of a dam. (Right) soniscope receiver against downstream face of a dam. (Adapted from Reference 32.)

ening process of cement paste, mortar, or concrete. The method was particularly useful for detecting changes during the first 36 h after adding water to the concrete mix. A very significant practical use of the method is the evaluation of the rate of setting, initial or final, for different types of cements or admixtures to be used for a given project. This aspect has been studied by many researchers.[19,38-40]

Whitehurst[39] has reported results of tests on 4 × 4 × 16 in. (102 × 102 × 406 mm) concrete prisms, using various types of cement. The concretes used had zero slump and immediately after casting the end plates of the forms were removed. Pulse velocity tests using the soniscope were made periodically, from shortly after the specimens were cast until 8 h or more had elapsed. Initial velocities of the order of 4000 ft/s (1220 m/s) were observed, and during the first few hours the velocities increased at a rapid rate. After periods varying from $4^1/_2$ to $8^1/_2$ h, the rate of increase suddenly changed and continued at a much slower pace. The point at which this occurred was taken as the time of set of concrete. The results of Whitehurst,[39] together with those reported by Cheesman,[40] are shown in Figure 8.

STUDIES ON DURABILITY OF CONCRETE

Aggressive environments will damage the structure of concrete and decrease the pulse velocity. Deterioration caused by freezing and thawing, sulfate exposure, corrosion of embedded items, etc. can be easily detected by the pulse velocity method and have been studied by various investigators.[5,19,36,41] Progressive deterioration of either a test specimen or *in situ* concrete can be monitored by conducting repetitive tests on the same concrete element. Deterioration of concrete due to fire exposure has also been investigated by the pulse velocity method.[26,42]

MEASUREMENT AND DETECTION OF CRACKS

This aspect of pulse velocity technique has been studied by various researchers.[5,43-46] As indicated earlier, the ultrasonic pulse transmits a very small amount of energy through air. Therefore, if a pulse traveling through the concrete comes upon an air-filled crack or

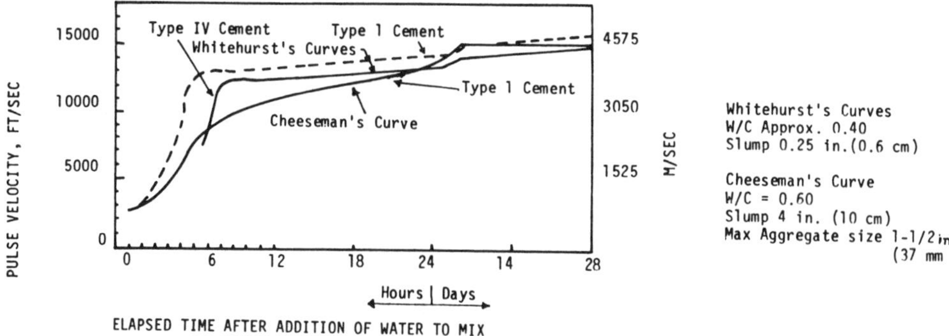

FIGURE 8. Relationship between pulse velocity and setting characteristics of concrete. (Adapted from References 39 and 40.)

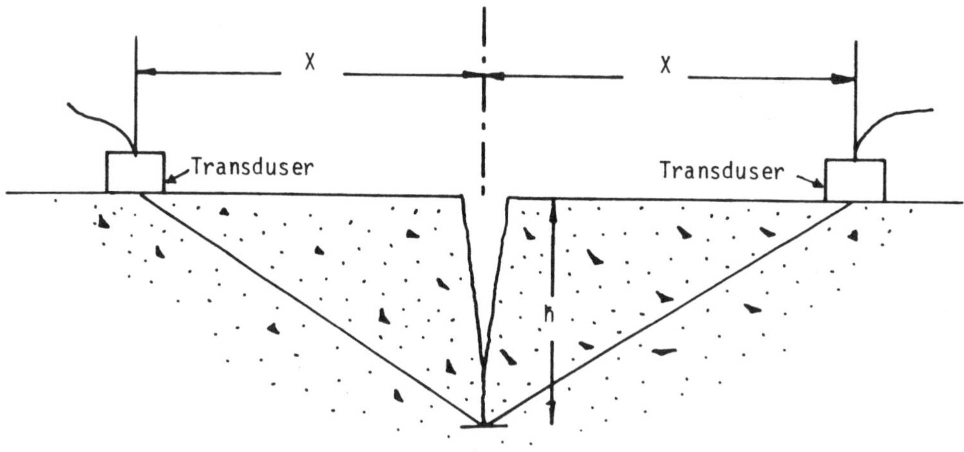

FIGURE 9. Measurement of crack depth.

a void whose projected area perpendicular to the path length is larger than the area of the transmitting transducer, the pulse will get diffracted around the defect. Thus, the pulse travel time will be greater than that through a similar concrete without any defect. The pulse velocity method, therefore, is effective in locating cracks, cavities, and other such defects. It should be pointed out that the application of this technique in locating flaws has serious limitations. For example, if cracks and flaws are small or if they are filled with water or other debris, thus allowing the wave to propagate through the flaw, the pulse velocity will not significantly decrease, implying that no flaw exists.

The depth of an air filled surface crack can be estimated by the pulse velocity method, Figure 9. The depth, h, is given by Equation 4.

$$h = (x/T_2)(T_1^2 - T_2^2)^{1/2} \tag{4}$$

Where x = distance of the transducer from the crack (note that both transducers must be placed equidistant from the crack); T_1 = transmit time around the crack; and T_2 = transit time along the surface of the same type of concrete without any crack (note that the surface path length for T_1 and T_2 must be equal).

It should be pointed out that for Equation 4 to be valid, the crack must be perpendicular to the concrete surface. A check should be made to determine if the crack is perpendicular

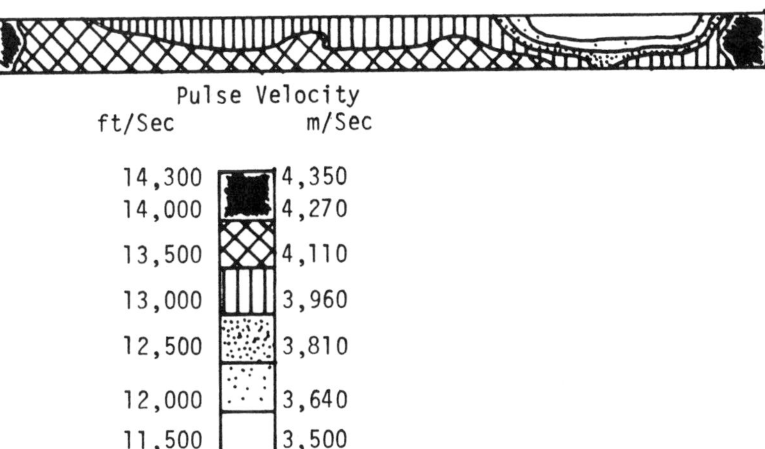

FIGURE 10. Variations of ultrasonic pulse velocity in reinforced concrete beam with selected weaker zones near an end. (Adapted from Reference 47.)

to the surface or not. This can be done as follows.[23] Place both the transducers equidistant from the crack and obtain the transit time. Move each transducer, in turn, away from the crack. If the transit time decreases, then the crack slopes towards the direction in which the transducer was moved.

INSPECTION OF REINFORCED CONCRETE FLEXURAL MEMBERS

Elvery and Din[47] have shown that pulse velocity measurements are more successful than crushing tests of companion specimens in determining the flexural strength of reinforced concrete beams. Data have been presented to indicate that pulse velocity measurement can clearly identify regions of low-strength concrete in reinforced concrete beams and the influence of these weak zones can be predicted with reasonable accuracy.

Figure 10 shows distribution of pulse velocity in a beam in which selected weak regions were made by inserting steel plates as temporary stop ends at the boundaries of the regions and placing concrete of much lower quality than that used in the rest of the beam within these boundaries. Then the pulse velocity method was used to properly identify the various types of concrete.

ADVANTAGES AND LIMITATIONS

The pulse velocity method is an excellent means for investigating the uniformity of concrete. The test procedure is simple and the available equipment in the market is easy to use in the laboratory as well as at a construction site.

The testing procedures have been standardized by ASTM and other organizations, and test equipment is available from several commercial sources. With the availability of small portable digital instruments, which are relatively inexpensive and easy to operate, ultrasonic testing should add a new dimension to the control of quality of concrete in the field.

Ultrasonic pulse velocity tests can be carried out on both laboratory-sized test specimens and concrete structures. This fact, combined with the knowledge that ultrasonic techniques provide the only available means of delineating both surface and internal cracks in concrete structures, enhances the usefulness of these tests.

Inasmuch as a large number of variables affect the relations between the strength parameters of concrete and its pulse velocity, the use of the latter to estimate the compressive and/or flexural strengths of concrete is not recommended unless previous correlating testing has been performed.

APPENDIX 1*

EFFECT OF REINFORCING BARS

The pulse velocity measured in reinforced concrete in the vicinity of reinforcing bars is often higher than in plain concrete of the same composition. This is because the pulse velocity in steel is 1.2 to 1.9 times the velocity in plain concrete and, under certain conditions, the first pulse to arrive at the receiving transducer travels partly in concrete and partly in steel. The apparent increase in pulse velocity depends, upon the proximity of the measurements to the reinforcing bar, the dimensions and number of the reinforcing bars, their orientation with respect to the propagation path, and the pulse velocity in the surrounding concrete.

AXIS OF REINFORCING BAR PERPENDICULAR TO DIRECTION OF PROPAGATION

The maximum influence of the presence of the reinforcing bars can be calculated, assuming that the pulse traverses the full diameter of each bar during its path. If there are *n* different bars of diameter Qi ($i = 1$ to n) directly in the path of the pulse, with their axes at right angles to the path of propagation (see Figure 11).

$$\frac{V_c}{V} = \frac{\left(1 - \frac{L_g}{L}\right)}{\left(1 - \frac{L_g V}{L v_a}\right)} \tag{5}$$

where V is the pulse velocity in the reinforced concrete, i.e., the measured pulse velocity
V_c is the pulse velocity in the plain concrete
V_s is the pulse velocity in the steel
L is he total path length
$L_s = \sum_{1}^{n} Q_i$ is the path length through steel

Values of $\frac{V_c}{V}$ are given in Table 3 for different amounts of steel in three types of concrete which could probably be rated as very poor, fair and very good materials.

In practice, $\frac{V_c}{V}$ is likely to be slightly higher than the values given in Table 5 because of misalignment of the reinforcing bars and because only a small fraction of the pulse energy will traverse the full diameter of each bar.

AXIS OF BAR PARALLEL TO DIRECTION OF PROPAGATION

If the edge of the bar is located at a distance ' a ' from the line joining the nearest points of the two transducers, and the path length between transducers is L, then the transit time T in either of the configurations of Figures 11B or 11C is

* From RILEM Recommendation NDTI, Paris, France, December 1972. With permission.

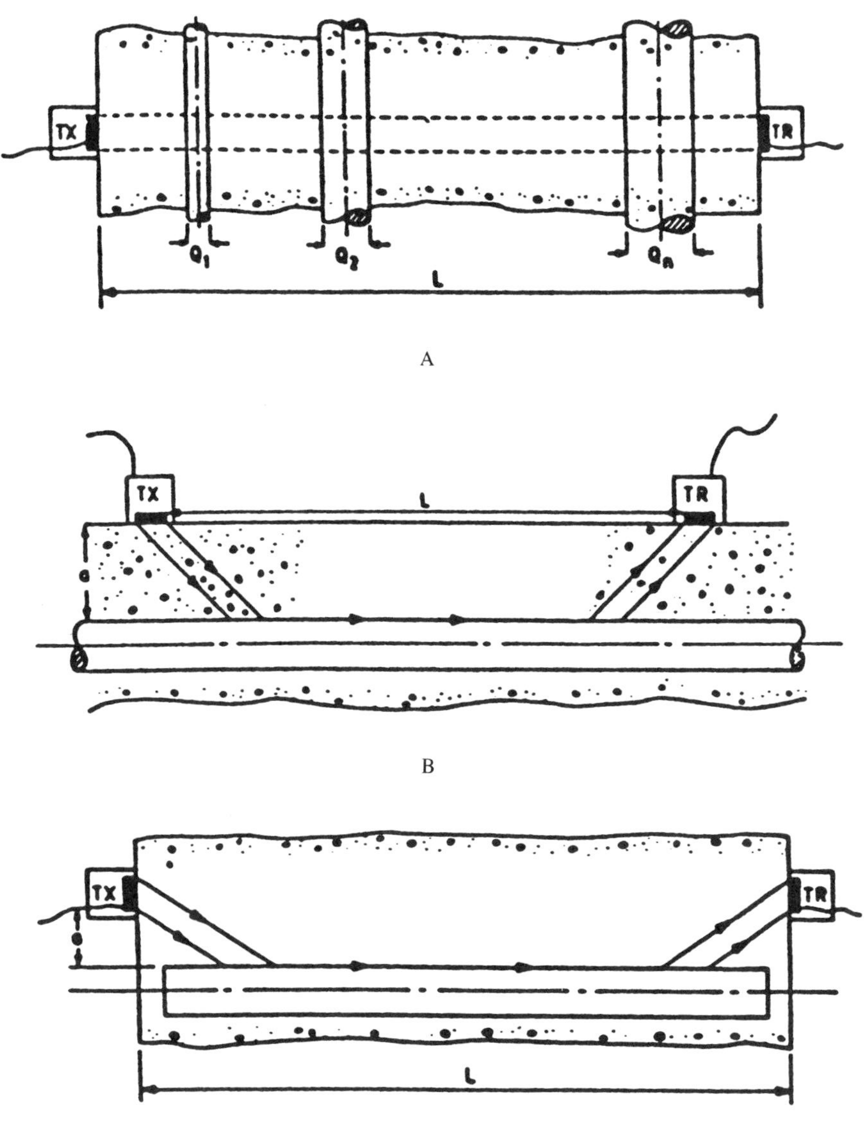

FIGURE 11. Measurements on reinforced concrete. (A) Reinforcing bars perpendicular to direction of propagation. (B) Reinforcing bar parallel to test surface. (C) Reinforcing bar parallel to direction of propagation.

$$T = \frac{L}{V_s} + 2a\sqrt{\frac{V_s^2 - V_c^2}{V_s V_c}}$$

for (6)

$$\frac{a}{L} < \frac{1}{2}\sqrt{\frac{V_s - V_c}{V_s + V_c}}$$

There is no influence of the steel when

$$\frac{a}{L} > \frac{1}{2}\sqrt{\frac{V_s - V_c}{V_s + V_c}}$$

TABLE 3
Influence of Steel Reinforcement-Line of Measurement Perpendicular to Axis of Bar

$$\frac{V_c}{V} = \frac{\text{pulse velocity in concrete}}{\text{measured pulse velocity}}$$

L_s/L	Very poor quality V_c = 3000 m/s	Fair quality V_c = 4000 m/s	Very good quality V_c = 5000 m/s
1/12	0.96	0.97	0.99
1/8	0.94	0.96	0.98
1/6	0.92	0.94	0.97
1/4	0.88	0.92	0.96
1/3	0.83	0.89	0.94
1/2	0.75	0.83	0.92

From Reference 23.

TABLE 4
Influence of Steel Reinforcement Line on Measurements Parallel to Axis of Bar

$$\frac{\text{True pulse velocity in concrete}}{\text{Measured pulse velocity in concrete}} = \frac{V_c}{V_s}$$

a/L	$\frac{V_c}{V_s} = 0.90$	$\frac{V_c}{V_s} = 0.80$	$\frac{V_c}{V_s} = 0.71$	$\frac{V_c}{V_s} = 0.60$
0	0.90	0.80	0.71	0.60
1/20	0.94	0.86	0.78	0.68
1/15	0.96	0.88	0.80	0.71
1/10	0.99	0.92	0.85	0.76
1/7	1.00	0.97	0.91	0.83
1/5	1.00	1.00	0.99	0.92
1/4	1.00	1.00	1.00	1.00

The difficulty of applying Equation (6) lies in deciding on the velocity (V_s) of propagation of the pulse along the steel bar. Propagation of the pulse is influenced by geometrical dispersion and the discussion in Section 4.3 is apposite. The value for V_s is thus likely to be between about 6000 m/s (i.e., the α velocity in the steel) and 5200 m/s (i.e., the bar velocity in the steel). A measure of this velocity can often be obtained by propagating along the axis of the embedded bar, and making allowance for any concrete cover at either end.

Corrections to the measured pulse velocity in the direction parallel to the reinforcement are given in Table 6. This table also indicates that, for bars which span most of the section, the lateral displacement of the line of measurement from the axis of the bar will usually be of the order of 0.2 to 0.25 L before the influence of the steel becomes negligible.

TWO-WAY REINFORCEMENT

Steel reinforcement in two or more directions complicates the interpretation of pulse velocity measurements. Corrections based on Tables 5 and 6 may be calculated for simple well-defined systems of reinforcement but it may become impossible to make any reliable corrections for more complicated heavily reinforced concrete.

REFERENCES

1. **Powers, T. C.,** Measuring Young's modulus of elasticity by means of sonic vibrations, *Proc. ASTM*, 38, Part II, 460, 1938.
2. **Obert, L.,** Sonic method of determining the modulus of elasticity of building materials under pressure, *Proc. ASTM*, 39, 987, 1939.
3. **Hornibrook, F. B.,** Application of sonic method to freezing and thawing studies of concrete, *ASTM Bull.*, 101, 5, 1939.
4. **Thomson, W. T.,** Measuring changes in physical properties of concrete by the dynamic method, *Proc. ASTM*, 40, 1113, 1940.
5. **Leslie, J. R. and Cheesman, W. J.,** An ultrasonic method of studying deterioration and cracking in concrete structures, *ACI J. Proc.*, 46 (1), 17, 1949.
6. **Jones, R.,** The Application of Ultrasonics to the Testing of Concrete, Research, London, England, May 1948, page 383.
7. **Whitehurst, E. A.,** Evaluation of Concrete Properties from Sonic Tests, ACI Monograph No. 2, ACI, Detroit, MI, 1966, 94 pages.
8. Nondestructive Methods of Testing Concrete, Review Rep., RR-2, Cement Research Institute of India, New Delhi, December 1969.
9. **Malhotra, V. M.,** Testing hardened concrete: nondestructive methods, ACI Monogr. No. 9, ACI, Detroit, MI, 1976.
10. **Naik, T.R.,** The Ultrasonic Testing of Concrete, published by ACI in *Experimental Methods in Concrete Structures for Practitioners,* Sabnis, G. M. and Fitzsimons, N., Eds., October 1979, (available from Department of Civil Engineering, the University of Wisconsin-Milwaukee).
11. ASTM Test Designation C597-83, Standard Test Method for Pulse Velocity Through Concrete, Annual Book of ASTM Standards, Vol 04.02, Philadelpha, 1987.
12. James Electronics, Inc., Instruction Manual for Model C-4899, V-Meter, Chicago, IL, May 1977.
13. **Jones, R.,** *Non-destructive Testing of Concrete,* Cambridge University Press, London, 1962.
14. **Facaoaru, I.,** Non-destructive testing of concrete in Romania, Symp. on Non-Destructive Testing of Concrete and Timber, Institution of Civil Engineers, London, 1970, 39.
15. **Bullock, R. E. and Whitehurst, E. A.,** Effect of certain variables on pulse velocities through concrete, *Highw. Res. Board Bull.*, 206, 37, 1959.
16. **Sturrup, V. R., Vecchio, F. J., and Caratin, H.,** Pulse velocity as a measure of concrete compressive strength, ACI SP 82-11, 1984, 201.
17. **Swamy, N. R. and Al-Hamed, A. H.,** The Use of Pulse Velocity Measurements to Estimate Strength of Air-Dried Cubes and Hence In situ Strength of Concrete, ACI SP 82, Malhotra, V. M., Ed., 1984, 247.
18. **Anderson, D. A. and Seals, R. K.,** Pulse velocity as a predictor of 28 and 90 day strength, *ACI J.*, 78, 116, 1981.
19. **Jones, R.,** Testing of concrete by an ultrasonic pulse technique, RILEM Int. Symp. on Nondestructive Testing of Materials and Structures, Paris, Vol. 1, Paper No. A-17, January 1954, 137. RILEM Bull. No. 19, 2nd part, November 1954.
20. **Kaplan, M. F.,** The effects of age and water to cement ratio upon the relation between ultrasonic pulse velocity and compressive strength of concrete, *Mag. Concr. Res.*, 11(32), 85, 1959.
21. **Kaplan, M. F.,** Compressive strength and ultrasonic pulse velocity relationships for concrete in columns, *ACI J.*, 29, No. 54-37, 675, 1958.
22. **Jones, R. and Facaoaru, I.,** Recommendations for testing concrete by the ultrasonic pulse method, Materials and Structures Research and Testing (Paris), 2 (19), 275, 1969.
23. RILEM Recommendation NDT 1, Testing of Concrete by the Ultrasonic Pulse Method, Paris, December 1972.
24. B.S. 4408, Part 5, Recommendations for Nondestructive Methods of Test for Concrete: Measurement of the Velocity of Ultrasonic Pulses in Concrete, British Standards Institution, London, February 1974.
25. **Chung, H. W.,** Effect of embedded steel bars upon ultrasonic testing of concrete, *Mag. Concr. Res. London,* 30 (102), 19, 1978.
26. **Chung, H. W. and Law, K. S.,** Diagnosing in situ concrete by ultrasonic pulse technique, *Concr. Int.*, October 1983, 42.
27. **Bungey, J. H.,** The Influence of Reinforcement on Ultrasonic Pulse Velocity Testing, ACI SP 82, Malhotra, V. M., Ed., 1984, 229.
28. **Philleo, R. E.,** Comparison of results of three methods for determining Young's modulus of elasticity of concrete, *ACI J.*, 26 (5), 461, 1955. Discussions, December 1955, 472-1.
29. **Goodell, C. E.,** Improved sonic apparatus for determining the dynamic modulus of concrete specimens, *ACI J.*, 27 (47-4), 53, 1950.
30. **Swamy, R. N.,** Dynamic Poisson's ratio of portland cement paste, mortar, and concrete, *Cement Concr. Res.*, 1, 559, 1971.

31. **Galan, A.,** Estimate of concrete strength by ultrasonic pulse velocity and damping constant, *ACI J.*, 64 (64-59), 678, 1967.
32. **Parker, W. E.,** Pulse velocity testing of concrete, *Proc. ASTM,* 53, 1033, 1953.
33. **Whitehurst, E. A.,** Soniscope tests concrete structures, *ACI J. Proc.*, 47(6), 433, 1951.
34. **Breuning, S. M. and Bone, A. J.,** Soniscope applied to maintenance of concrete structures, *Proc. Highw. Res. Board,* 33, 210, 1954.
35. **McHenry, D. and Oleson, C. C.,** Pulse velocity measurements on concrete dams, *Trans. 9th Int. Congr.* on Large Dams, Istanbul, 1967, Q34, R5, V. 3, 73.
36. **Chefdeville, J.,** Application of the Method Toward Estimating the Quality of Concrete, RILEM Bull. No. 15, Paris, Special Issue-Vibration Testing of Concrete, 2nd Part, August 1953, 59.
37. **Van der Winder, N. G. B. and Brant, A. W.,** Ultrasonic Testing for Fresh Mixes, Concrete, Cement and Concrete Association, Wexham Springs, England, December 1977, 25.
38. **Woods, K. B. and Mclaughlin, J. F.,** Application of pulse velocity tests to several laboratory studies of materials, *Highw. Res. Board,* Bulletin 206, 1959.
39. **Whitehurst, E. A.,** Use of Soniscope for Measuring Setting Time of Concrete, *Proc. ASTM,* 51, 1166, 1951.
40. **Cheesman, W. J.,** Dynamic testing of concrete with the soniscope apparatus, *Proc. Highw. Res. Board,* 29, 176, 1949.
41. **Neville, A. M.,** A study of deterioration of structural concrete made with high alumina concrete, Proc. Institution of Civil Engineers (London), 25, 287, 1963.
42. **Zoldners, N. G., Malhotra, V. M., and Wilson, H. S.,** High-temperature behavior of aluminous cement concretes containing different aggregates, *Proc. ASTM,* 63, 966, 1963.
43. **Sturrup, V. R.,** Evaluation of pulse velocity tests made by Ontario hydro, Bull. No. 206, Highway Research Board, 1, 1959.
44. **Jones, R.,** A method of studying the formation of cracks in a material subjected to stress, *Br. J. Appl. Physics (London),* 3, 229, 1952.
45. **Knab, L. J., Blessing, G. V., and Clifton, J. R.,** Laboratory evaluation of ultrasonics for crack detection in concrete, *ACI J.*, 80, 17, 1983.
46. **Rebic, M. P.,** The Distribution of Critical and Rupture Loads and Determination of the Factor of Crackability, ACI SP 82, Malhotra, V. M., Ed., 1984, 721.
47. **Elvery, R. H. and Din, N. M.,** Ultrasonic Inspection of Reinforced Concrete Flexural Members, Proc. Symp. on Nondestructive Testing of Concrete and Timber, Institution of Civil Engineers, London, June 1969, 23.
48. **Naik, T. R.,** Evaluation of an 80-year old Concrete Dam, a paper presented at the ACI Meeting in Seattle, WA, October 1987 (available from the Department of Civil Engineering, the University of Wisconsin-Milwaukee).

Chapter 8

COMBINED METHODS

A. Samarin

ABSTRACT

This chapter describes the theoretical and empirical based concepts as well as the history of development of combined nondestructive test methods for hardened concrete.

Of a number of purely nondestructive tests, the rebound (Schmidt) hammer and the ultrasonic pulse velocity combinations are the most commonly used. In the majority of cases, the need for *in situ* concrete strength evaluation arises as a result of suspect quality of concrete. By developing a prior correlation for a range of concrete grades and types, having only the source of coarse aggregate and a broad age group in common, it is possible to obtain good indication of the *in situ* strength of concrete, expressed as the value of a test result of a standard laboratory compressive specimen.

The quality of concrete, using combined nondestructive methods, is evaluated through the measurements and correlation of the surface hardness, density, elastic constants and the predicted compressive strength.

Use of combined methods is generally justifiable only if a reliable correlation for a particular type of concrete is developed prior to the evaluation of the subject quality concrete. The benefit of the small additional reliability of a combined test vs. a single nondestructive test should be assessed against the additional time, cost, and complexity of combined techniques.

INTRODUCTION

Although the need for nondestructive *in situ* testing of concrete has long been realized, it is seldom used in its own right for quality control and compliance purposes. In fact, the practice is to bring in a destructive (e.g., cores), a semi-destructive (e.g., pull-out of break off tests), or a nondestructive *in situ* test at the postmortem stage, either following noncompliance of a standard specimen test result or on observing signs of deterioration and distress in a structure.

Occasionally, nondestructive *in situ* tests are used to evaluate the "quality" of existing structural members for the purpose of subsequent modifications to that structure. The "quality" of concrete in practice is still commonly described in terms of its uniaxial compressive strength, evaluated statistically from the results of standard laboratory specimens, cast, compacted, cured, and tested under strictly prescribed conditions. There is of course a very good reason for maintaining this system. In those countries where ready-mixed concrete can be purchased from a pre-mixed concrete manufacturer, this method remains the only fair and reasonable way of evaluating the potential quality, i.e., strength of concrete as supplied to a building site.

However, with the present emphasis on the design of reinforced concrete structures according to the limit state concepts, there seems to be much greater need for better definition of the relationship between concrete quality and variability in actual structures and in standard, laboratory-cured test specimens of the same concrete. A combination of these *in situ* tests, if properly used, can improve some of these correlations. The extent to which the correlation can be improved should be balanced, however, by the additional time, cost, and resources required to use combined methods.

HISTORICAL DEVELOPMENT

In the preceding chapters several nondestructive and semi-destructive (or partially destructive) test methods have been described. These methods were developed to evaluate the strength, or strength related properties of concrete. In order to predict the strength of *in situ* concrete more accurately, a number of investigators have tried to apply more than one nondestructive method at the same time.

Some of the pioneering work in this field, carried out in the 1950s and 1960s, was reported by Kesler and Higuchi,[1] Skramtaev and Leshchinsky,[2] Wiebenga,[3] Facaoaru,[4] and McLeod.[5] Combined methods, as used in this chapter, refers to techniques in which one test improves the reliability and precision of the other in evaluating a property of concrete, e.g., strength or elastic modulus. When an additional test is used to provide new, but supporting information (e.g., location of reinforcing steel, using a covermeter) the tests are not considered combined, but supplementary. Of the reported combined methods, the following were considered promising by their investigators.

Dynamic modulus of elasticity and damping constant, as determined from resonance tests by Kesler and Higuchi,[1] were reported to correlate well with the compressive strength of concrete, regardless of its mix proportions, age, or moisture content. The accuracy of prediction was considered to be within 5%. Laboratory techniques were used, as the method was unsuitable for *in situ* measurements.

Wiebenga[3] used ultrasonic pulse velocity in place of dynamic modulus of elasticity and the damping constant as well as pulse attenuation in his laboratory tests. In each case, the use of either damping constant or pulse attenuation improved the accuracy of the predicted compressive strength. Galan[6] also used the combination of ultrasonic pulse velocity and the damping constant to estimate the *in situ* strength of concrete. The damping constant was determined by calibrating experimental curves of an oscillogram with the corresponding damped reverberated impulses. According to Galan, good correlation between the strength of concrete and the two acoustic characteristics, i.e., pulse velocity and damping constant, can be established. The pulse velocity expresses the elastic properties and the damping constant represents the inelastic behavior of concrete.

In the majority of cases, the differences between the estimated strength values and the values obtained by destructive testing was of the order of 5%, i.e., similar to the accuracy of the laboratory tests by Kesler and Higuchi. Most of the recent research work using the above technique has been conducted in the eastern European countries.

The new portable ultrasonic pulse velocity units currently used in the western countries have digital displays and are not equipped with oscilloscopes. Thus the application of pulse attenuation techniques combined with the ultrasonic pulse velocity would require additional equipment. It would also require highly skilled technical personnel, which can render this test method cost ineffective.

By far, the most popular combination, however, is the method based on the measurement of ultrasonic pulse velocity in conjunction with hardness measurements. This was confirmed by a number of surveys, such as Jones and Facaoaru[7] and Malhotra and Carette.[8] Historically, most of the credit for the development of this combined method should be attributed to Skramtaev and Leshchinsky,[2] Wiebenga,[3] MacLeod[5] and particularly to Facaoaru,[4,7,9] and for the promotion and popularization of the method to Malhotra.[8,10-12] Several isolated reports on the use of other types of combinations were published from time to time. For example, MacDonald and Ramakrishnan[13] considered pulse velocity and maturity, and Parsons and Naik[14] used the pullout test and maturity. However, the only recorded practical application at present seems to be the combination of ultrasonic pulse velocity and hardness measurements, and all other combinations, however theoretically promising, should thus be considered currently only as research and development.

COMBINED ULTRASONIC PULSE VELOCITY AND HARDNESS MEASUREMENT TECHNIQUES

THEORETICAL CONSIDERATIONS

Fundamental aspects of the hardness methods are given in Chapter 1 and those of ultrasonic pulse velocity in Chapter 5. However, a brief summary is given to highlight the major aspects of these tests.

Hardness scales are arbitrarily defined measures of the resistance of a material to indentation under static or dynamic load or resistance to scratch, abrasion, wear, cutting or drilling. Concrete test hammers evaluate surface hardness as a function of resiliency, i.e., the ability of a hammer to rebound or spring back. One of the original papers on the subject was published by Schmidt.[15] ASTM Method C 805 describes the test method for determining the rebound number of hardened concrete, and methods of hammer use and calibration are also given in the B.S.1881:Part 202:1986.

The propagation of a longitudinal or compressional disturbance and also a transverse or shear disturbance in a semi-infinite solid was explained by Poisson in 1828. Chree[16] and Lord Rayleigh[17] subsequently showed that by transmitting ultrasonic waves through blocks of different materials it was possible to measure the values of the elastic constants for these solids. The method was particularly useful for a material such as glass, for which the behavior under static loading causes considerable difficulty of interpretation.

The interpretation of the pulse velocity measurements in concrete is complicated by the heterogeneous and to some degree anisotropic nature of this material. The wave velocity is not determined directly, but is calculated from the time taken by a pulse to travel a measured distance. A piezoelectric transducer emitting vibrations at its fundamental frequency is placed in contact with the concrete surface so that the vibrations travel through the concrete and are received by another transducer, which is in contact with the opposite face of the test object.

In theory, for a semi-infinite elastic solid there is a unique relationship between the longitudinal wave velocity and the density, modulus of elasticity, and Poisson's ratio. The ultrasonic pulses produce relatively low stresses and strains and the behavior of concrete subjected to this test method can be considered elastic for all practical considerations (see Figure 1). Thus at least some of the theoretical correlations should retain a degree of validity in practical applications.

ASTM Method C 597 and B.S.1881:Part 203, 1986, describe the standard test methods for determination of pulse velocity through concrete.

DESCRIPTION OF TEST METHODS

The nondestructive testing of *in situ* concrete may be carried out for a number of reasons. The final objective of the evaluation, as well as the type and the extent of the available information, which is essential for interpreting the results will generally influence the selection of a particular method. Arbitrarily, the combined methods based on ultrasonic pulse velocity and rebound hammer techniques can be divided into two main groups.

In the first group, the prime objective is to determine the rate of strength gain in concrete and/or the variation in the strength within a group of concrete batches, mixed to the same proportions. All the ingredients of each mix (e.g., cement, aggregate, and admixture type) as well as their proportions are generally known in this case.

A classical example of this application is the SONREB method, developed largely due to the efforts of RILEM Technical Committees 7 NDT and 43 CND, and under the chairmanship of Facaoaru[18] has been adopted in Romania. A general relationship between compressive strength of concrete, rebound hammer number, and ultrasonic pulse velocity, in accordance with the tentative recommendations for "*In Situ* Concrete Strength Estimation

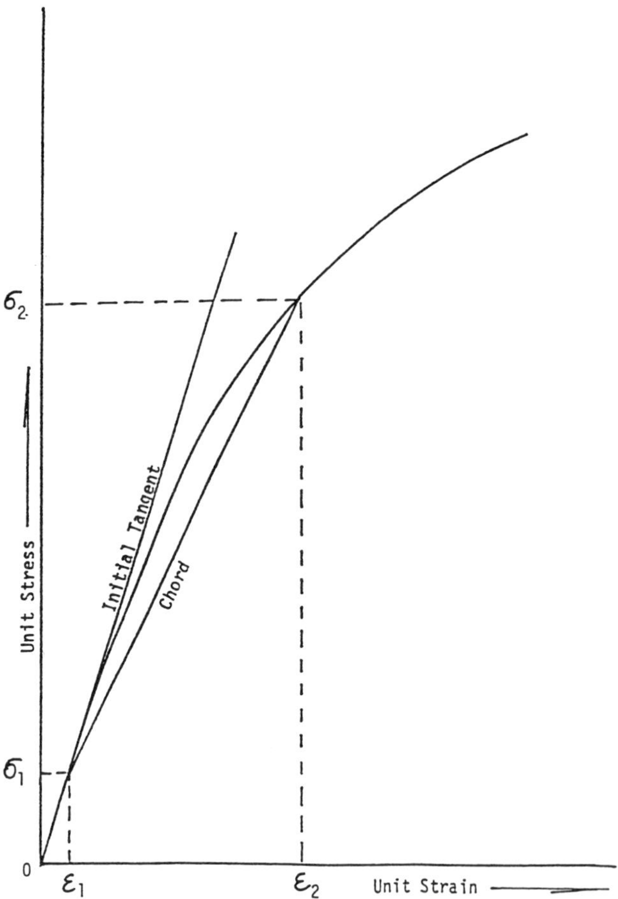

FIGURE 1. Typical relationship between the initial tangent and a chord modulus of elasticity for concrete. 0 to ϵ_1 is the range of potential disturbance in a static test, but also an elastic range in the dynamic test.

by Combined Non-Destructive Methods'', RILEM Committee TC 43 CND, 1983, forms the basis of SONREB technique. Figure 2 shows this relationship in the form of a nomogram. By knowing the rebound number and pulse velocity, the compressive strength is estimated.

A series of correction coefficients, developed for a specific concrete grade and type are then applied in order to improve the accuracy of prediction obtained from the nomogram. The following coefficients are used, according to Facaoaru:[18]

Cc = coefficient of influence of cement type
Cd = coefficient of influence of cement content
Ca = coefficient of influence of petrological aggregate type
Cg = coefficient of influence of aggregate fine fraction (less than 0.1 mm)
Co = coefficient of influence of maximum size of aggregate

The accuracy of the estimated strength (the range comprising 90% of all the results) is considered to be

1. 10% to 14% when the correlation relationship is developed with known strength values of cast specimens or cores and when the composition is known
2. 15% to 20% when only the composition is known

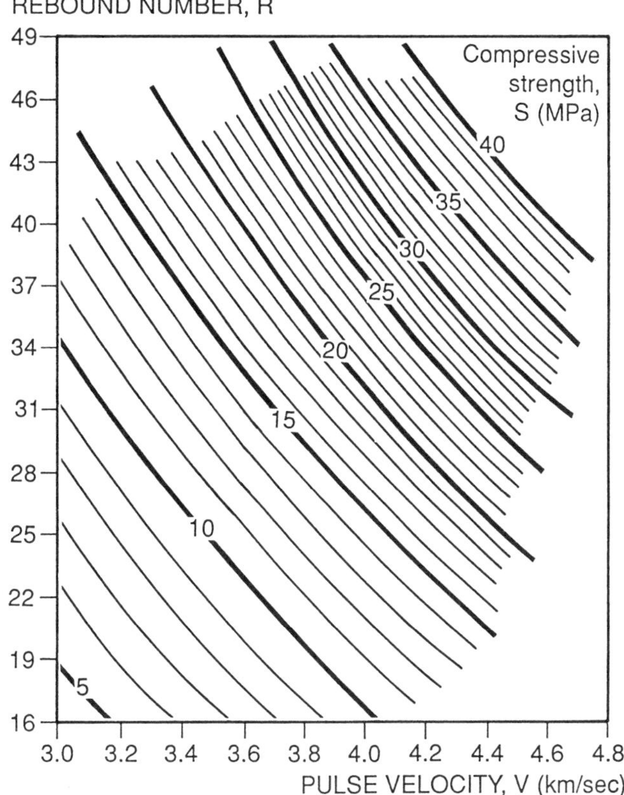

FIGURE 2. ISO-strength curves for reference concrete in SONREB-method.

In the second group, the prime objective is to determine the potential compressive strength of *in situ* concrete when its quality is considered to be suspect. Even if the intended mix ingredients and proportions are known (and quite often this is not the case), low compressive strength results, poor field performance, or even concrete appearance may lead to the belief that, due to either a mechanical malfunction or a human error, these ingredients and proportions differ from the designed values. Thus for the second group, we can say that the combined test method is used when concrete mix ingredients and proportions are NOT KNOWN.

The general approach for the second case is to develop a correlation relationship between pulse velocity and rebound hammer readings and compressive strength of standard laboratory specimens cast from the locally manufactured concretes. These forms of general correlation at least reduce the uncertainties due to variable ingredients and variations in mix proportions, as the sources of local raw materials are usually limited.

The accuracy (again shown as the range comprising 90% of all the results) when only specimens or cores are available is indicated by Facaoaru[18] to be between 12 and 16%. There are, however, ways in which this accuracy can be improved.

DEVELOPMENT OF TEST METHODS FOR PRACTICAL APPLICATIONS

As already mentioned, in the majority of cases the need for *in situ* concrete strength evaluation arises as a result of the suspect, and hence unknown, quality of the concrete in a structure.

In an ordinary concrete between 60 and 70% of the absolute volume is taken up by aggregate and the rest by cement paste, consisting of hydrated and unhydrated cement grains, chemically bound and free water, and entrained (small voids) or entrapped (larger voids) air. Subject to availability in a particular country or region, part of the cementitious material may be ground granulated blast furnace slag, fly ash, silica fume, or some other pozzolanic or reactive siliceous material. The paste may also contain chemical admixtures. The strength characteristics of a given cement paste, subjected to the influence of a particular environment, will be a function of *time,* and the strength characteristics of a given *aggregate,* for all practical purposes, can be considered time independent and a function of its petrological type only.

Thus, even in concrete of suspect quality and unknown composition, there are two variables which can be identified with a reasonable degree of accuracy, namely, *petrological type of the aggregate* and *approximate age of the concrete.* Aggregate can be identified by removing some of the matrix in an out-of-sight part of a structural member. Coarse aggregate is particularly easy to identify in this way and, in the majority of commercial grade concretes, coarse aggregate content is significantly higher than the fine.

Establishment of a series of specific correlations between the combination of rebound hammer number (R) and ultrasonic pulse velocity (V) and the compressive strength (S) of concretes, each containing a *particular aggregate type* and being of a *particular age group,* was completed in the early 1970s by Samarin and Smorchevsky.[19,20] Because the transit time of an ultrasonic pulse through concrete consists of the sum of transit times through aggregate and paste, the identification of *aggregate type* and *time* dependent properties of cement paste eliminates two major uncontrollable variables of the general correlation. The accuracy of the estimated compressive strength can thus be measurably improved. Yet another factor which can improve the accuracy of prediction, particularly over a wide range of concrete strength levels, is the provision for nonlinearity of some functions.

Work by Samarin[20] has shown that for Australian concretes the relationship between rebound hammer number (R) and the compressive strength (S) is nearly linear, and curve fitting analyses indicated that a fourth order function gives the best correlations between ultrasonic pulse velocity (V) and the compressive strength (S) for the same concretes. Thus the general equation for the rebound hammer correlation relationship is

$$S = a_0 + a_1 R \tag{1}$$

where a_0, a_1 are constants.
The general equation for the pulse velocity correlation relationship is of the form:

$$S = b_0 + b_1 V^4 \tag{2}$$

where b_0, b_1 are also constants.

It is worth mentioning that the almost universally accepted empirical relationship between the elastic modulus of concrete (E) and the compressive strength of concrete (S) is of the following general form:

$$E = A\,S^{05} \tag{3}$$

where A is a constant, depending on concrete density, statistical evaluation of strength and the selected system of measures. At the same time, the theory of propagation of stress waves through an elastic medium states that for a compression wave the following functional relationship is valid:

$$E = B\,V^2 \tag{4}$$

where B is a constant, depending on density and Poisson's ratio. Equating the right hand sides of Equations 3 and 4, we see that strength (S) is related to the fourth power of the pulse velocity (V), confirming the general validity of Equation 2.

Detailed consideration of the above derivation was given by Samarin and Meynink.[21] It was considered convenient in this work to divide concrete into three age groups, namely:

1. 7 days and younger
2. Over 7 days but less than 3 months
3. 3 months and older

It is known, for example as reported by Elvery and Ibrahim,[22] that the sensitivity of ultrasonic pulse velocity to concrete strength is very high in the first few days, but after about 5 to 7 days (depending on curing conditions) the results become considerably less reliable. Most of the concrete which is identified as being suspect is subsequently tested *in situ* at the age of between 1 and 3 months. The majority of the laboratory test data, for which the correlations have been developed, also fall into this period.

To develop a correlation relationship for each *age group* and *aggregate type,* concretes having a wide range of strength grades and mix composition are cast into standard laboratory (cylindrical) specimens. Compaction, curing, capping, etc. are carried out strictly according to the requirements of the relevant standards. Just prior to a compression strength test, each specimen is placed in a horizontal rig and the transit time of an ultrasonic pulse through the length of a capped cylinder is recorded. The pulse velocity is calculated as the ratio of the length over the transit time. The specimen is then placed in a compression machine and a load of approximately 1.4 MPa (200 psi) is applied, while 15 rebound hammer readings are taken around the circumference of the cylinder. A similar technique can be used with a cube or a prismatic specimen. The specimen is then tested in unconfined compression, using a test method which complies with the relevant standard requirements.

When the multiple correlation relationship for each *aggregate type* and each *age group* is developed, the results, as compared with correlations between compressive strength and ultrasonic pulse velocity alone, or between compressive strength and rebound hammer reading alone, indicate:

1. An increase in the multiple correlation coefficient above the correlation coefficient for rebound number alone for pulse velocity alone

   (Note that correlation coefficients for rebound number and strength are generally higher (better) than those for pulse velocity and strength.)

2. A decrease in the standard error of estimate for a multiple correlation relationship compared with relationships between rebound number and strength alone and between pulse velocity and strength alone

The degree of improvement due to the combined technique depends on a number of factors. Of these, the most significant (in the order of importance) appear to be

1. Grouping concretes for a particular multiple-regression analysis according to the petrological type of the coarse aggregate.
2. Use of least-squares curve fitting to establish the correct form of the relation between concrete strength and each independent variable separately. (For Australian concretes, the correlation between compressive strength and rebound hammer is very near linear, and pulse velocity has to be raised to the fourth power in order to produce near optimum curve-fitting in its functional relationship to strength.)

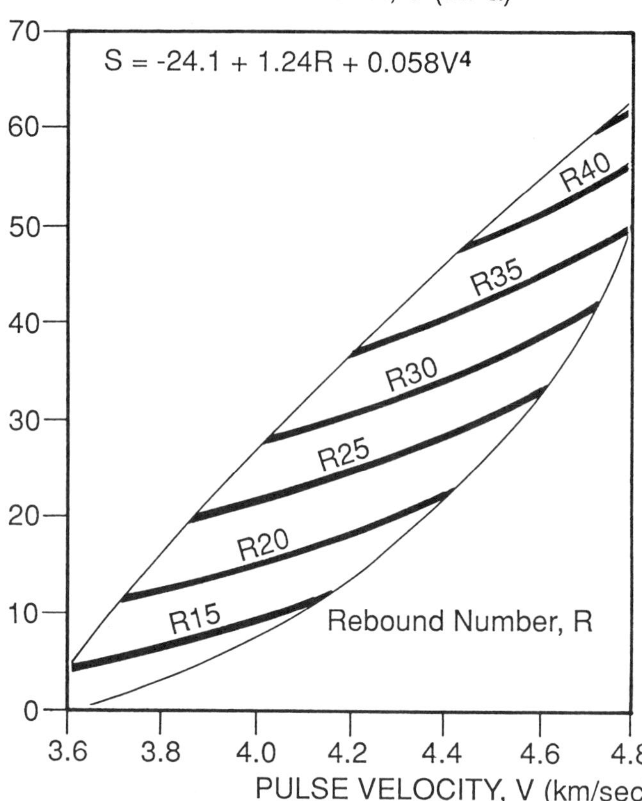

FIGURE 3. Nomogram for concrete of a particular aggregate type and age.

3. In establishing the multiple correlation relationship, a reasonably wide range of strength grades of concrete (say, from 20 to 50 MPa) all manufactured using identical coarse aggregate, should be used.

A typical multiple correlation relationship for Australian concrete in a form of a nomogram is shown in Figure 3. Early age effect puts greater emphasis on pulse velocity.

Development of a general multiple correlation equation for a particular country or even for a given district, for example as reported by Shah,[23] does not take into consideration the coarse aggregate effect and the benefit of a combined effect is reduced.

The curve fitting model used for each independent variable can have a significant effect on the relation to strength. Malhotra and Carette[8] in their analysis using multiple correlation methods compare their own test data with the correlations obtained by Samarin and Meynink[21] and Bellander.[24] In each case, when nonlinearity of pulse velocity vs. strength was not taken into account, the effect of the combined method was either of little significance, or there was no improvement at all due to multiple correlation.

A variety of nonlinear and linear multiple correlation equations were considered by different research workers and the results are compiled in Table 1. In every case, when nonlinearity between the pulse velocity (V) and strength (S) was taken into account the significance of the combined effect was enhanced.

LIMITATIONS AND ADVANTAGES OF COMBINED PULSE VELOCITY AND REBOUND NUMBER TECHNIQUE

First of all, most of the limitations which apply to the rebound number method by itself

TABLE 1
Various Multiple Regression Correlations Suggested by Different Researchers for Estimating Compressive Strength of Concrete

Researchers	Form of multi-regression equation	Significance of the combined effect
McLeod[5]	$S = k_o + k_1 R + k_2 V$	Significant in some cases, but not in others
Di Maio, et al.[25]		
Tanigawa, et al.[26]		
Knaze and Beno[27]	$S = a_0 + a_1 R + a_2 R^2$ $S = b_0 + b_1 V + b_2 V^2$	Use of nomogram:-"Curves of Equal Strength". Effect considered significant
Bellander[24]	$S = k_0 + k_1 R^3 + k_2 V$	Significant to a certain degree
Weibinga[3]	$\log_e S = k_0 + k_1 R + k_2 V$	Significant (within test conditions)
Shah[23]		
Tanigawa, et al.[26]		
Schickert[28]	$S = k_0 R^n V^m$	Some evidence of significant effect
Samarin et al.[19-21]	$S = k_0 + k_1 R + k_2 V^4$	Significant
Tanigawa et al.[26]	$S = V(a_0 + a_1 R + a_2 R^2 + a_3 R^3)$	Significant (but possibly too complex)

TABLE 2
Effect of Curing Conditions and Moisture Content of Concrete on the Reliability of Predictions Using Combined Techniques

Description of curing of cylinders	Pulse velocity m/sec	Rebound number	Compressive strength (MPa)	
			Predicted	Actual
1. 28 days moist	4.59	27.3	35.5	35.0
2. 28 days moist, dry surface by 65°C (2 h)	4.57	27.9	36.0	35.0
3. 26 days moist, dry surface by 50% RH (2 days)	4.56	29.1	37.0	38.0
4. 7 days moist, dry at 50% RH (21 days)	4.43	29.3	34.5	35.5
5. 7 days moist as for (4) but rewetted	4.41	27.2	31.5	31.5
6. 7 days moist/very dry in oven	4.10	28.9	28.0	39.0
7. 7 days moist as for (6) but rewetted	3.94	27.3	23.5	31.0

(Chapter 1) and also to the ultrasonic pulse velocity method by itself (Chapter 5) are likely to affect the reliability, sensitivity, and reproducibility of the results obtained by the combined technique. However, there are exceptions in those cases when a variation in properties of concrete produces opposite effects on the result of each component test. Most notable of these is the effect of variability of moisture content in concrete. An increase in the moisture content increases the ultrasonic pulse velocity but decreases the value of the rebound hammer number. This aspect of the combined method was reported by Bellander,[24] and it is also shown in Table 2 from unpublished research by Meynink and Samarin. In the work by Meynink and Samarin, the objective was to assess the influence of curing and the moisture condition of a specimen on the predicted vs. measured strength. The following method was employed:

Seven sets of cylinders (2 per set) cast from the same mix were subjected to the following curing regimes prior to testing in compression.

1. 28 days in water in 23°C
2. 28 days in water at 23°C, 2 h in oven at 65°C

3. 26 days in water at 23°C, 2 days in 50% RH, 23°C environment
4. 7 days in water at 23°C, 21 days in 50% RH, 23°C environment
5. As for Number 4 but prior to testing, rewetted by immersion in water for 2 h
6. 7 days in water at 23°C, 7 days in 50% RH, 23°C environment, 7 days in oven at 65°C, 7 days in oven at 110°C
7 As for Number 6 but prior to testing, rewetted by immersion in water for 2 h

The mix was typical of a concrete in Australia containing a natural normal weight aggregate and blended cement. With the exception of high temperature treatments 6 and 7, the strength calculated from a regression equation was within 1 MPa (about 145 psi) of measured strength for all curing and moisture conditions.

These results give further confidence in the combined technique developed by Samarin et al.[19-21] on the basis of prior correlations, when only the aggregate type and the age of concrete is known or can be identified. Results of heat treated specimens indicate that concrete exposed to fire may not be suitable for evaluation by the combined pulse velocity and hardness measurement technique. This view is confirmed by a detailed study of the effects of high temperatures on the reliability of strength estimates from combined nondestructive tests by Logothetis and Economou.[29]

Chung and Law[30] concluded that both the pulse velocity and the compressive strength of concrete are reduced by fire attack, but that the rate of reduction is not the same. It seems that, in order to evaluate the extent of fire damage, a prior multiple regression correlation relationship using the combined technique should be developed for concrete of a given composition subjected to a range of high temperatures.

The change in ultrasonic pulse velocity due to the presence of reinforcing bars in the direction of propogation of the pulse was investigated by a number of researchers, among whom the work by Chung[31] deserves particular mention. If the position of steel is known or can be located by a covermeter, corrections as given by Chung,[31] or as shown in B.S.1881:Part 203:1986, can be applied. However, complicated steel patterns in the direction of the ultrasonic pulse must be avoided.

The effect of incomplete consolidation of concrete on the accuracy of strength prediction using the combined method developed by Samarin et al.[19-21] was investigated by Samarin and Thomas.[32] Reduction in the concrete strength due to the lack of consolidation was correctly estimated by the combined nondestructive test. In superplasticized concrete, the estimated strength tended to be conservative, i.e., the actual results were slightly higher than predicted. The effect of concrete surface treatments, such as curing compounds and particularly surface treatments designed to improve abrasion resistance, was investigated by Sadegzadeh and Kettle.[33] The ultrasonic pulse velocity readings were not found to be significantly sensitive to surface treatments. The rebound hammer readings were affected by finishing techniques and curing regimes, but not by the *liquid* surface treatments. Hence, care must be taken in applying combined nondestructive test methods to concretes with abrasion resistant surface treatments. Similar considerations may apply to very old, surface carbonated concretes.

APPLICATION OF COMBINED TEST METHODS

In evaluating the *in situ* properties of concrete, one must take into account the potential differences between strengths in the lower and upper parts of structural members, and the extent to which this difference is affected by the size and the shape of a structural element. Wiebenga,[34] in his general report on the subject, states that the differences in strength between the upper and lower parts of walls and columns can often be 20 to 30% and in some cases the strength is up to 50% lower in the bottom parts of these members.

Destructive tests can be influenced by a number of factors, and these should be taken into consideration when comparison is made or correlations are established with the nondestructive tests. For example, Meynink and Samarin[35] found that the cores drilled in a horizontal direction generally give lower results than vertical cores taken at the same location.

Subject to the above considerations, combined methods for which prior correlations were developed for the local concrete materials have been successfully used for nondestructive *in situ* strength evaluation of concrete. Most of the reported cases of practical applications of the combined technique were in Europe and in Australia. The SONREB method, in which all concrete mix ingredients and proportions are usually known in advance, found practical use in Europe, as previously mentioned and reported by Facaoaru.[18] The application of this technique by Pohl[36] in the solution of a structural repair problem of a concrete silo resulted in considerable cost saving on the project. The main advantage of a nondestructive test is the possibility of obtaining a very large number of spot readings at a relatively low cost and without affecting the integrity of a structure.

Use of the combined nondestructive technique developed by Samarin et al.[19-21] became a routine method for evaluating the *in situ* quality of suspect concrete in many parts of Australia, and examples of its applications in a variety of projects involving office buildings, hospitals, and precast yards were given by Meynink and Samarin[35] and by Samarin and Dhir.[37] Some examples highlighting the practical use of this technique in Scotland were also reported.[37]

CONCLUSIONS

Combined nondestructive methods refer to techniques in which one test is used to improve the reliability of the *in situ* concrete strength estimated by means of another test alone.

The validity of a combined technique can be evaluated from the degree of improvement this additional test provides to the accuracy and reproducibility of predictions, vs. the additional cost and complexity of the combined method and the extent to which it is practicable to perform the additional tests *in situ*.

Of the various combinations proposed by different researchers and from the reported data it seems that only the combined techniques based on the ultrasonic pulse velocity and surface hardness measurement have been adopted in some parts of the world for practical evaluation of the *in situ* compressive strength of concrete. Presently, other combined techniques should be considered as being in the research and development phase.

The limitations of a combined method are usually those pertinent to the limitations of each component test, except when a variation in the properties of concrete affects the component test results in opposite directions. In this case, the errors can be self-correcting. Development of a prior correlation relationship is essential if the estimates from the combined tests are to be meaningful. The more information that can be obtained about the concrete ingredients, proportions, age, curing conditions etc., the more reliable the estimate is likely to be.

When testing suspect quality concrete of unknown composition, it is highly desirable to develop a prior correlation relationship in which factors such as aggregate type and approximate age of concrete are introduced as constants. For most *in situ* concretes an approximate age and petrological type of aggregate can be determined, thus reducing the number of uncontrollable variables in the equation by two.

The most important influences on the accuracy and reliability of strength estimates seem to be the coarse aggregate type in the concrete and the form of the multiple-regression equation. Nonlinear correlation relationships appear to provide more accurate estimates.

When a reliable prior correlation relationship exists for a particular concrete type, the use of combined nondestructive techniques provides a realistic alternative to destructive testing. It is often possible to perform a large and thus a representative number of tests at a reduced cost compared with coring, and without an adverse effect on the integrity of a structural element.

REFERENCES

1. **Kesler, C. E. and Higuchi, Y.,** Determination of compressive strength of concrete by using its sonic properties, *Proc. ASTM,* 53, 1044, 1953.
2. **Skramtaev, B. G. and Leshchinsky, M. Yu.,** Complex methods for non-destructive tests of concrete in constructions and structural works, RILEM, Bull., Paris, New Series No. 30, March 1966, 99.
3. **Wiebenga, J. G.,** A comparison between various combined non-destructive testing methods to derive the compressive strength of concrete, Rep. kBl-68-61/1418, Inst. TNO Veor Bouwmaterialen en Bouwconstructies, Delft, The Netherlands, 1968.
4. **Facaoaru, I.,** Non-destructive testing of concrete in Romania, Proc. Symp. on Non-destructive Testing of Concrete and Timber, London, June, 1969, Institute of Civil Engineers, London, 1970, 39.
5. **MacLeod, G.,** An assessment of two non-destructive techniques as a means of examining the quality and variability of concrete in structures, Rep. Cl/sfB/Eg/(A7q)UDC 666.972.017.620 179.1 42.454, Cement and Concrete Association, London, July 1971.
6. **Galan, A.,** Estimate of concrete strength by ultrasonic pulse velocity and damping constant, *ACI J.,* 64 (10), 678, 1967.
7. **Jones, R. and Facaoaru, I.,** Analysis of answers to a questionnaire on the ultrasonic pulse technique, RILEM, *Materials and Construction,* 1(5), 457, 1968.
8. **Malhotra, V. M. and Carette, G. C.,** In situ testing for concrete strength, CANMET publication: Progress in Concrete Technology, Malhotra, V. M., Ed., Canada, 1980, 749.
9. **Facaoaru, I.,** The correlation between direct and indirect testing methods for in-situ concrete strength determination, Proc. RILEM Conf., Quality Control of Concrete Situations, Stockholm, June 1979, Vol. 1, 147.
10. **Malhotra, V. M.,** Testing hardened concrete: non-destructive methods, ACI Monogr. No. 9, American Concrete Institute, Iowa State University Press, Detroit, 1979.
11. **Malhotra, V. M.,** In-situ strength evaluation of concrete, *Concrete International, Design and Construction,* Vol. 1, No. 9, September 1979, 40.
12. **Malhotra, V. M., Ed.,** *In Situ/Nondestructive testing of concrete,* ACI Spec. Publ., SP-82, American Concrete Institute, Detroit, 1984.
13. **MacDonald, C. N. and Ramakrishnan, V.,** Quality control of concrete using pulse velocity and maturity concepts, Proc. RILEM Conf., Quality Control of Concrete Structures, Stockholm, June 1979, Vol. 2, 113.
14. **Parsons, T. J. and Naik, T. R.,** Early age concrete strength determination by pullout testing and maturity, ACI, SP-82 (Ref. 12), Detroit, 1984, 177.
15. **Schmidt, E.,** Concrete testing hammer, *Schweiz. Bauz. (Zurich),* 68 (28), 378, 1958 (in German).
16. **Chree, C.,** On longitudinal vibrations, *Q. Math. J.,* 23, 317, 1889.
17. **Lord Rayleigh (Strutt, J. W.),** *The Theory of Sound,* Vol. 1, 2nd ed., Macmillan, London, 1894, chaps. 5, 7, and 8.
18. **Facaoaru, I.,** Romanian achievements in nondestructive strength testing of concrete, ACI, SP-82 (Ref. 12), Detroit, 1984, 35.
19. **Samarin, A. and Smorchevsky, G.,** The non-destructive testing of concrete, Central Research Laboratory, Internal Tech. Rep. No. 54, 1973.
20. **Samarin, A.,** Use of combined ultrasonic and rebound hammer method for determining strength of concrete structural members, Central Research Laboratory, Internal Tech. Rep., Ready Mixed Concrete Ltd. Tech. Bull., 1967.
21. **Samarin, A. and Meynink, P.,** Use of combined ultrasonic and rebound hammer method for determining strength of concrete structural members, *Concr. Int. Design Construction,* 3 (3), 25, 1981.
22. **Elvery, R. H. and Ibrahim, L. A. M.,** Ultrasonic assessment of concrete strength at early ages, *Mag. Concr. Res.,* 28 (97), 181, 1976.
23. **Shah, C. B.,** Estimation of strength of in situ concrete, *Indian Concr. J.,* 56 (11), 292, 1982.

24. **Bellander, U.,** Hallfasthet I fardig konstruktion - Del 3 - oforstorande metoder. Laboratorie - Och Faltforsok. (Concrete strength in finished structure - Part 3 - non-destructive testing methods.) Investigations in laboratory and in situ CBI Forskning Research, 3, 1977.
25. **Di Maio, A. A., Traversa, L. P., and Giovambattista, A.** Nondestructive combined methods applied to structural concrete members, *ASTM J. Cement Concrete Aggregates,* 7(2), Winter, 1985, 89.
26. **Tanigawa, Y., Baba, K., and Mori, H.,** Estimation of concrete strength by combined nondestructive testing method, ACI, SP-82, (Ref. 12), 57.
27. **Knaze, P. and Beno, P.,** The use of combined nondestructive testing methods to determine the compressive strength of concrete, RILEM, *Materials and Structures,* 18(105), 207, 1985.
28. **Schickert, G.,** Critical reflections on nondestructive testing of concrete, RILEM, *Materials and Structures,* 17(99), 217, 1984.
29. **Logothetis, L. L. and Economou, Chr.,** The influence of high temperatures on calibration of non-destructive testing for concrete, RILEM, Materials and Structures, 14(79), 1981.
30. **Chung, H. W. and Law, K. S.,** Diagnosing in situ concrete by ultrasonic pulse technique, *Concr. Int. Design Construction,* 7(2), 42, 1985.
31. **Chung, H. W.,** Effects of embedded steel bars upon ultrasonic testing of concrete, *Mag. Concr. Res.,* 30 (102), 19, 1978.
32. **Samarin, A. and Thomas, W. A.,** Strength evaluation of incompletely consolidated conventional and superplasticised concrete-combined non-destructive tests, Proc. 10th Austr. Road Research Board Conference, Sydney, Vol. 10, Part 3, 25—29 August 1980, 107.
33. **Sadegzadeh, M. and Kettle, R.,** Indirect and non-destructive method for assessing abrasion resistance of concrete, *Mag. Concr. Res.,* 38(137), 183, 1986.
34. **Wiebenga, J. G.,** Strength of concrete, General Rep. on Session 2.1, RILEM Symp., Quality Control of Concrete Structures, Stockholm, June 17—21, 1979, Proc., 115.
35. **Meynink, P. and Samarin, A.,** Assessment of compressive strength of concrete by cylinders, cores and non-destructive tests, RILEM Symp., Quality Control of Concrete Structures, Stockholm, June 17—21, 1979, Vol. 1, 127.
36. **Pohl, E.,** Combined non-destructive testing methods to assess the strength of in situ concrete for silo, RILEM Symp., Quality Control of Concrete Structures, Stockholm, June 17—21, 1979, Vol. 1, 151.
37. **Samarin, A. and Dhir, R. K.,** Determination of in situ concrete strength: rapidly and confidently by non-destructive testing, ACI SP-82, (Ref. 12), Detroit, 1984, 77.

Chapter 9

MAGNETIC/ELECTRICAL METHODS

K. R. Lauer

ABSTRACT

The initial portion of the chapter briefly describes the theory of magnetic induction, magnetic flux leakage, and nuclear magnetic resonance to facilitate an understanding of equipment used to locate reinforcement and determine the moisture content of concrete.

The remaining portion of the chapter discusses the electrical nature of concrete and the mechanism of reinforcement corrosion as a preliminary to understanding the use of electrical capacitance and resistance to measure moisture content, pavement thickness, and corrosion of reinforcement.

Where possible, the accuracy of current magnetic and electrical apparatus is indicated.

MAGNETIC/ELECTRICAL METHODS

Magnetic and electrical methods are being used in a number of ways to evaluate concrete structures. These methods are used (1) to locate reinforcement and measure member thickness by inductance; (2) measure the moisture content of concrete by means of its electrical properties and the nuclear magnetic resonance of hydrogen atoms; (3) measure the corrosion potential of reinforcement; (4) determine pavement thickness by electrical resistivity; and (5) locate defects and corrosion in reinforcement by measuring magnetic flux leakage.

Magnetic and electrical methods have received considerable attention in recent years. Their underlying principles range in complexity as do their practical applications in the field.

MAGNETIC METHODS

INTRODUCTION

Materials containing iron, nickel, and cobalt are strongly attracted to themselves and to each other when magnetized; they are called ferromagnetic materials. Other materials, such as oxygen, which are weakly attracted by magnetic field, are called paramagnetic materials.

In 1905, the magazine *Revue de Met* mentioned for the first time the possibility of detecting defects such as cracks, laminations, etc. in ferromagnetic materials by means of magnetic fields. In 1919, E. W. Hoke applied for the first patent in the U.S. on a magnetic inspection method which was granted in 1922.

Magnetic nondestructive testing techniques used in conjunction with concrete involve the magnetic properties of the reinforcement and the response of the hydrogen nuclei to such fields. Because of the need to control the magnetic field, electromagnets are used in most instances.

THEORY

At the present time, three different aspects of magnetic field phenomena are used in the nondestructive testing of reinforced concrete (1) alternating current excitation of conducting materials and their magnetic inductance; (2) direct current excitation resulting in magnetic flux leakage fields around defects in ferromagnetic materials; and (3) nuclear magnetic resonance.

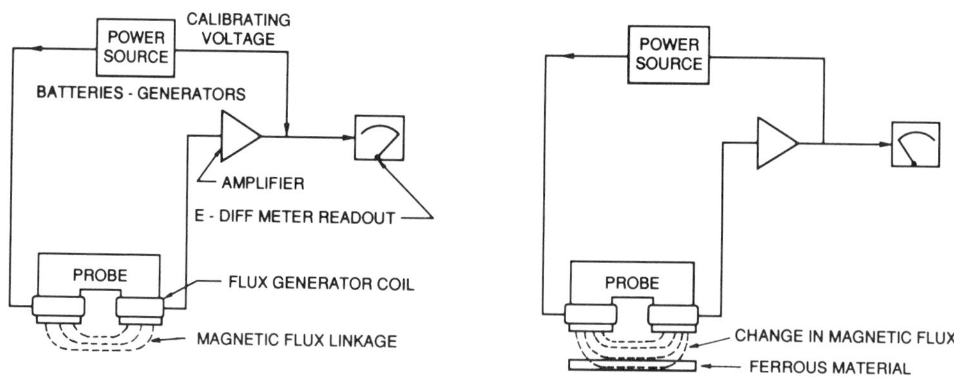

FIGURE 1. Principle of operation of induction meter used to locate reinforcement. (From Malhotra, V. M., Testing Hardened Concrete: Nondestructive Methods, ACI Monogr. No. 9, The Iowa State University Press, Ames, and American Concrete Institute, Detroit, 1976. With permission.)

Magnetic Induction

This technique is only applicable to ferromagnetic materials. Test equipment circuitry resembles a simple transformer in which the test object acts as a core (Figure 1). There is a primary coil, which is connected to a power supply delivering a low frequency (10 to 50 Hz) alternating current, and a secondary coil, which feeds into an amplifier circuit. In the absence of a test object, the primary coil induces a small voltage in the secondary coil, but when a ferromagnetic object is introduced near the coils, a much higher secondary voltage is induced. The amplitude of the induced signal in the secondary coil is a function of the magnetization characteristics, location, and geometry of the object.

The inductance of a coil can be reduced by bringing a conducting surface near the coil. It can be shown that the effect of a conducting plate on the coil is the same as the effect of a second coil, identical to the first, carrying a current equal and opposite to the coil current and located on the coil axis at a distance 2d from the original coil, where d is the coil-to-plate distance. The second coil is said to be the image of the first. The voltage induced in the first coil is seen to have two components. The first of these is due to the self-inductance of the coil in space, and the second is due to the mutual inductance between the coil and the plate. Thus the induced voltage is seen to be the sum of two components, one a constant and the other a function of coil-to plate spacing. As a result the inductance of the coil can be used to measure coil-to place distance, d, if the relationship between mutual inductance and d is known.

The probe unit consists of a highly permeable U shaped magnetic core on which two coils are mounted. An alternating current is passed through one of these coils and the current induced in the other coil is measured. The induced current depends upon the mutual inductance of the coil and upon the presence of the steel reinforcing bars. For a given probe the induced current is controlled by the distance between the reinforcement and the probe. This relatinship between induced current and distance from the probe to the reinforcement is not linear because the magnetic flux intensity decreases with the square of the distance. As a result, calibrated scales on commercial equipment are nonlinear. The magnetic permeability of concrete, even though low, will have some effect on the reading.

Flux Leakage Theory

Fundamentals of this theory have been explained in detail in a number of texts.[2-4] When ferromagnetic materials are magnetized, magnetic lines of force (or flux) flow through the material and complete a magnetic path between the poles. These magnetic lines of flux increase from zero at the center of the specimen and increase in density and strength toward

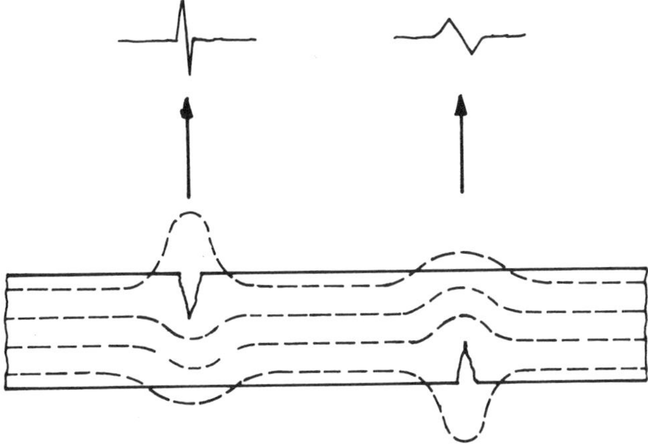

FIGURE 2. Effect of defects on flux pattern and measurement.

the outer surface. When the magnetic lines of flux are contained within the ferromagnetic object, it is difficult, if not impossible, to detect them in air space surrounding the member. However, if the surface is disrupted by a crack or defect, its magnetic permeability is drastically changed and leakage flux will emanate from the discontinuity. Measurement of the intensity of this leakage flux provides a basis for nondestructive identification of such discontinuities. Figure 2 illustrates how a notch or defect distorts the magnetic lines of flux causing leakage flux to exist in the surface of the ferromagnetic material.

Automatic flux leakage inspection systems use magnetic field sensors to detect and measure flux leakage signals. Flux leakage sensors usually have small diameters in order to have adequate sensitivity for detecting short length defects. Probes are typically spring loaded to provide constant lift-off (distance between probe and surface). Signals from probes are transmitted to the electronics unit where they can be filtered and analyzed by a continuous spectrum analyzer.

A majority of the sensors are inductive coil sensors or solid-state Hall effect sensors (electromotive forces developed as a result of the interaction of a steady current flowing in a steady magnetic field). Magnetic diodes and transistors, whose output current or gain change with magnetic field intensity, can also be used. To a lesser extent, magnetic tape systems have also been used.

The more highly magnetized the ferromagnetic object, the higher its leakage flux intensity from a given defect. The amount of leakage flux produced also depends on defect geometry. Broad, shallow defects will not produce a large outward component of leakage flux; neither will a defect whose long axis is parallel to the lines of flux. The latter are more easily detected with circular magnetic fields. Internal defects in thick parts may not be detected because the magnetic lines of flux nearly bypass the defect with little leakage. Defects oriented so that they are perpendicular to the surface and at right angles to the lines of flux will be more easily detected than defects lying at an angle with respect to the surface or flux lines. Defects lying at a shallow angle to the surface and oriented in the direction of the flux lines produce the weakest lines of leakage flux.

Nuclear Magnetic Resonance (NMR)

This technique is based on the interaction between nuclear magnetic dipole moments and a magnetic field. This interaction can be used as a basis for determining the amount of moisture present in a material by detection of a signal from the hydrogen nuclei in water molecules. The term resonance is used because the frequency of gyroscopic precession of the magnetic moments is detected in an applied magnetic field.

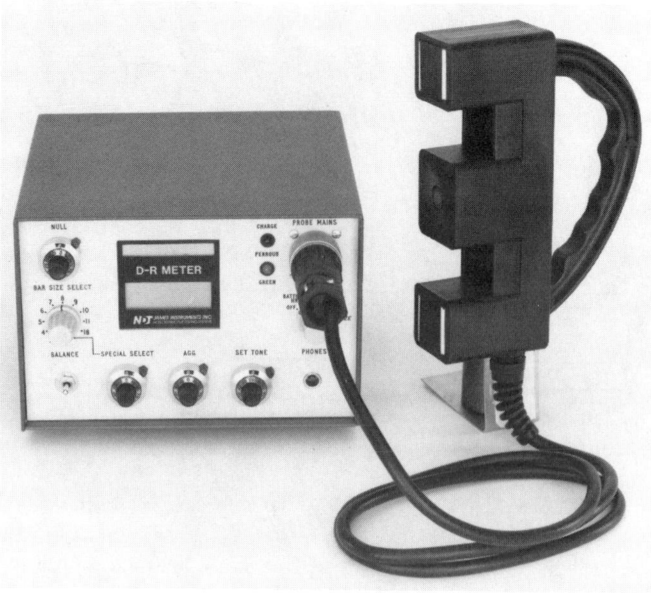

FIGURE 3. A meter used to locate reinforcement. Instrument includes a standard probe, a special probe for magnetic concrete, headphones and a spacer block for rebar measurement (Courtesy of NDT James Instrument Inc.)

Several methods of generating and detecting *NMR* signals are available. The method preferred for most practical applications is the transient or pulsed method because measurements can be made rapidly and the data obtained provide a maximum amount of information about the investigated species. Detailed information on this method is available in several texts.[5-7]

TEST METHODS

Depth of Concrete Cover

The induction principle resulted in the development of equipment for determining the location, sizes, and depth of reinforcement.[8,9] In 1951 an apparatus called the ''covermeter'' was developed in England by the Cement and Concrete Association in conjunction with the Cast Stone and Cast Concrete Products Industry.[8] Their reports indicate the effectiveness of this type of equipment for both plastic and hardened concrete. Refined versions are now available, using more sophisticated electronic circuits, which can detect reinforcement at depths of 12 in. A typical meter is shown in Figure 3.

Meters must be recalibrated for different probes. The probes are highly directional. A distinct maximum in induced current is observed when the long axis of the probe and reinforcement are aligned and when the probe is directly above the reinforcement. By using spacers of known thickness, the size of reinforcing bars between 3/8 and 2 in. (9.52 mm and 57.33 mm) can be estimated.

British Standard 4408 pt 1 suggests a basic calibration procedure involving a cube of concrete of given proportions with reinforcing bars at specified distances from the surface.[10]

These meters can be used to estimate the thickness of concrete members accessible from both sides. If a steel plate is aligned on one side with the probe on the other side, the measured induced current will indicate the thickness of the slab. The equipment must be especially calibrated for this use.

Commercial reinforcement bar locaters are portable, inexpensive instruments that can be easily used. Accuracy of ±2% or 0.1 in. up to depths of 6 in. for any bar size has been

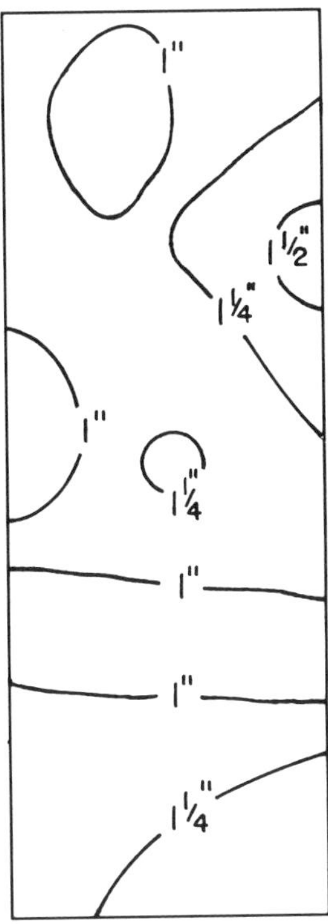

FIGURE 4. Depth of concrete cover on a reinforced concrete bridge deck.

claimed. A bar size accuracy of ± 10% to a depth of 8 in. is also indicated. Latest equipment utilize headphones which can detect by tone, a 3/4 in. bar at 12 in. depth. Tone trigger levels can be preset for depths less than 6 in. If cover determinations are carried out on a grid system over the concrete surface, equi-depth contours can be constructed which clearly illustrate the variability in depth of cover and any regions where it is less than satisfactory. An example of this type of map is illustrated in Figure 4.

The use of a hand-held instrument in mapping the cover of reinforcement in a bridge deck is very time consuming. The U.S. Federal Highway Administration developed a "Rolling Pachometer" which proved to be accurate, reliable, and capable of gathering data at a rate 20 times that of conventional hand-held methods.[11] The second generation system contains a modified hand-held meter, a battery operated, two channel, pressurized ink recorder; a speedometer; and associated electronics. The electronics include an amplifier, filters, voltage regulators, adjustable high and low reinforcing bar limit controls, a magnetic sensor, and counters for processing and displaying distance marks on the chart graph. A constant scan speed of 1 mph is required. The speedometer works in conjunction with the magnetic sensor which is located on one of the wheels. Pulses from the sensors, which are processed and used to indicate speed, are also fed into counters which trigger a distance mark on the left side of the chart paper (Figure 5). These distance marks occur every 18 in. of travel and, when used with the manual event marker switch, can be correlated to a particular area of interest on the bridge. The manual event mark is displayed on the left side of the chart paper.

Data, representing variations of the signal as the probe passes over the reinforcing bars, is displayed on the left channel of the chart paper. The sinusoidal nature of the recorded data represents the peaks associated with a reinforcing bar and, by measuring the distance of each peak to the edge of the chart paper graph, depth of cover can be determined from

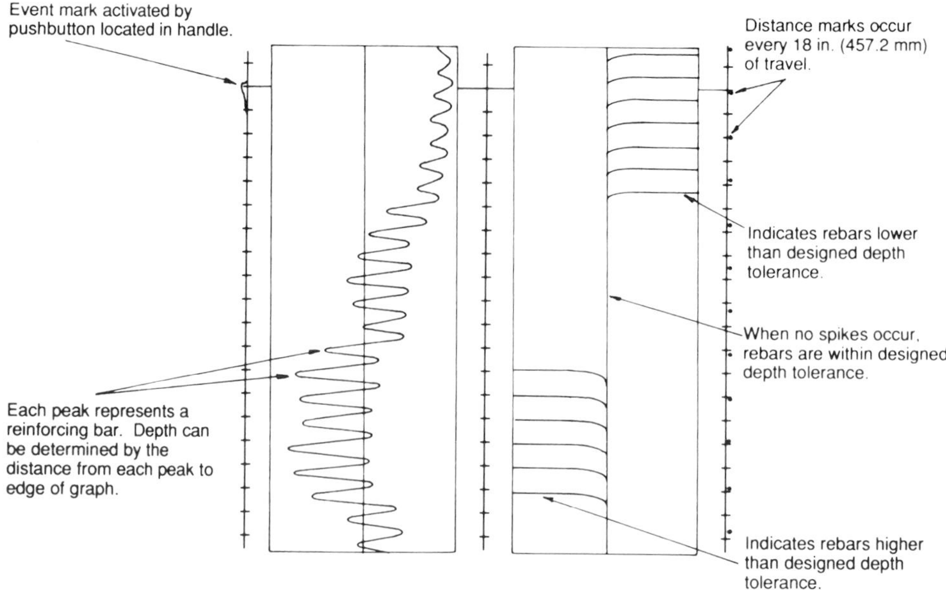

FIGURE 5. Typical chart recorder presentation of "Rolling Pachometer". (Adapted from Reference 11.)

a calibration curve. High and low limit controls permit identification of reinforcing bars either higher or lower than the allowed tolerances for the specified depth. This is indicated on the right hand channel. A spike to the left of center indicates a rebar that is lower than the preset limit. A spike to the right of center indicates a rebar that is higher than the preset limit.

Calibration curves for the apparatus are developed by constructing a test track with simulated bridge deck construction.

Repeatability, stability, and operation of the rolling instrument under various temperature extremes proved excellent. In comparison with other calibrated, hand-held meters, the results were also very good. A constant bias was, however, noted in both systems during the test of a particular bridge. The peaks indicated that the rebars were closer to the surface than when measured by coring operations. The error ranged from 1/8 to 1/4 in. depending upon the depth of the reinforcing bars. This bias effect was caused by the presence of magnetic aggregate particles. As a result, each bridge deck requires calibration with coring to correct for this kind of bias. The system is considered to be accurate within ± 0.25 in.

The accuracy of test systems using the induction principle are reduced by factors that affect the magnetic field within the range of the instrument. They include presence of more than one reinforcing bar — laps and second layers; metal tie wires and bar supports; and variations in the iron content of the aggregate and cement. As depth/spacing ratios increase, it becomes more difficult to discern individual bars. The operating temperature range of battery-operated models is generally relatively small, and such instruments will not function satisfactorily at temperatures below freezing.

Magnetic Field Method of Detecting Flaws in Reinforcement

Currently used inspection procedures rely heavily on rust staining, cracking, and spalling of concrete as an indication of problems with reinforcing steel. There is a great need for a practical nondestructive *in situ* method for detecting deterioration in the reinforcing steel of prestressed concrete highway bridge members. Of primary interest is the need to detect a 10% loss or greater of area due to corrosion and fracture of reinforcing bars and strands. Figure 6 illustrates regions of particular interest. A magnetic field method based on leakage

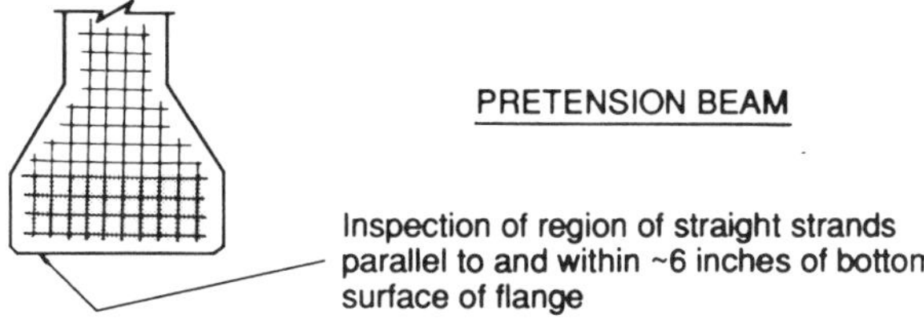

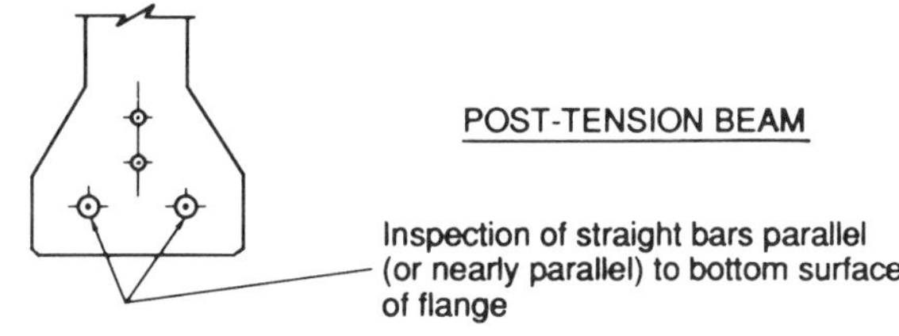

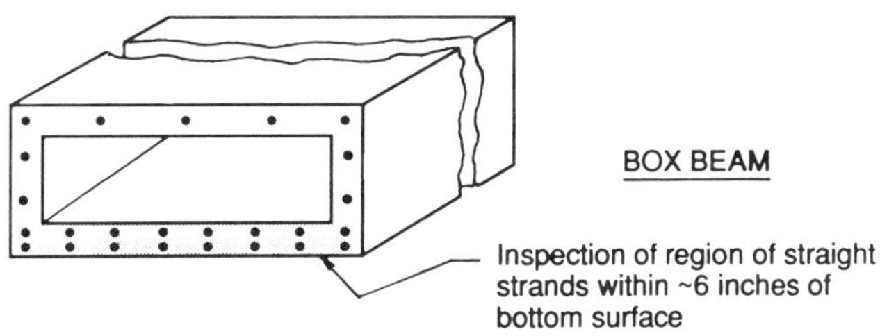

FIGURE 6. Critical reinforcement in structural bridge members. (Adapted from Reference 12.)

flux was considered to have the most promise for this application and was investigated by the U.S. Federal Highway Administration.[12]

This magnetic field method consists of applying a steady-state magnetic field to the beam under inspection and the use of a scanning magnetic field sensor to detect perturbances in the applied field caused by anomalies such as deterioration or cracks. The magnetizing field is produced by a dc excited electromagnet and a Hall-effect device is used as the magnetic field sensor. The experiments included varying degrees of deterioration, influence of adjacent unflawed steel elements, type of tendon duct, type of reinforcing steel, transverse rebar configuration, etc. Figure 7 illustrates typical magnetic responses for different steel configurations and degrees of deterioration. The results of the laboratory evaluation indicated good overall sensitivity to loss-of-section and excellent sensitivity to fracture with minimal degradation to signal response in the presence of a steel duct.

Field test results showed signatures with features similar to those observed in the laboratory as well as several prominent anomalous indications. From analyses of test data and field investigations it was determined that steel elements (chairs) were present in the structure which were completely unanticipated and were not indicated on the drawings. A method involving the "subtracting out" of configurational steel signatures so that signatures from deterioration and fracture could be recognized, proved successful. Additional work at the Ferguson Laboratory of the University of Texas at Austin confirmed the ability of the method to detect fractures in reinforcing strands as they developed from fatigue due to cyclic loadings.[13] The Federal Highway Administration is continuing to develop this system.

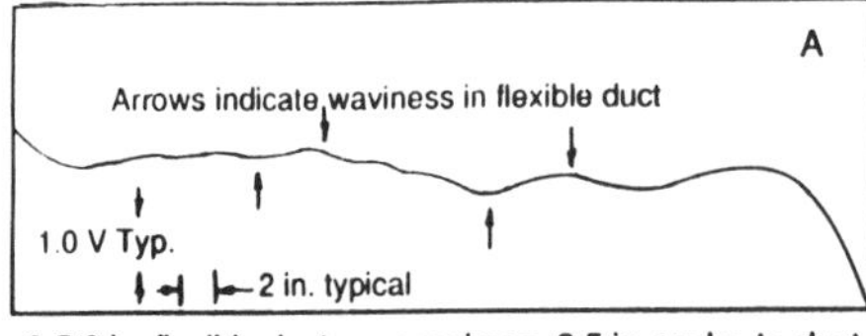

2-5/8 in. flexible duct, no specimen, 2.5 in. probe-to-duct spacing, 3 amp.

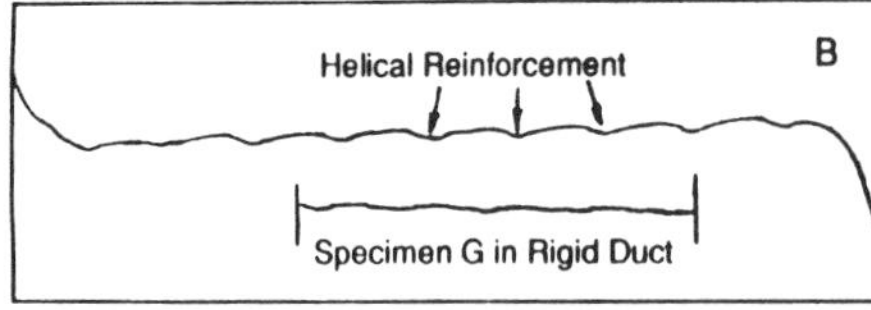

2-5/8 in. rigid duct, no specimen (insert with specimen), 2.5 in. probe-to-duct spacing, 3 amp.

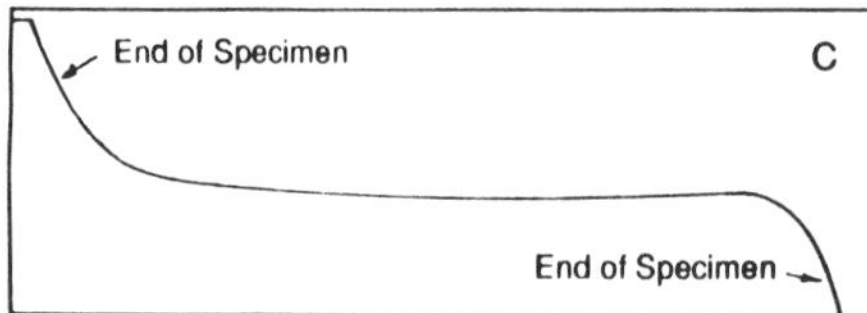

1 in. ϕ A722 Type II bar, no duct, 2.5 in. probe-to-bar spacing, 4 amp.

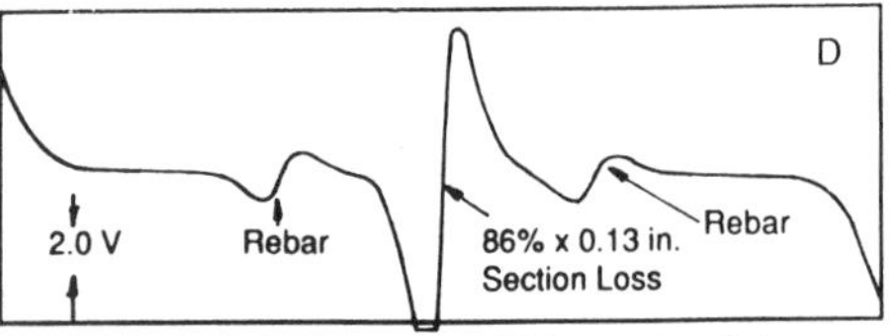

1/2 in. ϕ strand, 3/8 in. ϕ rebars 20 in. apart and above specimen, 1.5 in. probe-to-specimen spacing, 2 amp.

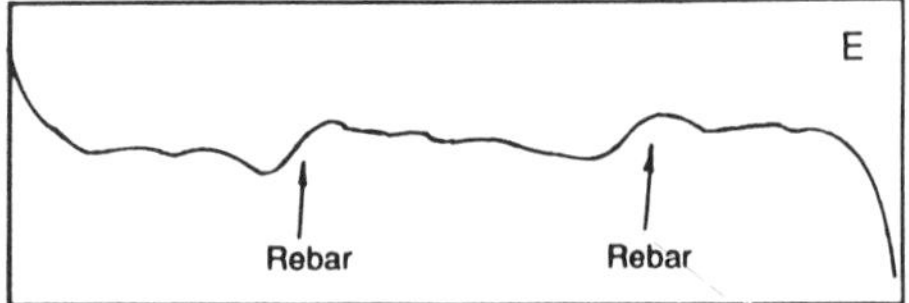

2-5/8 in. rigid duct, no specimen. 3/8 in. ϕ rebars 20 in. apart and above duct, 2.5 in. probe-to-duct spacing, 3 amp.

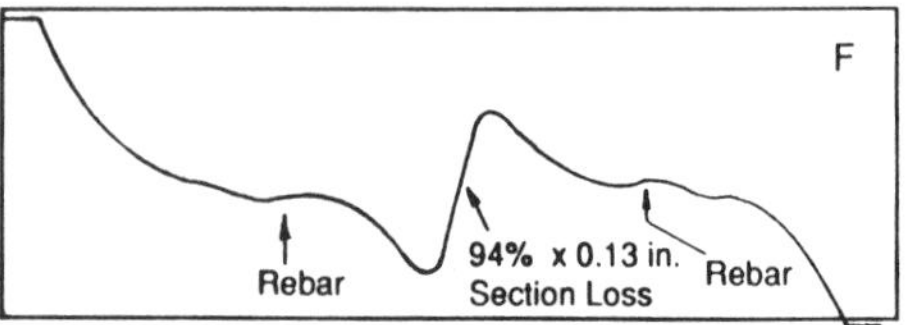

1 in. ϕ bar, rigid duct, 3/8 in. ϕ rebars 20 in. apart and above duct, 2.5 in. probe-to-specimen spacing, 3 amp.

FIGURE 7. Selected magnetic signatures from laboratory investigations of flux leakage instrument. (Adapted from Reference 12.)

Nuclear Magnetic Resonance Method of Determining Moisture Content

Corrosion of reinforcing steel in structural concrete is known to be caused by the combined action of chloride ions, moisture, and oxygen. Matzkanin et al. report on the development, fabrication, and evaluation of a nondestructive instrument for measuring the moisture content in reinforced concrete bridge decks.[14] This instrument is based on nuclear magnetic resonance (NMR) which is an electromagnetic method capable of determining the amount of moisture present in a material by detection of a signal from the hydrogen nuclei in water molecules. The transient (or pulsed) method is the most suitable one for practical applications.

The following discussion comes from the report by Matzkanin et al.[14] For transient NMR measurement, the material is exposed to a static magnetic field of intensity, H_o, and to a pulsed radiofrequency (RF) magnetic field, H_1 corresponding to the NMR frequency of the nuclei of interest. For hydrogen nuclei the NMR frequency is 2.1 MHz in a magnetic field of 494 O_ϵ(3.9 $\times$ 10^4 A/m). The H_1 field is generated by a transmitter which produces adequate power to cause the required RF current to flow in a tuned detection coil. Following each transmitter pulse (sequence) the NMR response of the excited nuclei induce a transient RF voltage in the detection coil. The prototype NMR Moisture Measurement System utilizes a two pulse sequence which provides the capability of distinguishing between NMR signals from free moisture and signals from bound hydrogen. For the application of NMR to moisture measurement in concrete bridge decks, a sensor assembly is required in which the test specimen is external to both the RF coil and magnet structure. A schematic illustration of the approach utilized in the prototype measurement system developed in this program is shown in Figure 8. A U-shaped magnet provides a magnetic bias field, H_o, extending outward from the open ends of the "U". The field direction is represented by the dashed line in Figure 8. The RF coil is a flat spiral located in the plane of the two magnet faces; two direction lines of the RF field, H_1, are shown in the figure. At a selected distance interval below the plane of the RF coil and magnet poles, the resonance relationship between the

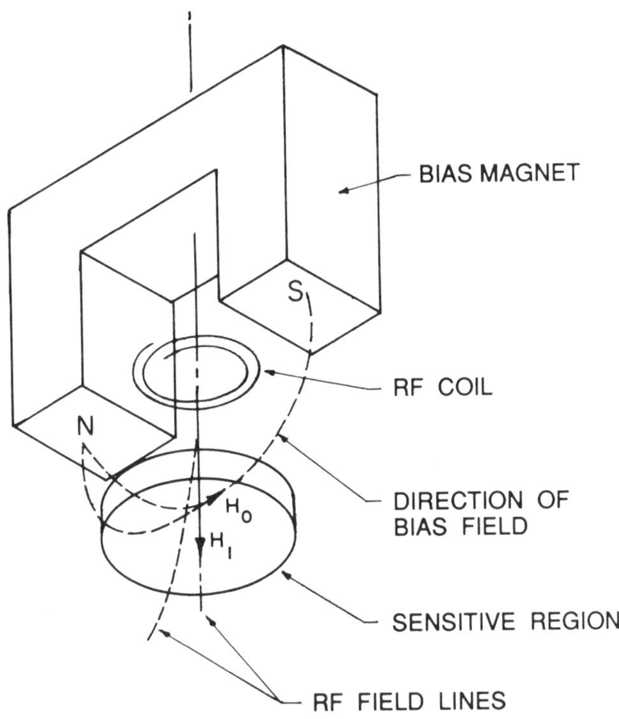

FIGURE 8. Schematic illustration of NMR detection for a specimen external to the RF coil and magnet structure. (Adapted from Reference 14.)

magnetic field intensity and RF frequency for hydrogen NMR can occur. The size, shape, and location of the sensitive region are determined by the magnitude and gradients of the magnetic bias field and the RF field.

The two assemblies comprising the prototype NMR Moisture Measurement System have exterior dimensions as shown in Figure 9. The electronics assembly contains the RF pulse generators operating at 2.1 MHz, amplifiers, signal processing components, and magnet power supply. The sensor assembly contains the electromagnet capable of producing the required hydrogen NMR field of 494 O_ϵ (3.9×10^4 A/m) at distances up to 6 in. (152 mm) away, the NMR detection coil and the tuning capacitor.

Specifications for the system indicate a moisture content range of from 1 to 6% by weight of the material with an accuracy of $\pm 0.2\%$ moisture for depths up to 2.75 in. and $\pm 0.4\%$ moisture for depths from 2.75 to 3.75 in. The measurement time per location was estimated at 2 min. A prototype instrument is in use at the Federal Highway Administration laboratories.

ELECTRICAL METHODS

INTRODUCTION

The changes in electrical properties of concrete have been investigated as a basis for understanding durability and the development of various nondestructive tests. The properties include electrical resistance, dielectric constant, and polarization resistance.

THEORY

Electrical Nature of Concrete

The evaporable water content of concrete varies with water-to-cement ratio, degree of

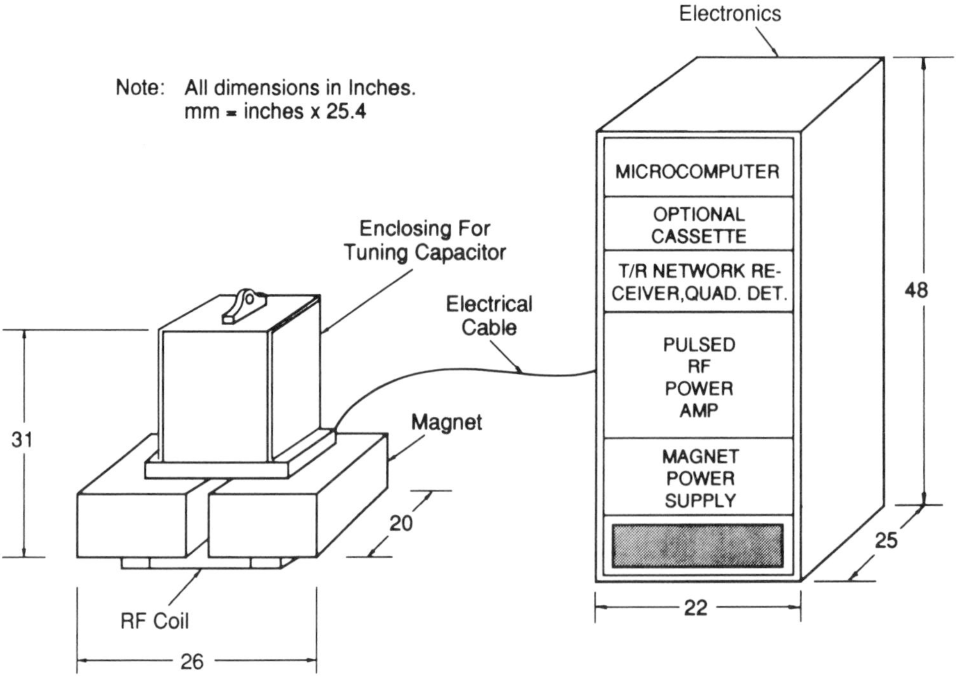

FIGURE 9. Diagram of sensor assembly and electronics assembly for NMR bridge deck moisture measurement system. (Adapted from Reference 14.)

hydration, and degree of saturation. This water contains ions, primarily Na^+, K^+, Ca^{++}, SO^{--}, and OH^-, whose concentrations vary with time. As a result the conduction of electricity by moist concrete could be expected to be essentially electrolytic as suggested by Nickkanin.[15] Tests by Hammond and Robson support this view.[16] Much of the following presentation of the theory is based on a paper by Monfore.[17]

The resistance of an electrolyte, or any other material, is directly proportional to the length and inversely proportional to the cross-sectional area. Thus

$$R = \rho\frac{L}{A} \tag{1}$$

where

R	=	resistance in ohms,
ρ	=	resistivity in ohm-cm
L	=	length in cm, and
A	=	cross-sectional area in sq cm.

Resistivity, ρ, is essentially constant for a given material under constant conditions and is numerically equal to the resistance of a 1-cm cube of the material.

Ohm's law states that the direct current through a metallic conductor is directly proportional to the potential applied and inversely proportional to the resistance of the conductor, i.e.,

$$I = \frac{E}{R} \tag{2}$$

where

I = current in amperes, and
E = potential in volts.

If the conductor is an electrolyte, the passage of direct current (movement of ions) will cause polarization and the establishment of a potential at the electrodes that opposes the applied potential. In such a case the current is

$$I = \frac{E_a - E_p}{R} \tag{3}$$

where

E_a = the applied potential in volts, and
E_p = the polarization potential in volts, or back emf, as it is frequently called.

Polarization potential (back emf) results from reactions that take place at the electrodes, reactions which depend upon the ions present and the materials of the electrodes. Thin films of oxygen, hydrogen, or other gases may be formed on the electrodes and may influence the potential created.

The use of alternating current does not avoid these polarization effects. Investigators such as Hammond and Robson suggest that the behavior of concrete can be modeled by a capacitor and resistor in parallel.[16] The current through such a combination is

$$I = \frac{E}{Z} \tag{4}$$

where

I = current in amperes,
E = potential in volts, and
Z = impedance in ohms.

For the particular case of a resistor and capacitor in parallel

$$Z = \frac{R}{(1 + \omega^2 C^2 R^2)^{1/2}} \tag{5}$$

where

ω = $2\pi f$, f being in hertz, Hz (cycles per second), and
C = capacitance in farads.

Another way of evaluating the model for alternating current is

$$I = j\omega FE\epsilon \tag{6}$$

where

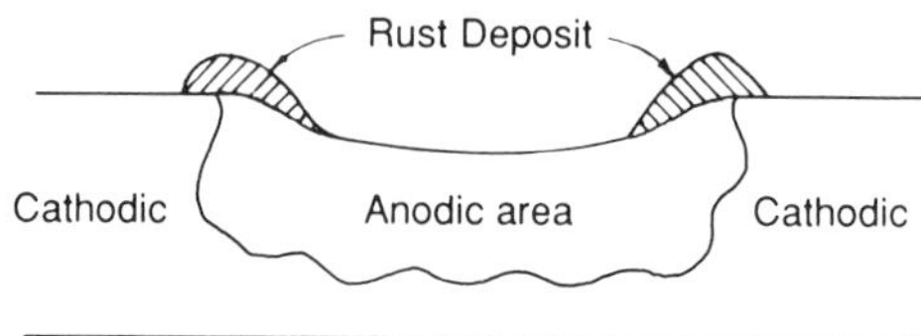

FIGURE 10. Rusting of an isolated steel bar.

F = a factor taking into account the geometry of the electrodes
E = the potential difference between the electrodes, volts
ω = $2\pi f$
and
ϵ = the complex dielectric constant defined as:

$$\epsilon = \epsilon' - j\epsilon'' \tag{7}$$

where

ϵ' = the dielectric constant of the material and

$$\epsilon'' = \frac{1}{R\omega} \tag{8}$$

The complex dielectric constant, ϵ, of a material is a measure of the extent and speed with which its molecular dipoles are aligned by an electric field. The real component, ϵ', is a measure of the extent of alignment, and the complete component ϵ'', is a measure of the speed of alignment, or frictional loss of electrical energy.

The ϵ-value of free water is about 40 times that of most solids, including cement and stone. It is about 8 times that of chemically bound water. The ϵ-value of moist concrete is a complex function of the dielectric properties of its components and the fact that its water exists in three states — unbound or free, chemically bound, and physically adsorbed.

Despite the complex relationship between a material's moisture content and its dielectric constant, this property has been used in the evaluation of moisture content.

Mechanism of Reinforcement Corrosion

The corrosion of steel in concrete is an electrochemical process that requires a flow of electrical current for the chemical corrosion reactions to proceed. It is similar to a corrosion cell of two dissimilar metals in which a current flow is established between two metals because of the difference in their electrical potential. A separate cathodic metal is not required for corrosion of steel. This is because different areas of the bar may develop ''active sites'' with higher electrochemical potentials, and thus set up anode-cathode pairs as illustrated in Figure 10, and corrosion occurs in localized anodic areas. The development of active sites can be caused by a variety of conditions, such as different impurity levels in the iron, different amounts of residual strain, or different concentrations of oxygen or electrolyte in contact with the steel.

For corrosion of steel embedded in concrete to occur, a number of conditions must be met:

1. The provision of anode-cathode sites
2. Maintenance of an electrical circuit
3. Presence of moisture and oxygen

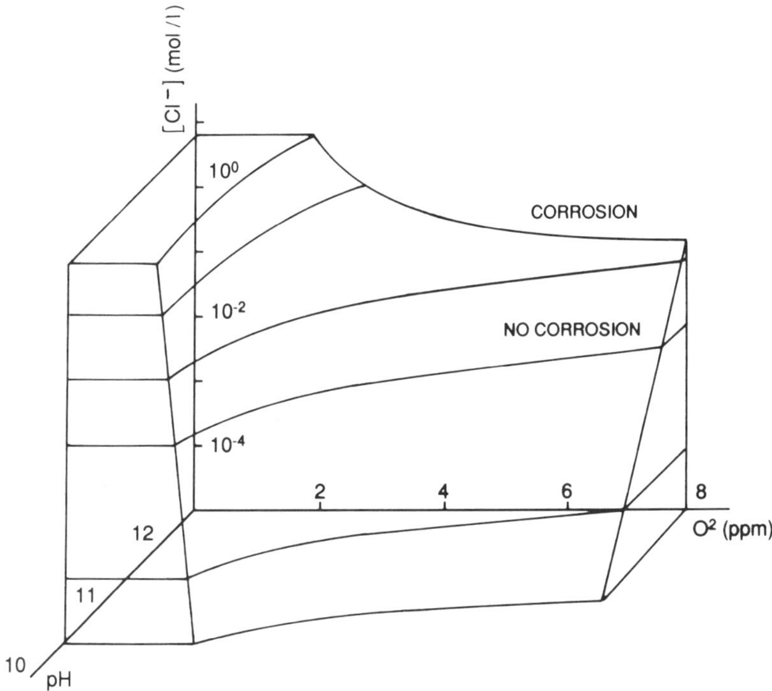

FIGURE 11. Regions of corrosion and no corrosion as a function of pH, chloride concentration, and oxygen concentration. (Adapted from Reference 18.)

Even with activity sites present reinforcement will not corrode if the normal highly alkaline conditions present in concrete prevail (a pH of about 12 to 12.5). At this level of alkalinity a passive oxide film forms on the surface of the reinforcement which prevents corrosion. This passive iron oxide layer is, however, destroyed when the pH is reduced to about 11.0 or below, and the normal porous oxide layer forms during corrosion. This critical reduction in pH occurs when calcium hydroxide, which maintains the high pH in cement paste, is converted to calcium carbonate by atmospheric carbonation. Chloride ions have the special ability to destroy the passive oxide film even at high alkalinities. The amount of chloride required to initiate corrosion depends on the pH of the solution in contact with the steel. Comparatively small quantities are needed to offset the alkalinity of portland cement. Concretes with a low permeability will limit the rate of corrosion by limiting the rate of diffusion of oxygen. Figure 11 is a summary of the relationships involving these variables.[18] If the values of pH, chloride content, and oxygen concentration result in a point below the curved surface, there is no corrosion.

The electrical potential differences, or voltage, between the anode and cathode may be measured by a voltmeter. This difference in measured voltage is not, however, very meaningful. It has been known for some time that the electrode potential of the steel in concrete is an indicator of corrosion feasibility. This electrode potential shows if a particular electrochemical reaction at the electrode is possible or impossible. For any chemical or electrochemical reaction, the thermodynamics and kinetics of the reaction determine the extent to which the reaction occurs. The thermodynamic factors determine if a reaction is possible and the kinetic factors determine the rate at which the reaction can proceed under a given set of conditions. In our case, the electrode potential is the thermodynamic factor, while concrete resistivity and availability of oxygen are the kinetic factors.

ELECTRICAL PROPERTIES OF CONCRETE

Little information exists in the literature relative to electrical properties of concrete. The

TABLE 1
Polarization Potential and Resistance under Varying D.C. Potential

Applied D.C. potential (volts)	Polarization potential, E_p (volts)	Resistance, R (ohms)
4 and 6	1.77	327
6 and 8	1.78	325
8 and 10	1.75	328

Note: Table 2 gives additional information on concrete of different ages.

From Monfore, G. E., *J. PCA Res. Dev. Lab.*, 10(2), 1968. With permission.

TABLE 2
Polarization Potential and Resistance of Concrete

Concrete	Age (days)	Applied D.C. potential (volts)	Polarization potential, E_p (volts)	Resistance, R (ohms)
A	7	4 and 6	1.84	412
A	28	4 and 6	1.84	514
A	90	4 and 6	1.85	642
B	7	4 and 6	1.80	308
B	28	4 and 6	1.76	435
B	90	4 and 6	1.77	598

From Monfore, G. E., *J. PCA Res. Dev. Lab.*, 10(2), 1968. With permission.

most comprehensive studies are those of Hammond and Robson[16] and Monfore.[17]

Moist concrete behaves essentially as an electrolyte with a resistivity in the order of 10^4 ohm-cm, a value in the range of semiconductors. Oven-dried concrete has a resistivity in the order of 10^{11} ohm-cm, a reasonably good insulator. The resistivity to direct current may be different since there is a greater polarizing effect.

Table 1 indicates essentially constant values of polarization potentials and resistance under varying applied D.C. voltages (see Equation 3). Table 2 gives additional information on concrete of different ages.

These data indicate an increase in resistance with age, probably as a result of a decrease in free water due to hydration. The polarization potential remained constant under different applied potentials.

Table 3 shows the effect of frequency to be minor in terms of resistance but substantial in terms of capacitance. The values of calculated impedance was essentially equal to the measured resistance.

The effect of potential on resistance or impedance was slight as shown in Table 4.

Since the resistivity of electrolytes decreases with increasing temperature, the resistivity of moist paste, mortar and concrete would also be expected to decrease with increasing temperature. Measurements of the resistivity of paste over a temperature range of 40 to 100°F indicated an average decrease in resistivity of 1% per degree increase in temperature.[17]

TEST METHODS

Capacitance Instruments for Measuring Moisture Content

Experimental investigations at frequencies up to 100 MHz have shown that both the real

TABLE 3
Effect of Frequency at Constant Potential

Frequency (hertz)	Resistance, R (ohms)	Capacitance, C (microfarads)	Impedance, Z (ohms)
100	571	0.07217	570.8
1000	561	0.00301	560.4
10000	551	0.00040	551.0

Note: The effect of potential on resistance or impedance was slight as shown in Table 4.

From Monfore, G. E., *J. PCA Res. Dev. Lab.*, 10(2), 1968. With permission.

TABLE 4
Effect of Potential at a Frequency of 1000 Hertz

Applied potential (volts)	Resistance, R (ohms)	Capacitance C (microfarads)	Impedance, Z (ohms)
2	964.4	0.00140	964.4
4	962.8	0.00207	962.7
6	962.0	0.00267	961.8
8	961.4	0.00295	961.3

From Monfore, G. E., *J. PCA Res. Dev. Lab.*, 10(2), 1968. With permission.

and imaginary components (see Equation 7) of the dielectric constant of building materials increase significantly with increasing moisture content.[16,18,19] The relationships are different for different materials. Bell et al. indicated that the moisture content of laboratory concrete specimens could be determined to ±0.25% for values less than 6% using a 10 MHz frequency.[18] Jones confirmed that high frequencies (10 to 100 MHz) minimize the influence of dissolved salts and faulty electrical contacts.[19]

Capacitance instruments are available to measure the moisture content of building materials. Various electrode configurations are available. The electrodes are attached to a constant frequency alternating current source and establish an electric field in the material to be tested. Current flow or power loss indicating moisture content is then measured. Most instruments are portable and easily operated. Figure 12 shows a moisture meter which works on this principle. A recent investigation by Knab et al. suggests that further study is required to establish the reliability of this method.[20] It is suggested that instruments be calibrated for a particular material. This would involve relating the measured power loss with the moisture content determined by a direct method such as heating in an oven.

Electrical Resistance Probe for Measuring Moisture Content

The resistance probe method involves measuring the electrical resistance of a material which decreases as the moisture content increases. Most instruments consist of two closely spaced probes and a meter-battery assembly enclosed in a housing. The probes are usually insulated except at the tips so that the region being measured lies between the tips of the probes. By having the probe penetrate soft materials the moisture content at various depths can be measured. This type of instrument should be calibrated for the particular material being tested.

FIGURE 12. Meter designed to measure moisture content of solids based on capacitance. (Courtesy of NDT James Instruments Inc.)

These simple, inexpensive instruments do not determine moisture contents precisely. However, a resistance-type meter has been used to trace moisture movement through concrete.[21]

Resistivity measurements can be used to estimate the probability of significant corrosion when half-cell potential tests show that corrosion is possible.[22] In general it has been shown that if:

1. The resistivity is greater than 12,000 ohm-cm, corrosion is unlikely to occur
2. The resistivity is in the range 5000 to 12,000 ohm-cm, corrosion will probably occur
3. The resistivity is less than 5000 ohm-cm, corrosion is almost certain to occur

Resistivity and Pavement Thickness

The resistance to the passage of electric current is different for different materials. Since concrete and subgrade have different characteristics the change in slope of a resistivity vs. depth curve will indicate pavement thickness.[22] A method for measuring resistivity, as illustrated in Figure 13, was suggested by Robertshaw and Brown.[23] Four electrodes are spaced equally. A current I is passed through the outer electrodes, C_1 and C_2 while the potential drop E between the inner electrodes, P_1 and P_2, is measured. The resistivity is calculated as:

$$\rho = \frac{2\pi AE}{I} \tag{9}$$

where A is the electrode spacing in cm.

When testing a concrete pavement the electrode system may be spaced at a 1- or 2-in. spacing for the initial readings and the system expanded in 1-in. increments for successive readings extending to a spacing equal to the pavement depth plus 3 to 6 in.

When plotting resistivity against electrode spacing or depth, a change in resistivity is normally encountered in the base layer that will produce a recognizable trend in the curve towards a higher or lower resistivity, signifying the presence of the underlying material.

The Moore Cumulative Curve Method of depth determination indicates a graphical treatment of data from this test procedure.[24] The assumption is made that equipotential hemispheres with a radius A are established around each current electrode (C_1 and C_2) (see Figure 13). Every point on the surface of the hemisphere has the same potential. By placing

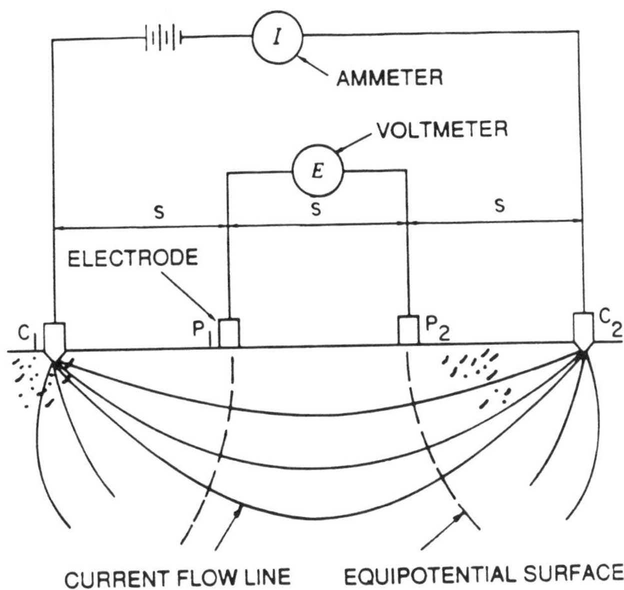

FIGURE 13. Vertical section along line of electrodes showing lines of current flow and equipotential surface in homogeneous materials. (From Robertshaw, J. and Brown, P. D., *Proc. Inst. Civil Engineers (London)*, 4(5), Part I, 645, 1955. With permission.)

the potential electrodes P_1 and P_2 at points on the surface where these hemispheres intersect the ground surface, it is possible to measure a potential drop that applies equally well at a depth A below the surface. As the electrode system is expanded to involve greater depth, the bottom of the hemispherical zones may involve a layer of differing electical resistivity, which produces a trend towards lower or higher resistivity and gives an indication of depth to the layer producing the resistivity change. The resistivity values are plotted against electrode spacing or depth as shown by the dashed-line curve of Figure 14. The solid line curve is a cumulative plotting of the data. The first point is the same value as the first point of the dashed-line curve plotted to a condensed scale. The second point is the sum of the resistivities for the first and second point, etc. Using a constant increment of depth throughout, the solid-line curve constitutes a graphical integration of the dashed-line curve. Straight lines drawn through the plotted points in the vicinity of a trend in the dashed-line curve intersect to give the depth to the subsurface layer producing the trend. Other intersections obtained in the cumulative curve were discounted as not being significant in the analysis, in the absence of additional recognizable trends in the dashed-line curve (see expanded scale dashed-line curve in Figure 14).

The Pennsylvania Department of Transportation built a modified version of this equipment.[25] The number of probes was increased from 4 to 48, spaced 1 in. apart. By use of a resistivity bridge read out device and push-button switches, spacing can be selected automatically, instead of physically moving the probes. As a result the instrument is capable of reading to a depth of 15 in. in approximately 5 min., one quarter the time required by the original instrument. An analysis of field data indicates at least 15 determinations are required per test area to provide a mean value no more than 1/4 in. greater than the average measured thickness.[25] Concern about the determination of the inflection point were allayed by a comparison among different operators. A comparison of three operators showed no significant difference at the 95% confidence level.

Electrical Resistance Probe for Reinforcement Corrosion

This technique is based on electrical resistance measurements on a thin section of *in*

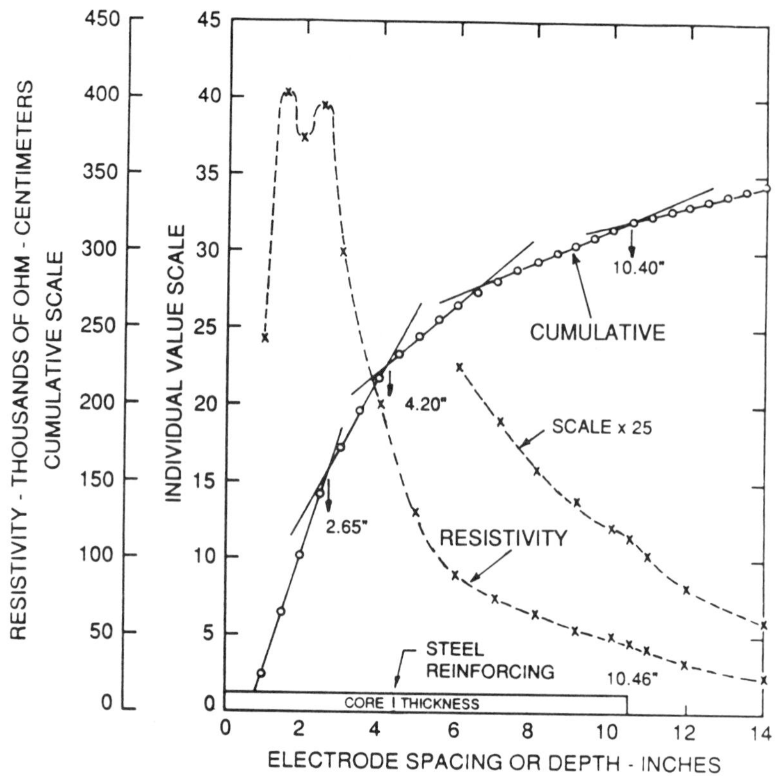

FIGURE 14. Earth resistivity test of reinforced concrete pavement on cement treated base. (Adapted from Reference 24.)

situ reinforcement.[22] Its resistance is inversely proportional to its thickness; as the thin slice is gradually consumed by corrosion it becomes thinner with a corresponding increase in resistance. As indicated in Figure 15, to facilitate measurements the probe is incorporated into a Wheatstone bridge network. One arm of the probe is protected from corrosion while the other arm is the inplace portion of the reinforcement. The measured resistance ratio can be used to monitor the corrosion rate.

$$R = \rho\frac{\ell}{A} = \rho\frac{\ell}{W}\left(\frac{1}{t}\right) = \frac{K}{t} \tag{10}$$

where

R	=	the resistance of the inplace specimen
ρ	=	the resistivity of the inplace specimen
ℓ	=	the length of the inplace specimen
W	=	the width of the inplace specimen
A	=	the cross-sectional area of the inplace specimen
t	=	the thickness of the inplace specimen and

K is a constant, therefore

$$\frac{R_T}{R_R} = \frac{K_T/t_T}{K_R/t_R} = \frac{K^*}{t_T} \tag{11}$$

where

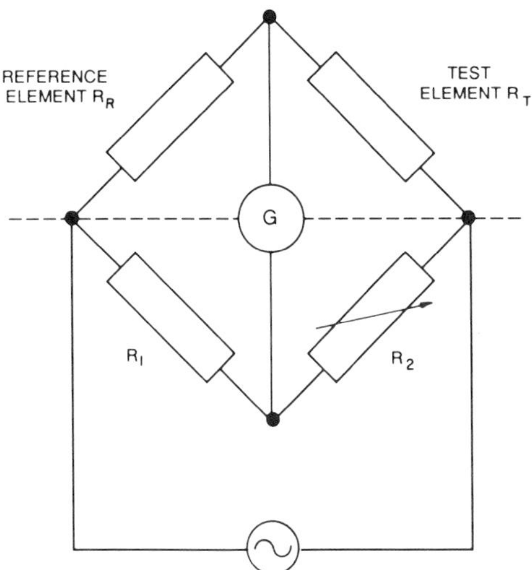

FIGURE 15. Basic circuit for electrical resistance probe technique.

R_T = the resistance of the exposed arm
R_R = the resistance of the protected arm
t_T = the thickness of the specimen in the exposed arm
t_R = the thickness of the specimen in the protected arm
K^*, K_T and K_R = constants
initially $t_T = t_R$, and

$$\text{initially } t_T = t_R \tag{12}$$

$$\frac{t_T - t_R^1}{a} = \text{rate of corrosion in } \mu\text{m year}^{-1}$$

during an exposure period of "a" years.
where

t_T = the initial thickness of the exposed arm at a = 0, and
t_t^1 = the thickness of exposed arm after exposure of "a" years.

Changes in temperature are automatically taken into account because both arms of the probe are located at the same place and are consequently exposed to the same temperature profile. In practice, the thickness of the specimens in the arms of the probe are in the range of 50 to 500 μm. The thinner the probe the shorter the life but the greater its sensitivity. The probes and AC bridge network can be made up in an electronics laboratory and are also available commercially.

The only significant disadvantages are the need for positioning the exposed arm of the probes during construction and the concerns of associated sampling techniques required for locating the probes in large structures subject to localized corrosion.

Half-Cell Potential Measurements

The measurement of electrode potential of steel reinforcement is made using the experimental arrangement shown in Figure 16. An electrical connection is made to the reinforcement at a convenient position enabling electrode potentials to be measured at any desired

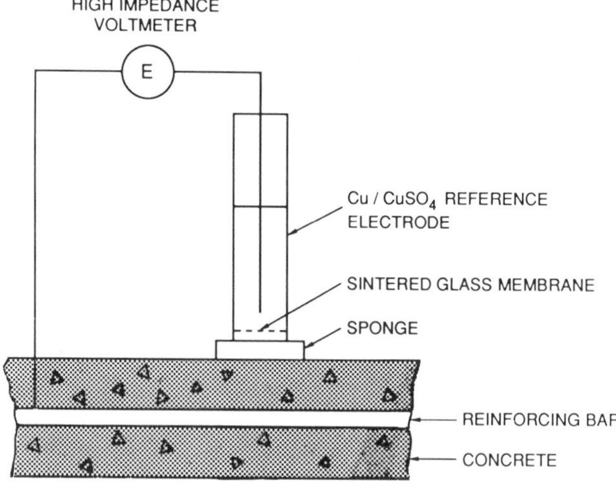

FIGURE 16. System for measuring electrode potential of reinforcement.

location by moving the half cell over the concrete surface in an orderly manner. The surface of the concrete being investigated is usually divided up into a grid system. The results can then be plotted in the form of an equipotential contour diagram (Figure 17).[26-29] The reinforcement in a structure is usually electrically continuous so that only one electrical connection is needed, but if there is a doubt over electrical continuity additional connections can be made or continuity tested. The potential difference between the reinforcement and the half cell is measured using a high impedance voltmeter. The two most commonly used half cells are the saturated calomel electrode and the saturated copper/copper sulfate electrode (CSE). The latter is more durable and has had considerable use.[26-29] It is described in detail in ASTM Test Method C876.[29]

The value of the potential measured is used to estimate the likelihood of corrosion but cannot indicate the corrosion rate.[23] The interpretations of these potentials varies with investigator and agency.[26-30] Generally accepted values representing corroding and noncorroding conditions are given in ASTM C876.[29]

1. If potentials over an area are more positive than -0.20 V CSE, there is a greater than 90% probability that no reinforcing steel corrosion is occurring.
2. If potentials over an area in the range of -0.20 to -0.35 V CSE, corrosion activity of the reinforcing steel in that area is uncertain.
3. If potentials over an area are more negative than -0.35 V CSE, there is a greater than 90% probability that reinforcing steel corrosion is occurring in that area.

Positive readings, if obtained, generally indicate a poor connection with the steel, insufficient moisture in the concrete, or the presence of stray currents and should not be considered valid.

According to ASTM C876, the difference between two half-cell readings taken at the same location with the same cell should not exceed 10 mV when the cell is disconnected and reconnected. The difference between two half-cell readings taken at the same location with two different cells should not exceed 20 mV.

The chief limitation is that it does not give information about the rate of corrosion.

Polarization Resistance and Reinforcement Corrosion

The National Bureau of Standards has developed a prototype portable system for meas-

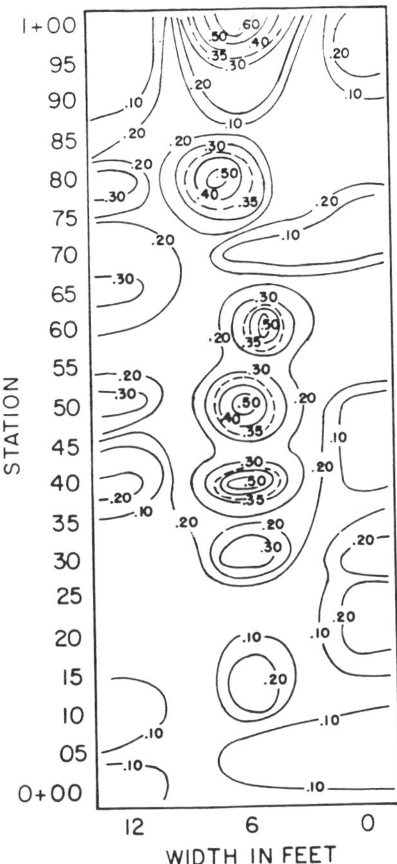

FIGURE 17. Equi-potential contour map. (From Van Deveer, J. R., *J. Am. Concr. Inst.*, Proc. V. 72, No. 12, 699, 1975. With permission.)

uring the corrosion rate of steel in concrete bridge decks.[18,31,32] A small, portable computer system is used to control the measurement of polarization resistance of steel in concrete. The polarization resistance is used to compute the corrosion current which indicates the rate at which the steel is corroding. The polarization technique employs a three electrode system having the steel specimen as one electrode, a voltage reference as a second electrode, and a counter electrode as a third electrode from which polarizing current is applied to the specimen. The circuit used for these measurements is that of Holler which incorporates a Wheatstone bridge for iR compensation as illustrated in Figure 18.[32] The iR error is an error which arises when potential measurements are made in the presence of an electric current in a resistive medium.

The NBS equipment was successfully used on three bridge decks. Ongoing studies are assessing the reliability and accuracy of the data obtained in the field.

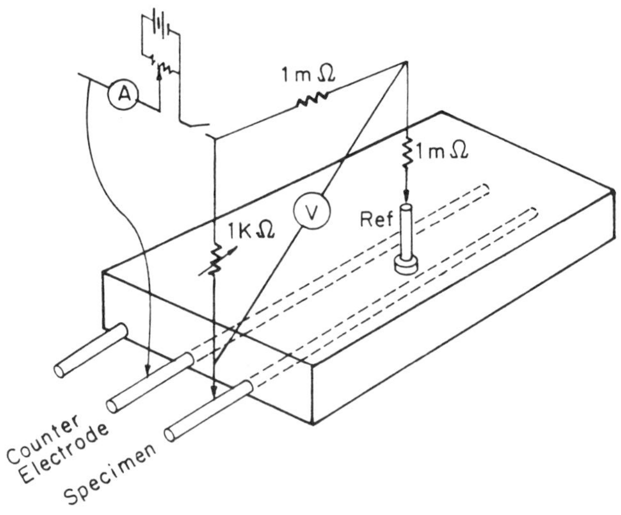

FIGURE 18. Holler's circuit for measuring iR compensation polarization resistance and compensation for IR error. (Adapted from Reference 31.)

REFERENCES

1. **Malhotra, V. M.,** Testing Hardened Concrete: Nondestructive Methods, ACI Monogr. No. 9, The Iowa State University Press, Ames, and American Concrete Institute, Detroit, MI, 1976.
2. **Mix, P. E.,** *Introduction to Nondestructive Testing,* John Wiley & Sons, New York, 1987.
3. **Lord, W.,** A survey of electromagnetic methods of nondestructive testing, in *Mechanics of Nondestructive Testing,* Stinchcomb, W. W., Ed., Plenum Press, New York, 1980, chap. 3.
4. **Lord, W., Ed.,** *Electromagnetic Methods of Nondestructive Testing,* Vol. 3, Gordon and Breach, New York, 1985.
5. **Abragam, A.,** *The Principles of Nuclear Magnetism,* Clarendon Press, Oxford, 1961.
6. **Sliehter, C. P.,** *Principles of Magnetic Resonance,* Harper & Row, New York, 1963.
7. **Andrew, E. R.,** *Nuclear Magnetic Resonance,* Cambridge University Press, Cambridge, England, 1955.
8. Determining the position of reinforcement in concrete, *Engineering (London),* 175(4555), 640, 1953.
9. **Rebut, P.,** Non-destructive apparatus for testing reinforced concrete: checking reinforcement with the "Pachometer" (in French), *Rev. Malenaux (Paris),* 556, 31, 1962.
10. British Standard 4408 p. 1, Non-destructive methods of test for concrete — electromagnetic cover measuring devices, British Standards Institution, London.
11. **Moore, K. R.,** Rapid measurement of concrete cover on bridge decks, *Public Roads,* 39(2) 1975.
12. **Kusenberger, F. N. and Barton, J. R.,** Detection of flaws in reinforcing steel in prestressed concrete bridge members, Rep. No. FHWA/AD-81/087, Federal Highway Administration, Washington, D.C., 1981.
13. **Barton, J. R.,** personal communication, 1987.
14. **Matzkanin, G. A., De Los Santos, A., and Whiting, D. A.,** Determination of moisture levels in structural concrete using pulsed NMR (Final Report), Southwest Research Institute, San Antonio, Texas, Federal Highway Administration, Washington, D.C., Report No. FHWA/RD-82-008, April 1982.
15. **Nikkanin, P.,** On the electrical properties of concrete and their applications, Valtion Teknillinen Tutkinuslaitos, Tiedotus, Sarja III, Rakennus 60 (1962), in French with English summary.
16. **Hammond, E. and Robson, T. D.,** Comparison of electrical properties of various cements and concretes, *The Engineer,* 199, 78 and 115, 1955.
17. **Monfore, G. E.,** The electrical resistivity of concrete, *J. PCA Res. Dev. Lab.,* 10(2), 1968.
18. **Bell, J. R., Leonards, G. A., and Dolch, W. L.,** Determination of moisture content of hardened concrete and its dielectric properties, *Proc. ASTM,* 63, 996, 1963.

19. **Jones, R.,** A review of the non-destructive testing of concrete, Proc. Symp. on Non-Destructive Testing of Concrete and Timber, Institution of Civil Engineers, London, June 1969, 1.
20. **Knab, R., Mathey, R. G., and Jenkins, D.,** Laboratory evaluation of nondestructive methods to measure moisture in built-up roofing systems, National Bureau of Standards Building Science Series 1311, 1981.
21. **Bracs, Gunars; Balint, Emergy; and Orchard, Dennis, F.,** Use of electrical resistance probes in tracing moisture permeation through concrete, *J. Am. Concr. Inst. Proc.,* 67(8), 642, 1970.
22. **Vassie, P. R.,** Evaluation of techniques for investigating the corrosion of steel in concrete, Department of the Environment, Department of Transport, TRRL Report SR397, Crowthorne, 1978.
23. **Robertshaw, J. and Brown, P. D.,** Geophysical methods of exploration and their application to civil engineering problems, Proc., The Institution of Civil Engineers (London), 4(5), Part I, 645, 1955.
24. **Moore, R. W.,** Earth resistivity tests applied as a rapid nondestructive procedure for determining thickness of concrete pavements, Highway Research Record, No. 218, 1968, 49, Transportation Research Board, Washington, D.C.
25. **Weber, W. G., Jr., Grey, R. L., and Cady, P. D.,** Rapid measurement of concrete pavement thickness and reinforcement location — field evaluation of nondestructive systems, National Cooperative Highway Research Program Report 168, 1976, Transportation Research Board, Washington, D.C.
26. **Van Deveer, J. R.,** Techniques for evaluating reinforced concrete bridge decks, *J. Am. Concr. Inst.,* December 1975, No. 12, Proc. 72, 697.
27. **Browne, R. D.,** The corrosion of concrete marine structures: the present situation, Concrete Structures — Proc. Int. Conf. Trondhiem, Taper, 1978, 177.
28. **Okada, K., Kobayashi, K., and Miyagawa, T.,** Corrosion monitoring method of reinforcing steel in offshore concrete structures, Malhotra, V. M., Ed., SP-82 of American Concrete Institute — *In Situ/Nondestructive Testing of Concrete,* 1984.
29. Standard test method for half cell potentials of reinforcing steel in concrete, ASTM C#876-80, 1987 Annual Book of ASTM Standards, V. 04.02, ASTM, Philadelphia, 557.
30. **Stratfull, R. F.,** Half cell potentials and the corrosion of steel in concrete, Highway Research Report, No. 433, 12, Transportation Research Board, Washington, D.C.
31. **Esalante, E., Cohen, M., and Kohn, A. H.,** Measuring the corrosion rate of reinforcing steel in concrete, NBSIR B4-2853, National Bureau of Standards, Washington, D.C.
32. **Esalante, E., Whitenton, E., and Qiu, F.,** Measuring the rate of corrosion of reinforcing steel in concrete — final report, NBSIR 86-3456, National Bureau of Standards, Washington, D.C.
33. **Holler, H. D.,** Studies of galvanic couples, *J. Electrochem. Soc.,* 97, 271, 1950.

Chapter 10

RADIOACTIVE/NUCLEAR METHODS

Terry M. Mitchell

ABSTRACT

Radioactive and nuclear methods can be useful analytic or diagnostic tools, but, with an exception or two, are not widely used in concrete testing currently. The methods are based on directing radiation from sources such as radioisotopes and X-ray generators against or through fresh or hardened concrete samples. The radiation collected after interaction with the concrete provides information about physical characteristics such as composition, density, and structural integrity.

Gamma radiometry is the most widely used method, primarily for density determinations on roller-compacted and bridge deck concretes. Radiography is used occasionally in concrete laboratories for studying microstructure and in the field for confirming the integrity of structural concrete. Infrequently used, neutron-gamma techniques provide composition information on fresh or hardened concrete. The radioactive and nuclear methods are fast and accurate, but their use has been limited by the often complex technology involved, high initial costs, and training and licensing requirements.

INTRODUCTION

Radioactive and nuclear methods for testing concrete have been the subject of numerous research studies but, with an exception or two, are not widely used. The methods are generally fast and accurate and often provide information not available to any other means. On the other hand, their limited use is likely due to often complex technology, high initial costs, and training and licensing requirements. Another factor which limits the effectiveness of these, as well as many other nondestructive methods, is the heterogeneous nature of concrete itself, whether within a small sample, across a construction site, or from one project to another. Nevertheless, many of the radioactive and nuclear methods can be very useful analytic or diagnostic tools, and the disadvantages cited should not be overstressed.

The methods available use radiation produced by radioisotope sources, X-ray generators, and nuclear reactors to bombard fresh or hardened concrete samples. The radiation which is transmitted through, attenuated by, or emitted by the concrete is then collected and analyzed. The collected radiation can provide information about physical characteristics such as composition, density, and structural integrity.

GENERAL PRINCIPLES

Although "radioactive" and "nuclear" have specific and distinct meanings, they are often used interchangeably in nondestructive testing contexts to refer to test methods which use the interaction of wave or particle radiation with matter to supply analytic or diagnostic information about the material. (In this chapter, the methods will be referred to, generically, as "nuclear methods.")

The nuclear methods used to test concrete can be separated into three categories: (1) radiometry; (2) radiography; and (3) neutron-gamma techniques. Radiometry describes techniques in which a radiation source and a detector are placed on the same or opposite sides of a concrete sample; a portion of the radiation from the source passes through the concrete and reaches the detector where it produces a series of electrical pulses. When these pulses are counted, the resulting count or count rate is a measure of the dimensions or physical characteristics, e.g., density or composition, of the concrete sample. Radiography describes techniques in which a radiation source and photographic film (the radiation detector) are placed on opposites sides of a concrete sample. After exposing the film, the result is a photographic image of the sample's interior, which is primarily used to locate defects in the concrete. Neutron-gamma techniques, rarely used in the concrete industry, are those in which a concrete sample is irradiated with neutrons, one type of radiation, and gamma rays, a second type, are emitted and detected. The result is a series of counts which are a measure of the composition of the concrete.

SOURCE, INTERACTION, AND DETECTION

Each nuclear testing method is a system composed of a radiation source, a mode of interaction with the concrete, and a radiation detector. A general description of each of these three components is needed before focussing on individual nuclear techniques.

Sources generate two types of radiation, electromagnetic waves and particles. The electromagnetic waves employed in nondestructive testing of concrete are gamma rays and X-rays. Wave radiation is characterized by the energy it carries, usually expressed in units of electron volts, eV (or kilo electron volts, keV, or mega electron volts, MeV). Gamma rays are emitted from reactions inside an atomic nucleus and typically carry energies from a few keV to several MeV. X-rays are emitted from interactions outside the nucleus among orbital and free electrons. They typically have energies from a few eV up to 100 keV, although much higher energies can be produced in X-ray tubes. Neutrons are the only particles of interest in concrete testing. They are uncharged particles which are also characterized by the energy they carry. Neutrons with energies greater than 10 keV are described as "fast", between 0.5 eV and 10 keV as "epithermal," and less than 0.5 eV as "slow".

The interaction of gamma and X-rays with concrete can be characterized as penetration with attenuation. That is, if a beam of gamma rays strikes a sample of concrete, some of the radiation will pass through the sample; a portion will be removed from the beam by absorption; and another portion will be removed by being scattered out of the beam (when gamma rays scatter, they lose energy and change direction). If the rays are travelling in a narrow beam, the intensity I of the beam falls off exponentially according to the relationship:

$$I = I_0 e^{-\mu x} \qquad (1)$$

where

I_o is the intensity of the incident beam
x is the distance from the surface where the beam strikes
μ is the linear absorption coefficient.

For the gamma and X-ray energies common in nuclear instruments used to test concrete, the absorption coefficient includes contributions from a scattering reaction, called Compton scattering, and an absorption reaction, called photoelectric absorption. In Compton scattering, a gamma or X-ray loses energy and is deflected into a new direction by a collision with a free electron. In photoelectric absorption, a gamma or X-ray is completely absorbed by an atom, which then emits a previously bound electron. The relative contributions of Compton scattering and photoelectric absorption are a function of the energy of the incident gamma or X-rays. In concrete, Compton scattering is the dominant process for gamma or X-ray energies in the range from 60 keV to 15 MeV, while photoelectric absorption dominates below 60 keV.

The amount of Compton scattering which occurs at a given gamma or X-ray energy is a function of the density of the sample being irradiated. The amount of photoelectric absorption which occurs is chiefly a function of the chemical composition of the sample, and increases as the fourth power of the atomic number of the elements present. (The latter effect is one of the reasons lead, with an atomic number of 82, is such a good absorber of X-rays.)

The interaction of neutrons with concrete can also be characterized as penetration with attenuation, although the reactions are different than those for gamma and X-rays. Except for nuclear reactors, most sources used to test concrete generate fast neutrons. Neutrons are scattered by collisions with atoms of the various elements in a concrete sample, losing energy and changing direction in each collision. Hydrogen atoms are the most effective scatterers, by far, and collisions with hydrogen atoms rapidly change neutrons from fast to slow. A measurement of the number of slow neutrons present therefore serves as an indicator of how much hydrogen is present in a sample, and since the only hydrogen present in concrete typically is in water molecules, slow neutron detection can be used as a measure of water content.

Neutrons are also absorbed in concrete, and, in general, slow neutrons are more likely to be absorbed than fast neutrons. Unlike the photoelectric absorption process for gamma and X-rays, which increases as atomic number increases, the absorption process for slow neutrons shows no such regular relationship. Although atoms of all elements absorb neutrons to some extent, certain elements, e.g., boron and cadmium, are much stronger absorbers than others. Absorption plays a minor role in neutron-based test methods, except in the neutron-gamma methods.

When neutrons are absorbed, many of the capturing atoms become radioactive. The newly formed atoms are usually unstable isotopes which subsequently decay, emitting gamma rays (or other types of radiation) with energies characteristic of the isotope. Some of these gamma rays are emitted almost instantly, others are emitted over a period of time ranging from seconds to years, depending on the isotope. If the gamma rays of specific energies are counted during specified time periods, the quantities of various elements present in a concrete sample can be established. Counting the gamma rays emitted almost instantly is the basis for neutron capture gamma ray analysis, while counting those emitted after some delay is the basis for neutron activation analysis.

Radiation detectors are the third component of a nuclear testing method. Having interacted with the concrete sample, the radiation carries information about sample properties or structure. The task of the detector and the associated electronics is to collect, process, and analyze that information to put it in a form useful to the engineer.

Radiography methods generally employ photographic film as detectors (although photosensitive paper, fluorescent screens, and electronic detectors are also used), while radiometry and neutron-gamma methods employ gas-filled tubes or scintillation crystals (crystals which transform incident radiation into flashes of light). In X- and gamma radiography, the rays strike the film emulsion and expose it much the same as light exposes conventional photographic film. Neutron radiography is slightly more complex since neutrons will not expose the film emulsion directly; the neutrons are forced to strike a screen of a material which absorbs them and then emits secondary radiation which exposes the film.

The detectors for radiometry and neutron-gamma methods absorb a portion of the radiation and turn it into electrical pulses or currents which can be counted or analyzed. In Geiger-Müller tubes, for example, gamma or X-rays ionize some of the gas in the tube; when the amount of ionization is then multiplied by a high voltage applied across the tube, it produces an electrical pulse which indicates radiation has interacted in the tube. Geiger-Müller tubes are used widely in nuclear density gauges. Neutrons do not ionize the gas in a gas-filled tube directly, but they are absorbed by boron trifluoride or ^{3}He in a tube; the latter gases emit secondary radiation which ionizes the gas in the tube and produces electrical pulses. Gas-filled neutron detectors are widely used in moisture gauges in agriculture and civil engineering applications.

Scintillation crystals are used primarily in the neutron-gamma methods, where knowing the energy of the detected gamma rays is useful. When gamma rays are absorbed in a scintillation crystal (sodium iodide is a typical crystal material), they produce tiny flashes of light. A photomultiplier tube turns the light into electrical pulses, the amplitudes of which are directly proportional to the energies of the gamma rays absorbed. Pulse height analyzers allow counts to be made over a single gamma ray energy range or over many such ranges to provide a distribution of counts as a function of energy. Scintillation crystal detection in neutron-gamma methods has uses such as establishing the amount of chloride ion present in concrete bridge decks or the amount of cement in freshly mixed or cured concrete samples.

After the brief discussion of radiation safety that follows, the remainder of this chapter will be devoted to examining radiometry, radiography, and neutron-gamma methods in more detail. The discussion of each of the three methods includes historical background, descriptions of the test equipment and procedures, case histories, and advantages and disadvantages. Appropriate references are also provided.

RADIATION SAFETY

With few exceptions, anyone who owns radioactive sources or who performs any of the nuclear testing procedures discussed in this chapter must hold a license from the appropriate governmental authority. Licensing requirements are imposed to insure that neither employees nor the general public are unnecessarily exposed to radiation and that an appropriate radiation safety program is in place. License applications typically require a listing of the individuals responsible for the radiation safety program (and the training they have received), descriptions of the training required for individuals using any of the sources, a description of the facilities and of the radiation monitoring equipment available, and an outline of the radiation safety program. Licensing requirements should not discourage potential users of nuclear testing methods, but should be regarded as promoting and insuring safe usage. More information about radiation safety for the nuclear testing methods is available in References 1 to 3.

RADIOMETRY

HISTORICAL BACKGROUND

Gamma radiometry is widely used in highway construction for density determinations on soils, soil-aggregates, and asphalt concrete, in the paper and pulp industry for thickness monitoring, and in other industries, but is just beginning to gain acceptance for testing portland cement concretes (PCC). Neutron radiometry, likewise, is widely used in highway construction (on soils and asphalt concrete), in well-logging, and in roofing rehabilitation, but is rarely used in testing PCC.

The development of gamma radiometry techniques did not really begin until radioisotope sources became widely available with the advent of nuclear reactors (after World War II). Malhotra[4] reported Smith and Whiffin as the first users of gamma radiometry on concrete in 1952; they made direct transmission measurements using a ^{60}Co source* inserted in a hole in a concrete block and a Geiger-Müller tube detector external to the block. The apparatus allowed measurements of variations in density with depth in order to evaluate the effectiveness of an experimental surface-vibrating machine.

In his 1976 survey, Malhotra[4] reported gamma radiometry had been used for measuring the *in situ* density of structural concrete members, the thickness of concrete slabs, and the

* Isotope radiation sources such as ^{60}Co are identified by the chemical symbol for the element (Co for cobalt) preceded by the isotope's atomic weight (60) in a superscript.

density variations in drilled cores from concrete road slabs. With the possible exception of its application in Eastern Europe for monitoring density in precast concrete units, radiometry was still an experimental nondestructive testing tool for concrete at that time. Density monitoring applications increased in the highway industry after a 1972 report by Clear and Hay[5] showed the importance of consolidation in increasing the resistance of concrete to penetration by chloride ions. A number of U.S. state and Canadian province highway agencies began to use commercially available nuclear gauges to evaluate the density achieved in bridge deck overlays, particularly overlays employing low slump, low water-cement ratio mixes. In 1979, The American Association of State Highway and Transportation Officials (AASHTO) adopted a standard method, T 271, for the ''Density of Plastic and Hardened Portland Cement Concrete in Place by Nuclear Methods,''[6] and, in 1984, the American Society for Testing and Materials (ASTM) followed with a slightly different version, Test Method C 1040.[7]

Most recently, Whiting et al.[8] showed the strong influence of consolidation on several critical properties of concrete including strength, bond to reinforcing steel, and resistance to chloride ion penetration. They also evaluated several existing nuclear (gamma radiometric) gauges and strongly recommended their use for monitoring consolidation during construction.[9] They pointed out the value of density monitoring in evaluating the quality of concrete construction itself, rather than just the quality of the materials being delivered to the job site. Currently no procedures are in standard use to measure the in-place quality of concrete immediately after placement; that quality is not assessed until measurements such as strength, penetration resistance, and/or smoothness can be made after the concrete has hardened.

Gamma radiometry is also being used extensively for monitoring the density of roller-compacted concretes.[10,11] Densification is critical to strength development in these mixtures of cement (and pozzolans), aggregates, and a minimal amount of water. After placement, the concrete is compacted by rollers, much the same as asphalt concrete pavements. Commerically available nuclear gauges have become standard tools for insuring these concretes are adequately compacted.

Gamma radiometry has found limited application in composition determinations on PCC. When radioisotope sources emit low energy (below 60 keV) gamma rays, photoelectric absorption is the predominant attenuation mechanism, rather than Compton scattering. Since the absorption per atom increases as the fourth power of the atomic number Z, it is most sensitive to the highest Z element present in a sample. Noting that calcium in portland cement is the highest Z element present in significant quantities in PCC (in mixtures containing non-calcareous aggregates), Berry[12] used ^{241}Am (60 keV gammas) in a prototype backscatter device for measuring the cement content of fresh concrete. Mitchell[13] refined the technique, and the resulting instrument was reported to measure cement contents to within $\pm 8\%$ (at a 95% confidence level) for siliceous aggregate mixtures and to within $\pm 11\%$ for calcareous aggregate mixtures. Because of the sensitivity of photoelectric absorption to Z, the cement content procedure required calibration on a series of mixtures of different cement contents for a given aggregate source. This sensitivity to aggregate composition remains a barrier to further application of the technique.

A shortlived but interesting application of gamma radiometry was in pavement thickness determinations.[14] Gamma ray absorption is a function of the thickness of a specimen (see Equation 1). Therefore, a source could be placed beneath a PCC pavement, and, if a detector were positioned directly over the source, the count recorded by the detector would be a function of the pavement thickness. Researchers placed thumbtack-shaped ^{46}Sc sources on a pavement subbase before a PCC pavement was placed. The sources were difficult to locate after the concrete was placed, however, and the technique was abandoned albeit with a recommendation that it deserved further research.

Neutron radiometry has seen few applications for testing PCC although neutron moisture gauges are widely used in other industries. In 1976, Malhotra's[4] state-of-the-practice included only a single reference to the use of neutron gauges on concrete, by Bhargava in 1969. Bhargava[15] used a neutron moisture gauge to measure the water content at three locations (top, middle, and bottom) of 6 in. × 12 in. × 5 ft (150 mm × 305 mm × 1.5 m) mortar and concrete columns. In another application Lepper and Rodgers[16] used commercially available neutron moisture gauges with the probe placed at the center of 1/2 and 1 ft^3 (14 and 28 dm^3) volumes (unit weight measures) of fresh PCC. The latter researchers found that the gauges could establish water contents to within ±3 to 6% of the actual value at a 95% confidence level; the accuracy depended on the gauge model and the sample volume.

TEST EQUIPMENT AND PROCEDURES

All gamma radiometry systems are composed of (1) a radioisotope source of gamma rays; (2) the object (concrete) being examined; and (3) a radiation detector and counter. Measurements are made in either of two modes, direct transmission (Figure 1) or backscatter (Figure 2).

In direct transmission, the specimen, or at least a portion of it, is positioned between the source and the detector. The source and detector may both be external to the concrete sample (Figure 1A), e.g., in making density scans on cores or thickness determinations on pavements; the source may be inside the concrete and the detector outside (Figure 1B), e.g., in determining the density of a newly placed pavement or bridge deck; or the source and detector may both be inside the concrete (Figure 1C), e.g., in determining the density of a particular stratum in a newly placed pavement or in a hardened cast concrete pile.

In direct transmission, the gamma rays of interest are those that travel in a straight (or nearly straight) line from the source to the detector. Gamma rays that are scattered through sharp angles, or are scattered more than once, generally do not reach the detector. The fraction of the originally emitted radiation that reaches the detector is primarily a function of the density of the concrete, and of the shortest distance between the source and the detector through the concrete, as shown in Equation 1. Typical gamma ray paths are shown in Figure 1. The actual volume of the concrete through which gamma rays reach the detector, i.e, the volume which contributes to the measurement being made, is usually ellipsoidal in shape (see, e.g., Figure 1B), with one end of the volume at the source, the other at the detector. Sources typically used in direct transmission devices allow measurements to be made through 2 to 12 in. (50 to 300 mm) of concrete.

In backscatter measurement, the source and the detector are next to each other, although separated by radiation shielding. No portion of the concrete sample lies on a direct path between the source and detector. The source and detector may both be external to the concrete (Figure 2A), e.g., in determining the density of a newly placed pavement or bridge deck from the top surface of the concrete; or they may both be in a probe which is inserted in the concrete (Figure 2B), e.g., in a borehole in a cast pile.

In backscatter, only gamma rays that have been scattered one or more times within the concrete can reach the detector. The shielding prevents radiation from traveling directly from the source to the detector. Examples of gamma ray paths are shown in Figure 2A. Each time a gamma ray is scattered it changes direction and loses some of its energy. As its energy decreases, the gamma ray becomes increasingly susceptible to photoelectric absorption. Consequently, backscatter measurements are more sensitive to the chemical composition of the concrete sample than are direct transmission measurements in which unscattered gamma rays form the bulk of the detected radiation.

Backscatter measurements made from the surface are usually easier to perform than direct transmission measurements which require access to the interior or opposite side of the concrete. However, backscatter has another shortcoming besides sensitivity to chemical composition: the concrete closest to the source and detector contributes more to the radiation

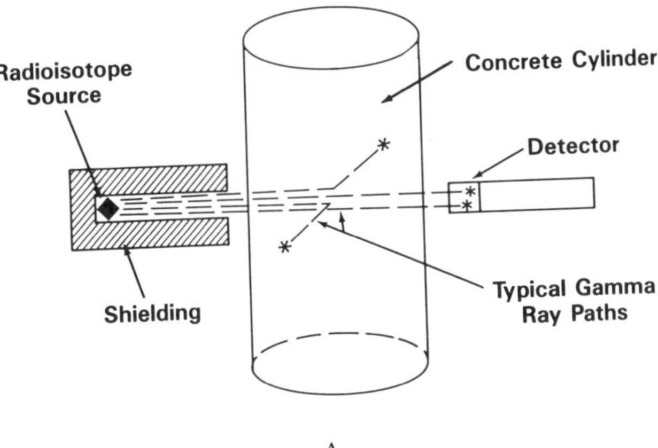

A

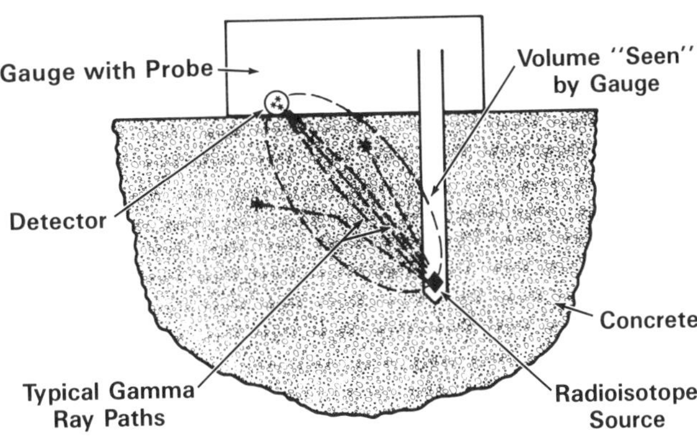

B

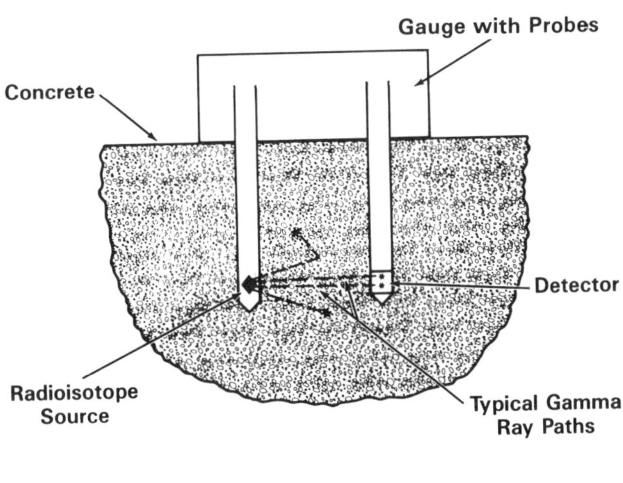

C

FIGURE 1. Gamma radiometry schematics — direct transmission mode: (A) source and detector both external to concrete; (B) source internal, detector external; and (C) source and detector both internal.

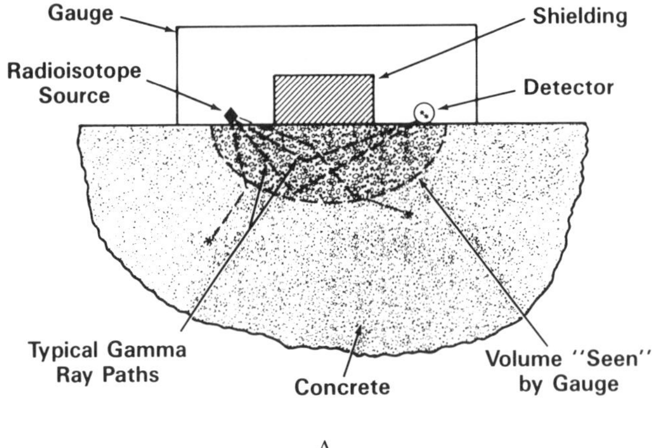

A

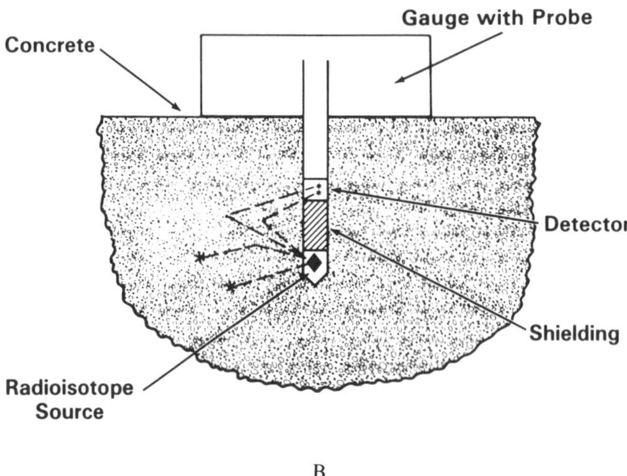

B

FIGURE 2. Gamma radiometry schematics — backscatter mode: (A) source and detector both external to concrete; and (B) both in probe internal to concrete.

count than does material further away. For typical commercially available backscatter density gauges, the top 1 in. (25 mm) of a concrete sample yields 50 to 70% of the density reading, the top 2 in. (50 mm) yield 80 to 95%, and there is almost no contribution from below 3 in. (75 mm). The actual volume of the concrete through which gamma rays reach the detector is roughly half of an ellipsoid (see Figure 2A), with one end of the ellipsoid at the source, the other at the detector.

The source, detector, and shielding arrangement can be modified to somewhat increase the depth to which a backscatter gauge will be sensitive. One of the most depth sensitive gauges is a prototype device built for mounting on the back of a slipform paver for continuous density monitoring;[17] slightly over 70% of the device's reading comes from the top 2 in. (50 mm) of concrete and about 5% comes from below $3^1/_2$ in. (90 mm). Using a gauge model with minimal depth sensitivity may be desirable for applications such as measuring density of a thin (1- to 2-in. [25 to 50 mm] thick) overlay on a bridge deck. (Backscatter measurement has a third disadvantage, sensitivity to surface roughness, but this is rarely a concern for measurements on concrete.)

Gamma radiometry systems for monitoring density generally use ^{137}Cs (662 keV) sources, but ^{226}Ra (a wide range of gamma ray energies, which can be treated as equivalent, on the average, to a 750 keV emission) and ^{60}Co (1.173 and 1.332 MeV) are employed in some. These sources are among the few that have the right combination of long half-life* and sufficiently high initial gamma ray energy for density measurements. The half-life of ^{137}Cs, for example, is 30 years.

Most commercially available density gauges employ gas-filled Geiger-Müller (G-M) tubes as gamma ray detectors because of their ruggedness and reliability. Some prototype devices have employed sodium iodide scintillation crystals as detectors. The crystals are more efficient capturers of gamma rays than G-M tubes. They also can energy discriminate among the gamma rays they capture, a feature which can be used to minimize chemical composition effects in backscatter mode operation. However, the crystals are temperature and shock sensitive and, unless carefully packaged, they are less suitable for field applications than the G-M detectors.

Portable gauges for gamma radiometry density determinations are widely available. A typical gauge is able to make both direct transmission and backscatter measurements, as shown in Figures 1B and 2A, respectively. The gamma ray source, usually 8 to 10 mCi** of ^{137}Cs, is located in the tip of a retractable (into the gauge case) stainless steel rod. The movable source rod allows direct transmission measurements to be made at depths up to 8 or 12 in. (200 or 300 mm), or backscatter measurements when the rod is retracted into the gauge case. The typical gauge would have one or two G-M tubes inside the gauge cause about 10 in. (250 mm) from the source rod. With the source rod inserted 6 in. (150 mm) deep into the concrete, the direct transmission source-to-detector distance would be about 11 in. (280 mm).

Detailed procedures for both direct transmission and backscatter measurements are given in ASTM Standard Test Method C 1040.[7] Density measurements require establishment of calibration curves (count rate vs. sample density) prior to conducting a test on a concrete sample. Calibration curves are created using fixed density blocks, typically of granite, limestone, aluminum, and/or magnesium. Method C 1040 encourages users to adjust the calibration curves for local materials by preparing fresh concrete samples in fixed volume containers (the containers must be at least 18 × 18 × 6 in. [450 × 450 × 150 mm] for backscatter measurements). The nuclear gauge readings on the concrete in such a container are compared with the density established gravimetrically, i.e., from the weight and volume of the sample, and the calibration curve is shifted accordingly.

In-place tests on concrete are straightforward. For a direct transmission measurement, the most common configuration is that shown in Figure 1B; the gauge is seated with the source rod inserted into a hole which has been formed by a steel auger or pin. For a backscatter measurement, the most common configuration is shown in Figure 2A, with the gauge seated on the fresh or hardened concrete at the test location. Care must be taken to insure reinforcing steel is not present in the volume "seen" by the gauge; reinforcing steel can produce a misleadingly high reading on the gauge display. Counts are accumulated, typically over a 1- or 4-min period, and the density is determined from the calibration curve (or read directly off a gauge in which the calibration curve has been internally programmed).

Tests with other gamma radiometry configurations (Figures 1A, 1C, and 2B) employ the same types of sources and detectors. Various shielding designs are used around both sources and detectors in order to collimate the gamma rays into a beam and focus it into a

* The rate at which radioisotope sources emit radiation decreases with time; the half-life is the period of time in which the rate decreases by a factor of two.

** The curie (Ci) has traditionally been the unit for measuring activity of a radioisotope source; it is defined as exactly 3.7×10^{10} disintegrations/second. The curie is being replaced by an SI unit, the becquerel (Bq), which is equivalent to 2.703×10^{-11} Ci.

specific area of a sample. The two probe direct transmission technique (Figure 2C) needs additional development but has considerable potential for monitoring consolidation at particular depths, for example, below the reinforcing steel in reinforced concrete pavements. Iddings and Melancon used a commercially available gauge with a 5 mCi ^{137}Cs source and a 1-in. dia. × 1-in. (25 mm × 25 mm) sodium iodide scintillation crystal, respectively, in two probes which were separated by 12 in. (300 mm) of concrete.[18] They reported the effective vertical layer thickness for density measurements with this system to be about 1 in. (25 mm).

Neutron radiometric procedures usually employ a source/detector configuration similar to that used in gamma backscatter probes, as in Figure 2B. The probe might contain a 100 mCi fast neutron source ($^{241}Am/Be$) and a gas-filled BF_3 or ^3He detector. Because the detector is almost totally insensitive to fast neutrons, no shielding is employed between it and the source. The response of a neutron radiometry gauge arises from a much larger volume of concrete than does that of a gamma backscatter gauge. For example, a neutron radiometry probe completely surrounded by concrete with a water content of 250 lb/yd^3 (150 kg/m^3) will effectively be seeing the concrete up to 14 in. (350 mm) away; a gamma backscatter probe in the same concrete will be seeing the concrete no more than 4 in. (100 mm) away.

CASE HISTORIES

As noted previously, Malhotra[4] reported examples of radiometric techniques for measuring *in situ* density of structural concrete members, thickness of concrete slabs, and density variations in drilled cores. The principal applications of gamma radiometry recently have been in monitoring the density of conventional PCC bridge decks and pavements and of roller-compacted concrete (RCC) pavements and other structures during construction. Examples are presented here of what can be achieved with these techniques:

Static Gamma Radiometry on Conventional PCC AND RCC Pavements

Tayabji and Whiting[9] reported on two field tests using commercially available nuclear density gauges on PCC pavement. A typical data collection effort is shown in Figure 3A, where measurements are being taken during slipform paving operations from a platform that was part of a dowel inserter.[19] Readings were taken in the direct transmission mode, with a 10 mCi ^{137}Cs source inserted 8 in. (200 mm) into the pavement. The technician made a 15-sec count at every third stop of the inserter unit, i.e., at about 42-ft (13-m) intervals. Measurements were made at eight locations in each of eight lots. The average consolidation in the eight lots ranged from 98.9 to 100.2% of the rodded unit weight, with standard deviations within the lots ranging from 0.5 to 1.3%.

Figure 3B shows a similar density gauge being used on a newly placed RCC pavement. Density monitoring is critical on RCC projects since high density is needed to develop adequate flexural strength. On a pavement project such as the one shown in the figure, the concrete behind the paver typically has a density of 95% of the laboratory maximum, but will reach 98% after additional roller compaction.

Ozyildirim[20] cautions that static gauges are not suitable for exact determinations of degree of consolidation in the field, because variations in component proportions or air content within acceptable ranges can cause variations in the maximum density attainable.

Dynamic Gamma Radiometry on Pavements

Figure 4 is a photograph of the Consolidation Monitoring Device (CMD) in use over a newly placed, conventional PCC pavement. The CMD is a non-contact backscatter density gauge, shown here with the source/detector unit mounted on a track on the back of a slipform paver. The unit rides back and forth transversely while the paver moves forward, thus

A

FIGURE 3. Static nuclear density gauges in use during construction of (A) conventional PCC pavement[19] and (B) roller-compacted concrete.[11] (A, Photo courtesy of FHWA; B, Photo courtesy of Kirby T. Meyer, P. E., Austin, TX. Reprinted with permission from Malisch, W. R., *Conc. Constr.*, 33, 13, 1988.)

monitoring the density of a significant portion of the pavement. The CMD uses a 500 mCi ^{137}Cs source and a 1-3/4 in. dia × 4 in. long (45 × 102 mm) sodium iodide scintillation crystal. Results indicate that the device is capable of duplicating core density measurements within a ± 2-1/4 lbs/ft^3 (±36 kg/m^3) range at a 95% confidence level.[17] The CMD appears to be effective for tasks such as establishing proper vibrator operation, alerting vibrator malfunctions, and detecting significant changes in mixture composition, i.e., too little or too much air entrainment.

ADVANTAGES AND LIMITATIONS

Gamma radiometry offers engineers a means for rapidly assessing the density, and therefore the potential quality, of concrete immediately after placement. Table 1 summarizes the advantages and limitations of backscatter and direct transmission gamma radiometry techniques and of neutron radiometry as well.

FIGURE 3B

Direct transmission gamma radiometry has been used for density measurements on hardened concrete, but its speed, accuracy, and need for internal access make it most suitable for quality control measurements before newly placed concrete undergoes setting. Backscatter gamma radiometry is limited by its inability to respond to portions of the concrete much below the surface, but it can be used over both fresh and hardened concrete and can be used, in non-contact devices, to continuously monitor density over large areas. Gamma radiometry techniques have gained some acceptance in density monitoring of bridge deck concrete and fairly widespread acceptance for density monitoring of roller-compacted concrete pavement and structures.

Although widely used in other industries, neutron radiometry techniques have been only minimally developed for water content measurements in concrete.

RADIOGRAPHY

HISTORICAL BACKGROUND

X-radiography techniques were developed around the turn of the century and came to be used extensively in the 1920s. As with gamma radiometry, the development of gamma radiography techniques followed the ready availability of radioisotope sources after World War II.

X- and gamma radiography procedures are used primarily for examining welded products and castings for defects, e.g., for slag inclusions or porosity in welds and for gas cavities, blowholes, or cracks in castings. In the mid-1970s, Barton stated that X- and gamma radiography had become a $500 million a year business in the U.S. alone.[21]

Usage of X- and gamma radiography on concrete has developed very slowly. Malhotra[4] reported Mullins and Pearson making the first published field application in 1949; they used X-rays to show variations in density and to locate reinforcing bars. The same author reported finding only one other field application of X-rays up to the time of his writing (1976): a

FIGURE 4. Dynamic nuclear density gauge (Consolidation Monitoring Device) in use during construction of conventional PCC pavement.[17] (Photo courtesy of FHWA.)

TABLE 1
Radiometry

Technique	Advantages	Limitations
Gamma radiometry for density	Technology well-developed Rapid Simple, rugged, and portable equipment Moderate initial cost Minimal operator skills	Requires license to operate Requires radiation safety program
Backscatter mode	Suitable for fresh or hardened concrete Can scan large volumes of concrete continuously	Limited depth sensitivity Sensitive to concrete's chemical composition and surface roughness
Direct transmission mode	Very accurate Suitable primarily for fresh concrete Low chemical sensitivity	Requires access to inside or opposite side of concrete
Neutron radiometry for water content	Rapid Accurate Large sample size minimizes effects of concrete's heterogeneity	Technology needs further development Requires calibration on local concrete materials Requires license to operate Requires radiation safety program

study on bond stress in prestressed concrete beams. The only field X-ray work on concrete recently has been done with the Scorpion II system developed in France.[22,23] The prototype Scorpion II includes a linear accelerator X-ray source mounted on a movable crane; the resulting system is designed for evaluation of prestressed concrete bridges. The equipment has been used for examining the quality of grouting and of concrete and for establishing the location and condition of prestressing cables.

Gamma radiography has enjoyed somewhat greater success in field applications, particularly in the U.K. Malhotra[4] summarized applications including determining the position and condition of reinforcing steel, establishing the location and extent of voids in concrete and in the grouting of post-tensioned concrete, and the detection of variable consolidation in concrete. The first generation Scorpion device used gamma ray sources for many of the same applications as those discussed previously for its successor (the higher energy X-rays generated by Scorpion II's linear accelerator double the thickness of the concrete which can be examined).[23] The British Standards Institution has adopted a standard for gamma radiography, BS 1881: Part 205, which indicates the usefulness of radiography for locating and identifying steel and voids in structural concrete.[24] The British standard includes a number of recommendations for investigators considering radiographic examination of concrete.

Both X- and gamma radiography have also been used in the laboratory to study the internal structure of concrete. Malhotra[4] summarized applications prior to 1976, which included studies of microcracking before and during loading and of crack propagation. The only laboratory application subsequent to 1976 appears to be by Slate, whose work included further studies of microcracking.[25]

Neutron radiography is also a very young field, only widely used since the mid-1960s. Typically more expensive than X- and gamma radiography, it finds its principal uses in applications where its interactions with particular elements in a sample are stronger than those of X- or gamma rays, or in hostile environments where intense gamma radiation is already present. Hawkesworth[26] cites the following five areas where neutron radiography is advantageous:

1. Highly radioactive objects, such as nuclear reactor fuel rods
2. Hydrogenous materials, such as seals, gaskets, and explosives
3. Materials containing high neutron absorbers such as boron and cadmium
4. Samples in which hydrogenous material or high neutron absorbers are embedded in heavy metals, such as copper, brass, lead, or steel, which are low neutron absorbers
5. Materials where the correct balance among different isotopes is important, e.g., new reactor fuel

Another author[27] says " . . . three applications, inspection of explosive devices, inspection of aircraft gas-cooled turbine blades for residual core material, and inspection of nuclear fuel pins, make up well over 90% of today's (1976) neutron radiography market".

The only applications of neutron radiography for testing concrete have been the recent work by Najjar et al.[28] in the U.S., by Reijonen and Pihlajavaara[29] in Finland, and by Mo Da-Wei et al.[30] in China. The first group used the technique to study microcracking, the second used it to determine the thickness of the carbonated layer, and the third used it to study the water permeability of concrete samples.

TEST EQUIPMENT AND PROCEDURES

As shown in Figure 5, all radiography systems are composed of (1) a beam from a radiation source; (2) the object being examined; and (3) an image collector. In X-radiography, the radiation source is an X-ray tube. When high voltage is applied to the tube, X-rays are

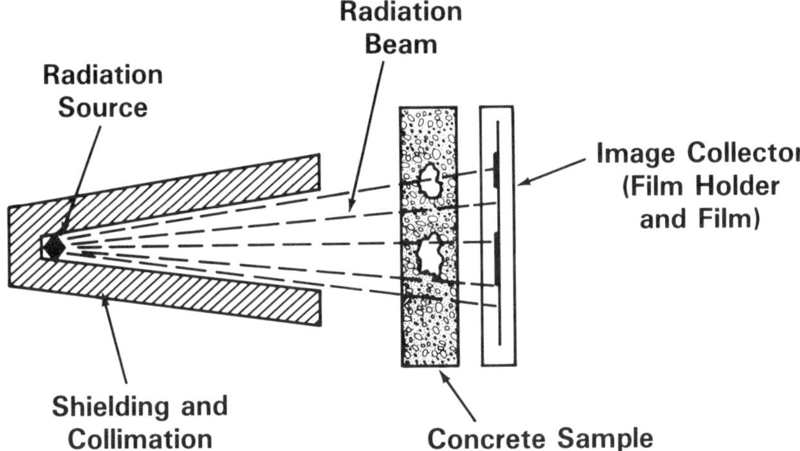

FIGURE 5. Radiography schematic.

emitted with energies proportional to the voltage. According to Malhotra,[4] X-ray sources used on concrete in laboratory studies typically produce maximum energies ranging from 30 to 125 keV. The French Scorpion II, designed for field applications, generates X-rays up to 4.5 MeV. In gamma radiography, a radioisotope source emits gamma rays continuously. The commonly used sources include ^{60}Co (which emits 1.17 and 1.33 MeV gammas), ^{137}Cs (0.67 MeV gammas), and ^{192}Ir (gammas at a number of energies between 0.21 and 0.61 MeV). Source selection depends on the planned application: the energy, half-life, exposure time required, and cost may all be considerations.

Once generated, the X- or gamma rays are collimated and the resulting beam is directed against the concrete sample. The radiation is attenuated, primarily by scattering out of the beam; the amount of attenuation depends on the thickness and density of the concrete and the initial energy of the radiation. The thickness of a typical concrete which will reduce the intensity of a radiation beam by 1/2 is 1.9 in. (48 mm) for ^{192}Ir, 2.1 in. (53 mm) for ^{137}Cs, 2.7 in. (69 mm) for ^{60}Co, and 5.3 in. (135 mm) for 4.5 MeV X-rays from an accelerator. The effect is multiplicative, so that, for a ^{137}Cs source, 4.2 in. (107 mm) of concrete reduces the beam intensity to 1/4 of its original value, 6.3 in. (160 mm) to 1/8 of its original value, etc. Beyond a certain thickness the beam intensity is reduced to a point that clear images cannot be obtained and collection times are impractically long. Figure 6 shows typical concrete thicknesses which can be examined in the field for three of the radiation sources. In laboratory studies, e.g., of microcracking, concrete test specimens are less than 1 in. (25 mm) thick and X-rays in the 100 keV range are appropriate.

After the radiation beam traverses the sample, it is collected and turned into an image on photographic film. Steel attenuates gamma and X-rays much more than concrete does (reducing the intensity of a ^{60}Co beam by 1/2 requires 0.87 in. [22 mm] of steel or 2.7 in. [69 mm] of concrete), while air attenuates much less effectively than concrete. These differences in attenuation therefore can yield a photographic image of the internal structure of the concrete. Exposure times can range from a few seconds to an hour or longer, depending on the thickness of the concrete sample. Film selection depends on the range of thicknesses and densities being analyzed. Intensifying screens, usually of lead foil, can be used adjacent to the film to reduce exposure times and to minimize blurring due to secondary radiation.

Radioscopy, in which a special fluorescent screen collects the image and converts it, in reat time, into a television image, has been used to a limited extent on concrete, e.g., in a modification of the Scorpion system.[23] The maximum concrete thickness which can practically be viewed by a radioscopy system is only about 3/4 of that possible with a radiography system.

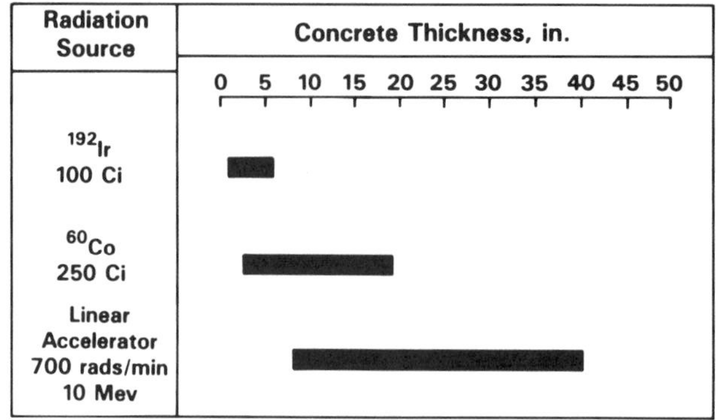

FIGURE 6. Concrete thickness ranges for three typical gamma radiography sources (Adapted from Dufay[23]). (Original figure courtesy of Laboratoire Régional des Ponts-et-Chausées de BLOIS.)

X-ray tomography also has potential applications as a radiographic technique on concrete.[31,32] Tomographic equipment, now widely used in medicine, collects radiographic information along a single plane through the object, repeats the process from various perspectives around it, and then uses a computer to construct an image of the internal structure. Tomography systems are very expensive and, up to now, have only been used in a limited way in research studies on concrete.

In neutron radiography, three types of sources are used, depending on the application. For laboratory studies, neutron beams from nuclear research reactors are the most effective sources. Accelerators are used in field applications, as are radioisotopes such as ^{124}Sb/Be and ^{252}Cf. Nuclear reactor beams produce the clearest images in the shortest exposure times. Accelerators produce more intense beams and hence have shorter exposure times than do radioisotopes; the latter are the most portable and require the least operator skill.

The laboratory studies on concrete have involved irradiating samples ranging from slices 0.15 in. (3.8 mm) thick to cylinders 2 in. (50 mm) in diameter.[28-30] The neutron beam is attenuated by (1) hydrogen-produced scattering, in the case of moisture studies; or (2) absorption by a high neutron absorber such as gadolinium, in the case of cracking studies. In the latter case, the concrete sample is impregnated, prior to irradiation, with a compound containing the absorbing element; the compound acts as a contrasting agent.

Photographic film is again used for image collection but cannot be directly exposed by neutrons. The film must instead be wrapped in a converter, gadolinium foil for example, which turns the neutrons into light or charged particles which will then expose the film. Exposure times again depend on the thickness of the sample, and typically are an hour or longer. Najjar[33] provides useful information on both experimental equipment and sample preparation.

CASE HISTORIES

The principal applications of radiography to concrete to date are principally in two cateogries: (1) X-, gamma, and neutron radiography in laboratory studies of internal microstructure, particularly of microcracking, and (2) X- and gamma radiography in field studies of macrostructure, e.g., the location of reinforcing steel and of voids or areas of inadequate consolidation. Two examples are presented here of what can be achieved with these techniques.

X- and Neutron Radiography for Microcracking

Malhotra[4] summarized much of the pre-1976 work in X- and gamma radiography of the microstructure of concrete. Two recent articles by Najjar et al.[28,34] illustrate the effectiveness of the two techniques in microcracking studies. Figures 7A and B show X- and neutron radiographs, respectively, of a 0.15 in. (3.8 mm) thick slice of a 4 in. (102 mm) diameter concrete sample that had been loaded to just beyond the peak compressive stress. Figures 8A, B, and C show, respectively, a conventional photograph, a conventional X-radiograph, and a contrasting-agent X-radiograph of a 0.30 in. (7 mm) thick slice of a 4 in. (102 mm) diameter sample, also loaded to just beyond the peak stress. For both figures, specimens of the desired thickness were sawed from concrete cylinders that had been loaded to the desired level. The specimens were then cleaned, ground, and polished.

The X-radiography in Figure 7A was accomplished with a 150-kV industrial X-ray machine operated at 40 kV, 5-mA current, and 2 min. of exposure time on a fine-grained Kodak Type T X-ray film. That in Figure 8B and 8C was accomplished with a 150-kV veterinary medicine diagnostic X-ray machine operated at 44 kV, 150-mA current, and 0.33 sec of exposure time on Kodak Type X-OMATG film. The neutron radiography was accomplished with a beam of slow neutrons generated from the TRIGA Mark II nuclear reactor at Cornell University. The neutron flux was approximately 10^6 neutrons/cm^2 s at the specimen. The slow neutrons passed through the specimen and the film, and then a portion were absorbed by a 25 μm thick gadolinium foil; the secondary radiation from this absorption exposed the film.

Figures 7B and 8C show the quality of images of microcracking which can be obtained from radiography when contrasting agents are employed. In 7B a gadolinium nitrate solution was applied to the polished specimen surface and allowed to penetrate into the cracks and voids; in 8C the surface was treated similarly with a lead nitrate solution. The two agents are efficient absorbers, respectively, of neutrons and X-rays, thus highlighting the microcracks in the radiographs. It should be noted that the radiographs are not images of the surface but are images of the full thickness of the specimen, so care must be taken in interpreting and establishing the effective width and depth of the cracks.

X- and Gamma Radiography for Macrostructure

Figure 9 is a photograph of the French-built Scorpion II system in use on a prestressed concrete bridge structure. The handling arm, shown extended to beneath the bridge, includes a miniaturized linear accelerator which can generate X-rays with energies up to 4.5 Mev. The detector, either a film cassette for radiography or an image converter and camera for radioscopy, is positioned in the interior of the concrete girder.

Figure 10A is a photograph of Scorpion II's radioscopic image of a prestressing cable anchorage in concrete; Figure 10B is a schematic of the cable anchorage and interpretation of the radiograph. The radioscopic image was observed through 14 in. (360 mm) of concrete in real time; a sensitive electronic camera (25 images/s and 10^{-4} lux) captured the image of the fluorescent screen and displayed it on a television monitor. The radioscopic image is not sharp but does clearly show the large voids ("heterogeneity") behind the anchorage and the smaller voids ("bad contact") on the front side.

It should be noted that a radiograph of a thick concrete member is based on the entire thickness; therefore, thin cracks or planar defects perpendicular to the radiation beam will have little effect on the beam and may not be detectable in the radiograph.

ADVANTAGES AND LIMITATIONS

X- and gamma radiography offer researchers a means for visually examining the internal microstructure of concrete specimens and offer engineers a means for inspecting the quality of construction and materials in concrete structures. Table 2 summarizes the advantages and limitations of the two techniques and of neutron radiography as well. X-ray techniques are fast; capable of viewing interiors of concrete sections up to 3 ft (1 m) thick; can generate radiation at the optimum energy level for a given specimen thickness; and require minimal

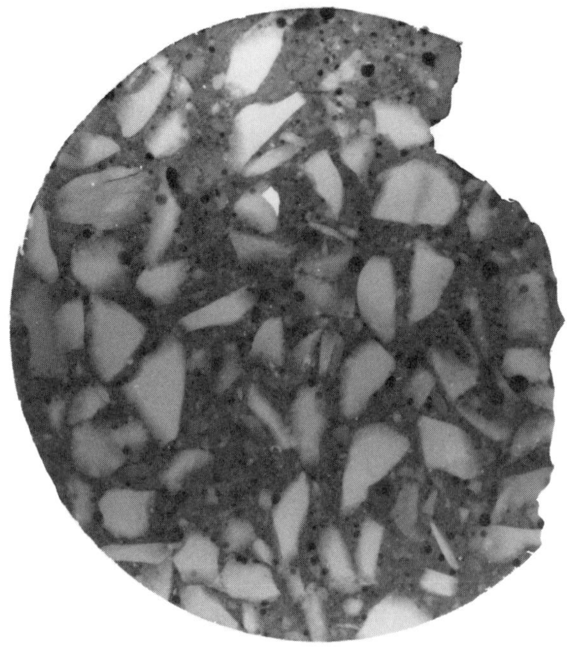

A

B

FIGURE 7. (A) X-radiograph of 4 in. dia., 0.15 in. thick, (102 × 3.8 mm) slice of concrete after loading to peak stress; (B) neutron radiograph of same specimen.[28] (Copyright ASTM. Reprinted with permission.)

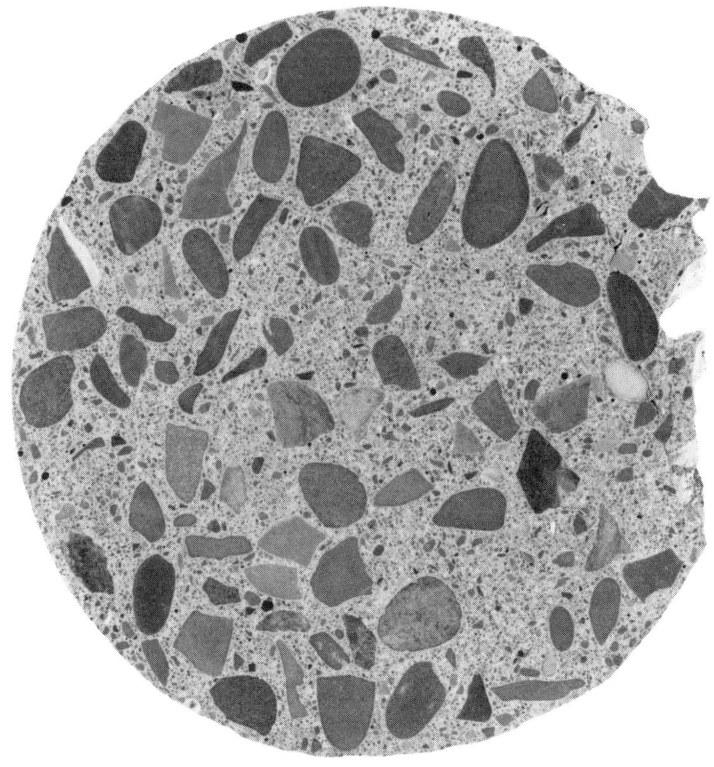

A

FIGURE 8. (A) Conventional photograph of 4 in. dia., 0.30 in. thick (102 × 7 mm) slice of concrete after loading to peak stress; (B) conventional X-radiograph of same specimen; and (C) X-radiograph of same specimen after treatment with contrasting agent.[34] (Copyright ASTM. Reprinted with permission.)

shielding since radiation emission can be terminated by turning off the electrical power supply. On the other hand, X-ray equipment is expensive and must be operated carefully because of high voltages and high radiation levels. Until recent miniaturization of X-ray generators such as in the Scorpion II system, gamma radiography equipment has generally been more portable. Gamma ray equipment is less expensive and is easier to operate because electrical power is not necessary. The continuous radioactivity from the gamma ray sources requires more shielding and additional safety interlocks to prevent accidental exposures.

Neutron radiography is currently a tool for researchers to use in examining the internal microstructure of concrete specimens. It is most useful where specimens contain elements that interact more readily with neutrons than with X- or gamma radiation.

NEUTRON-GAMMA TECHNIQUES

HISTORICAL BACKGROUND

Neutron-gamma techniques, which include neutron activation analysis (NAA) and neutron capture-prompt gamma ray analysis (NCGA), are important tools in criminal forensics, authentication of art objects, and establishment of trace elements in pollutants; as yet they have only served as a research aid, for composition measurements, on concrete. Malhotra[4] summarized the pre-1976 work on concrete, stating that the neutron-gamma techniques were "largely undeveloped for application to concrete and little published data were available on their laboratory and field use for this application".

FIGURE 8B

In the decade following Malhotra's statement, applications to concrete have continued at a very limited pace. Iddings and Arman[35] reported construction of field NAA equipment for measuring the cement content of fresh concrete, of hardened concrete in place, and of concrete cores. Taylor[36] developed a multielement analyzer which used both NAA and NCGA to establish cement, water, and fine and coarse aggregate contents of fresh concrete. And Rhodes et al.[37] constructed a mobile device for nondestructively measuring the chloride content of concrete bridge decks.

TEST EQUIPMENT, PROCEDURES, AND CASE HISTORY

Neutron-gamma systems are composed of three components: (1) a source of neutrons; (2) the sample being examined; and (3) a gamma ray collection and counting system. When samples can be brought into the laboratory, research nuclear reactors are the optimum neutron source. However, all of the devices named in the previous section were designed for testing concrete in the field and employed radioisotope sources, either ^{239}Pu/Be (^{239}Pu emits alpha particles which are captured by the Be to produce neutrons) or ^{252}Cf (^{252}Cf emits neutrons directly as a result of spontaneous fission). ^{252}Cf source sizes typically are 200 to 500 μg, which generate 5 to 12 $\times$ 10^8 neutrons/s.

Neutron-gamma systems work best with high neutron levels from the sources. Because this means sources are highly radioactive and because neutrons are very penetrating radiation, the mobile neutron-gamma systems require considerable shielding. For example, the shield for Iddings' field system was a 36 in. (90 mm) diameter sphere of borated water extended polyester, along with a small lead shield; the total assembly weighed some 800 lb (350 kg).[35]

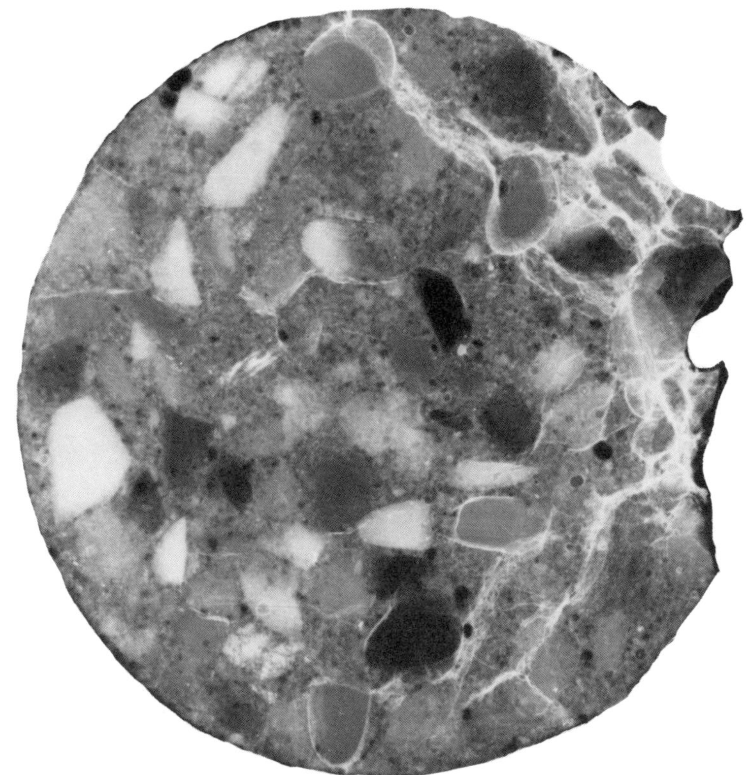

FIGURE 8C

FIGURE 9. Scorpion II X-radiography system on prestressed concrete bridge structure. (Photo courtesy of Laboratoire Régional des Ponts-et-Chausées de BLOIS.)

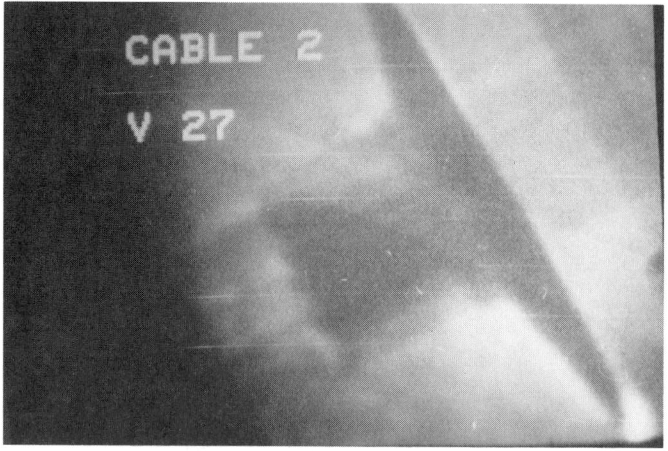

A

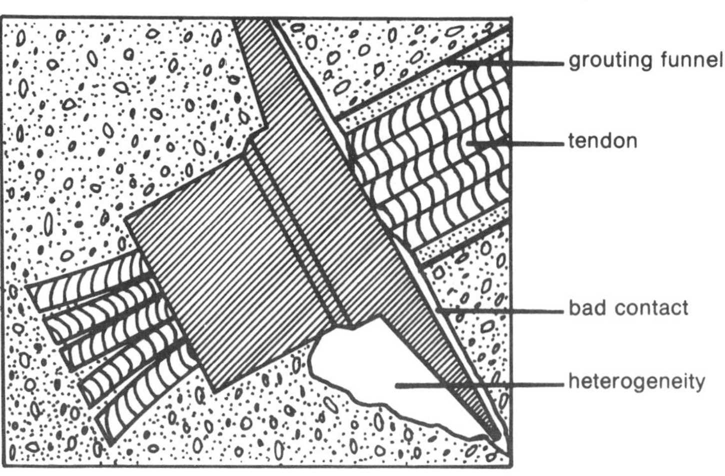

B

FIGURE 10. Scorpion X-radioscopic image of prestressing cable anchorage in concrete: (A) photograph of radioscopic image; (B) schematic of cable anchorage and interpretation of photograph. (Photo courtesy of Laboratoire Régional des Ponts-et-Chausées de BLOIS.)

Sample sizes can be relatively large, e.g., up to 10 in. diameter by 8 in. high (250 mm × 200 mm) in the Iddings' experiments; this can be helpful in minimizing the effects of concrete's characteristic heterogeneity. Because of the high radiation levels at the sources, test procedures require remote controls to transfer the sample into the vicinity of the source or to bring the source out of its shield into the vicinity of the sample.

The neutrons interact with the various elements present in the sample in three principal ways: (a) absorption by atoms and prompt (almost immediate) emission of characteristic gamma rays; (b) inelastic scattering by atoms, also with prompt emission of gamma rays; and (c) absorption by atoms which decay and emit characteristic gamma rays over a longer period of time (seconds, minutes, or longer, after the irradiation occurs). The first two interactions are the basis for NCGA, the third for NAA.

TABLE 2
Radiography

Technique	Advantages	Limitations
All radiography techniques		Expensive Require skilled operators Require license to operate Require radiation safety program Thin cracks or planar defects perpendicular to radiation beam may be difficult to detect
X-ray	Useful for examining internal macrostructure, e.g., steel location and voids, or microstructure Fast View through concrete up to 3 ft (1 m) thick Minimal shielding Energy adjustable for different applications	Primarily a research tool Very expensive initially Potential high voltage hazard
Gamma ray	Useful for examining internal macrostructure, e.g., steel location and voids, or microstructure More portable than X-ray Less expensive than X-ray No electrical power required	View through concrete up to 18 in. (0.5 m) thick Long exposure times
Neutron	Useful for examining microstructure	Primarily a research tool Very expensive initially Little advantage over other radiography methods for most applications on concrete

Each element present in a sample has a different probability of capturing (or scattering) neutrons. When neutrons are absorbed, the resulting atom is in an excited state, and the atom can emit radiation of various types to return to a more stable state. When gamma rays are emitted, either as prompt or activation gammas, they have an energy or energies characteristic of the element. The number of gamma rays of a specific energy emitted in a specified period of time can then be related to the amount of the element present in the sample. Scintillation crystals are generally used in the detecting and counting system because of their absorption efficiency and their capability for resolving gamma ray energies.

The distinction between the two neutron-gamma techniques is based on the time at which the radiation detection and count is made: the NCGA measurements require the irradiation and the counting to be done simultaneously; in NAA the source and sample are separated after irradiation and the sample and detector are brought together, after some delay period (to allow potentially interfering gamma ray emissions to die down), for the counting.

The Iddings NAA system for measuring the cement content of fresh concrete is an example of a typical prototype neutron-gamma system.[35] The device employed 140 μg of ^{252}Cf and a 5 in. dia × 5 in. long (127 mm × 127 mm) sodium iodide scintillation crystal and determined the cement content of a 5 in. dia × 5 in. high container of concrete. The sample was brought adjacent to the source, irradiated for 4 min, and then removed. After a 1 min radioactive decay period, the 3.09 and 4.05 Mev gamma rays (8.5 min half-life) characteristic of ^{49}Ca were counted for 4 min. In concretes containing siliceous coarse and fine aggregates, the cement is the only component containing a significant quantity of calcium. The system required initial calibration with samples having known cement contents and using the same cement and aggregates as would be used in subsequent "unknown" samples. Results indicated the Iddings NAA system could establish cement contents in typical concretes with siliceous aggregates within 10% of the actual value (at a 95% confidence level).

Other irradiation, decay, and counting period combinations and detection over other gamma ray energy levels allow other elements to be measured quantitatively. The Rhodes instrument, for example, is based on the detection and counting of the capture and activation gamma rays emitted by ^{35}Cl and ^{37}Cl, respectively, for detecting the chloride ion level in bridge deck concrete.[37] Techniques based on adding small quantities of an easily irradiated and detected element, such as indium, to concrete are considered periodically as a quality control tool: concrete samples from construction projects could be irradiated to verify their cement content.[38]

ADVANTAGES AND LIMITATIONS

Neutron-gamma techniques have some potential for field composition measurements on both fresh and hardened concrete because of their speed, accuracy, and large sample sizes. However, no systems have been commercially developed for concrete because (1) heavy shielding is required; (2) electronics are complex and expensive for a system with a limited commercial market; and (3) interferences from elements other than the one(s) of interest force frequent recalibrations and limit accuracy with some component compositions, e.g., calcareous aggregates.

REFERENCES

1. *Metals Handbook*, 8th ed., Vol. 11, *Nondestructive Inspection and Quality Control*, Boyer, H. E., Ed., American Society for Metals, Metals Park, Ohio, 1976, 114.
2. Code of Federal Regulations, Title 10, Energy, Parts 0 to 50, U.S. Government Printing Office, Washington, D.C., 1988.
3. **Kerr, G. W.,** Regulatory control for neutron radiography, in *Practical Applications of Neutron Radiography and Gaging*, ASTM STP 586, Berger, H., Ed., American Society for Testing and Materials, Philadelphia, 1976, 93.
4. **Malhotra, V. M.,** Testing hardened concrete: nondestructive methods, ACI Monogr. No. 9, American Concrete Institute, Detroit, 1976, 109.
5. **Clear, K. C. and Hay, R. E.,** Time-to-Corrosion of Reinforcing Steel in Concrete Slabs, Vol. 1, Effect of Mix Design and Construction Parameters, Rep. FHWA-RD-73-32, U.S. Federal Highway Administration, Washington, D.C., 1973.
6. Standard Method of Test for Density of Plastic and Hardened Portland Cement Concrete in Place by Nuclear Methods, AASHTO Designation T271-83, *Methods of Sampling and Testing*, 14th ed., American Association of State Highway and Transportation Officials, Washington, D.C., 1986.
7. Standard Test Methods for Density of Hardened and Unhardened Concrete in Place by Nuclear Methods, ASTM Designation C1040-84, 1989 Annual Book of ASTM Standards, Vol. 04.02, Concrete and Aggregates, American Society for Testing and Materials, Philadelphia, 1989.
8. **Whiting, D., Seegebrecht, G. W., and Tayabji, S.,** Effect of degree of consolidation on some important properties of concrete, in *Consolidation of Concrete*, ACI SP-96, Gebler, S. H., Ed., American Concrete Institute, Detroit, MI, 1987, 125.
9. **Tayabji, S. D. and Whiting, D.,** Field evaluation of concrete pavement consolidation, *Transp. Res. Rec.*, 1110, 90, 1988.
10. **Schrader, E. K.,** Compaction of roller compacted concrete, in *Consolidation of Concrete*, ACI SP-96, Gebler, S. H., Ed., American Concrete Institute, Detroit, MI, 1987, 77.
11. **Malisch, W. R.,** Roller compacted concrete pavements, *Concr. Construction*, 33, 13, 1988.

12. **Berry, P. F.,** Radioisotope X- and Gamma-Ray Methods for Field Analysis of Wet Concrete Quality: Phase II — Instrument Design and Operation, Rep. ORO-3842-2, U.S. Atomic Energy Commission, Washington, D.C., 1970.
13. **Mitchell, T. M.,** Measurement of cement content by using nuclear backscatter-and-absorption gauge, *Transp. Res. Rec.*, 692, 34, 1978.
14. **Weber, W. G., Jr., Grey, R. L., and Cady, P. D.,** Rapid measurement of concrete pavement thickness and reinforcement location, Rep. NCHRP 168, Transportation Research Board, Washington, D.C., 1976.
15. **Bhargava, J.,** Nuclear and radiographic methods for the study of concrete, Civil Engineering and Building Construction Series 60, ACTA Polytechnica Scandinavia Publishing Office, Stockholm, 1969.
16. **Lepper, H. A., Jr. and Rodgers, R. B.,** Nuclear methods for determining the water content and unit weight of fresh concrete, *J. Mater.*, JMLSA, 6(4), 826, 1971.
17. **Mitchell, T. M., Lee, P. L., and Eggert, G. J.,** The CMD: a device for continuous monitoring of the consolidation of plastic concrete, *Public Roads,* 42, 148, 1979.
18. **Iddings, F. A. and Melancon, J. L.,** Feasibility of Development of a Nuclear Density Gage for Determining the Density of Plastic Concrete at a Particular Stratum, Rep. FHWA/LA-81/149, Louisiana Department of Transportation and Development, Baton Rouge, LA, 1981.
19. **Whiting, D. A. and Tayabji, S. D.,** Relationship of Consolidation to Performance of Concrete Pavements, Rep. FHWA/RD-87/095, U.S. Federal Highway Administration, Washington, D.C., 1988.
20. **Ozyildirim, H. C.,** Evaluation of a Nuclear Gage for Controlling the Consolidation of Fresh Concrete, Rep. FHWA/VA-81/41, Virginia Department of Highways and Transportation, Richmond, VA, 1981.
21. **Barton, J. P.,** Neutron radiography — an overview, in *Practical Applications of Neutron Radiography and Gaging,* ASTM STP 586, Berger, H., Ed., American Society for Testing and Materials, Philadelphia, 1976, 5.
22. **Champion, M. and Dufay, J.-C.,** Naissance du SCORPION: système de radioscopie télévisée par rayonnement pour l'inspection des ouvrages en béton, *Rev. Gen. Routes Aerodromes,* 589, 7, 1982.
23. **Dufay, J.-C. and Piccardi, J.,** Scorpion: premier système de radioscopie télévisée haute énergie pour le contrôle non destructif des ouvrages d'art en béton précontraint, *Bull. Liaison Lab. P. & Ch.*, 139, 77, 1985.
24. Recommendations for radiography of concrete, BS 1881: Part 205, British Standards Institute, London, 1986.
25. **Slate, F. O.,** Microscopic observation of cracks in concrete, with emphasis on methods developed and used at Cornell University / X-ray technique for studying cracks in concrete with emphasis on methods developed and used at Cornell University, in *Fracture Mechanics of Concrete,* Witmann, F. H., Ed., Elsevier, 1984, chap. 3, sections 1 and 2.
26. **Hawkesworth, M. R.,** Neutron radiography in industry, in *Recent Developments in Nondestructive Testing,* The Welding Institute, Cambridge, Great Britain, 1978, 13.
27. **Underhill, P. E. and Newacheck, R. L.,** Miscellaneous applications of neutron radiography, in *Practical Applications of Neutron Radiography and Gaging,* ASTM STP 586, Berger, H., Ed., American Society for Testing and Materials, Philadelphia, 1976, 252.
28. **Najjar, W. S., Aderhold, H. C., and Hover, K. C.,** The application of neutron radiography to the study of microcracking in concrete, *Cement Concr. Aggregates,* 8(2), 103, 1986.
29. **Reijonen, H. and Pihlajavaara, S. E.,** On the determination by neutron radiography of the thickness of the carbonated layer of concrete based upon changes in water content, *Cement Concr. Res.*, 2, 607, 1972.
30. **Da-Wei, M., Chao-Zong, Z., Zhi-Ping, G., Yi-Si, L., Fu-Lin, A., Qi-Tian, M., Zhi-Min, W., and Huiz-Han, L.,** The application of neutron radiography to the measurement of the water permeability of concrete, in *Proc. 2nd World Conf. on Neutron Radiography,* Barton, J. P., Farny, G., Person, J.-L., and Rottger, H., Eds., D. Reidel, Boston, 1986, 255.
31. **Gilboy, W. B. and Foster, J.,** Industrial applications of computerized tomography with X- and gamma-radiation, in *Research Techniques in Nondestructive Testing,* Vol. 6, Sharpe, R. S., Ed., Academic Press, London, 1985, 255.
32. **Morgan, I. L., Ellinger, H., Klinksiek, R., and Thompson, J. N.,** Examination of concrete by computerized tomography, *J. Am. Concr. Inst.*, Proc., 77(1), 23, 1980.
33. **Najjar, W. S.,** The Development and Application of Neutron Radiography to Study Concrete, with Emphasis on Microcracking, Ph.D. thesis, Cornell University, Ithaca, NY, 1987.
34. **Najjar, W. S. and Hover, K. C.,** Modification of the X-radiography technique to include a contrast agent for identifying and studying microcracking in concrete, *Cement Conc. Aggregates,* 10(1), 15, 1988.
35. **Iddings, F. A. and Arman, A.,** Determination of Cement Content in Soil-Cement Mixtures and Concrete, Rep. FHWA-LA-LSU-NS-2, Louisiana Department of Highways, Baton Rouge, 1976.
36. **Howdyshell, P. A.,** A Comparative Evaluation of the Neutron/Gamma and Kelly-Vail Techniques for Determining Water and Cement Content of Fresh Concrete, Rep. CERL SP-M-216, Construction Engineering Research Laboratory, Champaign, IL, 1977.

37. **Rhodes, J. R., Stout, J. A., Sieberg, R. D., and Schindler, J. S.,** In Situ Determination of the Chloride Content of Portland Cement Concrete Bridge Decks, Rep. FHWA/RD-80/030, U.S. Federal Highway Administration, Washington, D.C., 1980.
38. **Frevert, E.,** Zerstörungsfreie feststellung des zementgehaltes von beton, in *Radiation and Isotope Techniques in Civil Engineering,* Proc. Conf. of the EURISOTOP Office, Brussels, 1970, 717.

Chapter 11

SHORT-PULSE RADAR METHODS

Gerardo G. Clemeña

ABSTRACT

Short-pulse radar is a powerful scientific tool with a wide range of applications in the testing of concrete. It is gaining acceptance as a useful and rapid technique for nondestructive detection of delaminations and other types of defects in bare or overlaid reinforced concrete decks. It also shows potential for other applications — such as monitoring of cement hydration or strength development in concrete, study of the effect of various admixtures on curing of concrete, determination of water content in fresh concrete, and measurement of the thickness of concrete members.

To facilitate the understanding of these applications, the physical principles behind short-pulse radar are presented. The advantages and limitations of radar in these applications are also discussed.

INTRODUCTION

Experiments in the early 1900s proved the feasibility of transmitting electromagnetic (EM) waves through space as a beam and receiving the reflected signal from an airborne object in the path of the beam. Because of the military significance of this technology, it experienced a rapid advancement during World War II. Subsequent refinement in microwave* sources and detection circuits made it possible to accurately locate planes, ships, clouds, land forms, and, in fact, any object capable of reflecting EM waves. These applications were made possible by the realization that different objects have their own characteristic scattering and reflection properties toward EM waves and that EM waves travel through free space with a constant speed, equivalent to that of light. Such detection systems were eventually called radar, which is an acronym for **ra**dio **d**etection **a**nd **r**anging.

Since much of this technology was also applicable to transmission of EM signals through solids, experiments were conducted in the early 1950s using radar as a tool for probing solids, such as rock and soil. It was quickly recognized that the speed of microwave and its amplitude as a function of distance traveled in a solid could vary significantly from one material to another, and that these properties can be used to identify and profile subsurface geological features. This led to the development in the late 1960s of several types of radar systems, which were called ground-probing radars (GPR) because of their original intended applications.

Since then, GPR has been put to a variety of uses, including determining the thickness and structure of glaciers, locating ice in permafrost,[1] finding sewer lines and buried cables,[2] measuring the thickness of sea ice,[3] profiling the bottom of lakes and rivers,[4] examining the subsurface of the moon,[5] detecting buried containerized hazardous waste,[6] and measuring scouring around bridge foundations.[7]

The earliest study on the use of GPR in areas related to civil engineering, but not to concrete itself, was that reported in 1974 by Bertram et al., which dealt with the inspection of airfield for voids underneath pavements.[8] This study was followed by other studies, which

* An EM wave which has a wavelength between about 0.3 and 30 cm, corresponding to frequencies on the order of 1 to 100 GHz.

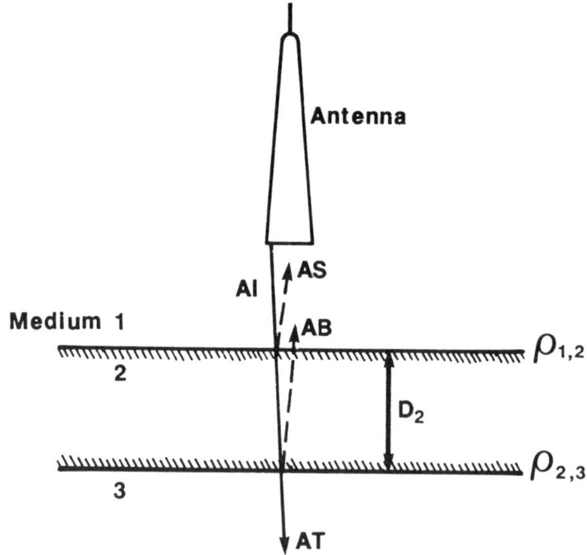

FIGURE 1. Propagation of EM energy through dielectric boundaries.

included locating undermining underneath concrete sidewalks[9] and pavements.[10,11] As the following discussion will show, GPR can be used to test concrete for other purposes. Because of the nature of the microwave pulses that are employed by the radar systems used and because the applications are no longer limited to the probing of sursurface geological features, GPR is more appropriately called short-pulse radar.

PRINCIPLE OF SHORT-PULSE RADAR

Short-pulse radar is the electromagnetic analog of sonic and ultrasonic pulse-echo methods. It is governed by the processes involved in the propagation of electromagnetic energy through materials of different dielectric constants.

BEHAVIOR OF MICROWAVE AT THE INTERFACE OF TWO DIFFERENT MATERIALS

Consider the behavior of a beam of EM energy (such as microwave) as it strikes an interface, or boundary, between two materials of different dielectric constants (see Figure 1). A portion of the energy is reflected, and the remainder penetrates through the interface into the second material. The intensity of the reflected energy, AR, is related to the intensity of the incident energy, AI, by the following relationship

$$\rho_{1,2} = \frac{AR}{AI} = \frac{\eta_2 - \eta_1}{\eta_2 + \eta_1} \tag{1}$$

where

$\rho_{1,2}$ = the reflection coefficient at the interface, and
η_1, η_2 = the wave impedances of materials 1 and 2, respectively, in ohms.

For any nonmetallic material, such as concrete or soil, the wave impedance is given by

$$\eta = \sqrt{\frac{\mu_0}{\epsilon}} \tag{2}$$

where

μ_0 = the magnetic permeability of air, which is $4\pi \times 10^{-7}$ henry/meter, and
ϵ = the dielectric constant of the material in farad/meter.

(Metals are perfect reflectors of EM waves, because the wave impedances for metals are zero.) Since the wave impedance of air, η_0 is

$$\eta_0 = \sqrt{\frac{\mu_0}{\epsilon_0}} \tag{3}$$

and if we define the relative dielectric constant, ϵ_r, of a material as

$$\epsilon_r = \frac{\epsilon}{\epsilon_0} \tag{4}$$

where ϵ_0 = the dielectric constant of air, which is 8.85×10^{-12} farad/meter. Then, we may rewrite Equation 2 as

$$\eta = \frac{\eta_0}{\sqrt{\epsilon_r}} \tag{5}$$

and Equation 1 as

$$\rho_{1,2} = \frac{\sqrt{\epsilon_{r1}} - \sqrt{\epsilon_{r2}}}{\sqrt{\epsilon_{r1}} + \sqrt{\epsilon_{r2}}} \tag{6}$$

where ϵ_{ri} and ϵ_{r2} are the relative dielectric constants of materials 1 and 2, respectively.

Equation 6 indicates that when a beam of microwave strikes the interface between two materials, the amount of reflection ($\rho_{1,2}$) is dictated by the values of the relative dielectric constants of the two materials. If material 2 has a larger relative dielectric constant than material 1, then $\rho_{1,2}$ would have a negative value; i.e., with the absolute value indicating the relative strength of the reflected energy and the negative sign indicating that the polarity of the reflected energy is opposite of that arbitrarily set for the incident energy.

PROPAGATION OF MICROWAVE ENERGY THROUGH A MATERIAL

After penetrating the interface and into material 2, the wave propagates through material 2 with a speed, V_2, given by

$$V_2 = \frac{C}{\sqrt{\epsilon_{r2}}} \tag{7}$$

where C is the propagation speed of EM waves through air, which is equivalent to the speed of light, or 1 ft/ns (0.3 m/ns). As the wave propagates through material 2, its energy is attenuated as follows:

$$A = 12.863 \times 10^{-8} f\sqrt{\epsilon_{r2}} \{\sqrt{1 + \tan^2\delta} - 1\}^{1/2} \tag{8}$$

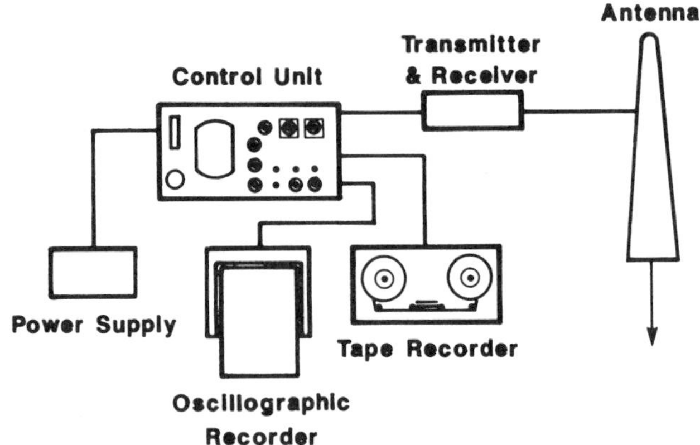

FIGURE 2. Components of a typical short-pulse radar system.

where

A = attenuation, in decibel/meter*,
f = wave frequency, in Hz,

and the loss tangent (or dissipation factor) is related to σ, the electrical conductivity (in mho/meter) of the material by:

$$\tan \delta = 1.80 \times 10^{10} \frac{\sigma}{f\epsilon_{r2}} \tag{9}$$

When the remaining microwave energy reaches another interface, a portion will be reflected back through material 2 as given by Equation 6. The resulting two-way transit time (t_2) of the microwave energy through material 2 can be expressed as

$$t_2 = \frac{2D_2}{V_2} = \frac{2D_2\sqrt{\epsilon_{r2}}}{C} \tag{10}$$

where D_2 is the thickness of material 2.

We shall discuss later how some of these basic principles serve as the basis for the various applications in which short-pulse radar is already being used routinely or has shown promise.

INSTRUMENTATION

Short-pulse radar systems are used in applications related to inspection of concrete. These radar systems operate by transmitting a single pulse that is followed by a "dead time" in which reflected signals are returned to the receiver. A basic radar system consists of a control unit, a monostatic antenna (i.e., an antenna that is used for both transmitting and receiving), an oscillographic recorder, and a power converter for DC operation (see Figure 2). In the inspections of sizeable concrete structures, such as bridge decks and pavements, a multi-channel instrumentation tape recorder is mandatory due to the relatively fast rate at

* Decibel is a unit for describing the ratio of two intensities or powers; if I_1 and I_2 are two amounts of intensity, the first is said to be n decibels greater, where $n = 10 \log (I_1/I_2)$.

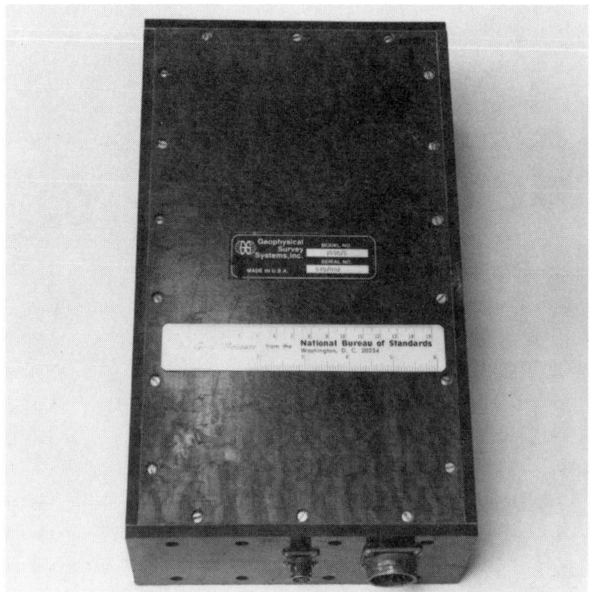

A

FIGURE 3. (A) A "bowtie" radar antenna, and (B) A control unit and an oscillographic recorder.

which the inspection has to be carried out. In addition, tape-recorded radar signals are amenable to treatment by signal processing techniques that facilitate data analysis and interpretation.

In the inspection of concrete, it is desirable to use a radar antenna with relatively high resolution or short pulse width, such as 1 ns or preferrably less. The two commercially available antennas that are often cited in the literature have pulse widths of approximately 1 ns. With such antennas the thinnest layers of a concrete that can be distinguished or accurately measured, if it is assumed that the concrete has a dielectric constant of 6, would be 1.2 to 2.4 in. (3.1 to 6.1 cm), depending on the radar system.

Figure 3 shows one of these antennas, which is referred to as a "bowtie" antenna because of the physical shape of its transmitting and receiving element within the housing. This particular antenna has a center frequency of 900 MHz and a characteristic pulse width of 1.1 ns. At 12 in. (0.3 m) away from an object, it provides a coverage of approximately 12 in. $\times$ 15 in. (0.30 m $\times$ 0.38 m). The other commercially available antenna is called a "horn" antenna, because of its outer appearance. It is usually used in a noncontact manner, as it is scanned over the surface of the concrete.

It is quite difficult to estimate a radar system's depth of penetration before the inspection is actually performed, especially the penetration into reinforced concrete, since penetration is affected by the moisture content of the concrete and the amount and type of reinforcement. However, for dry and unreinforced concrete the penetration can be as much as 24 in. (0.61 m).

In operation, a circuit within the radar control unit that is shown in Figure 3B generates a trigger pulse signal at a rate of 50 kHz, i.e., a pulse at every 20 sec. Each trigger pulse, in turn, causes a solid-state impulse generator in the transmitting antenna to produce a pulse with a very fast rise time, which is then electrically discharged from the antenna in the transducer as a short burst of electromagnetic energy. The resulting pulse is then radiated into the material being examined.

As the radiated pulses travel through the material, different reflections will occur at

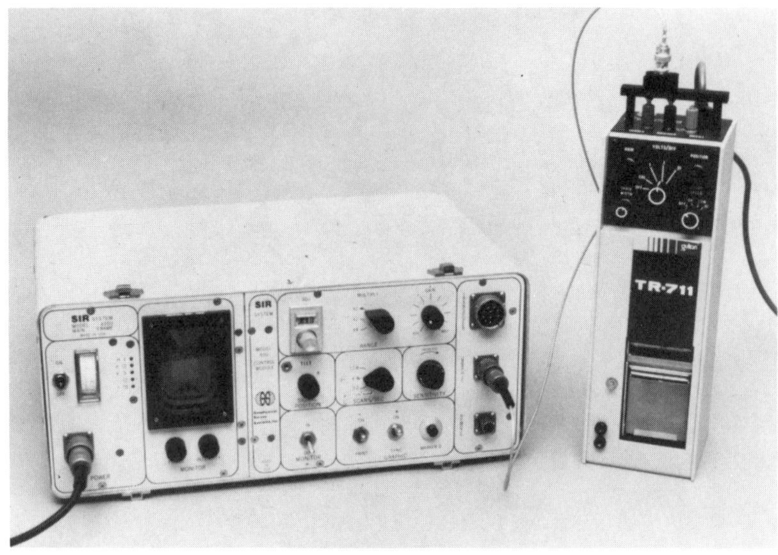

FIGURE 3B

interfaces that represent changing dielectric properties. Each reflected electromagnetic pulse arrives back at the receiving antenna at a different time that is governed by the depth of the corresponding reflecting interface and the dielectric constant of the intervening material (see Equation 10). A receiver circuit reconstructs the reflected pulses at an expanded time scale by a time-domain sampling technique. The resulting replicas of the received radar signals are amplified and further conditioned in the control unit before they are fed to an output.

The analog output can be displayed on an oscilloscope, an oscillographic recorder, or a facsimile gray-scale graphic recorder. It can also be recorded on magnetic tape for future processing or analysis. On an oscilloscope or an oscillographic recorder, the received radar signals may appear similar to the waveform illustrated in Figure 4A, depending on the radar system used. For the particular contact radar system from which this waveform was obtained, the received signal consists of three basic components. At the top is the transmitted pulse, or the feed-through of the transmitted pulse into the receiver section, that serves as a time reference. Immediately following the transmitted pulse is a strong surface reflection, the shape of which is indicative of the shape of the radar pulse transmitted by the antenna. Then, at a later time equal to the pulse travel time from the surface to an interface and back to the antenna, the interface reflection appears. The vertical scale is the time scale, which can be calibrated by a pulse generator that produces pulses at equally spaced time durations. If the wave speed in the material is known the time scale can be converted to a corresponding depth scale.

A facsimile graphic recorder will print the portions of the waveforms that exceed a selected threshold amplitude. As the antenna is scanned across the surface, the positive and negative peaks in the waveforms are displayed as bands of varying gray tones, whereas the zero crossings between peaks are displayed as narrow white lines, as shown in Figure 4B. Notice that the main feature throughout the profile at varying times (or depths) is a group of three closely related bands across the chart, each of which represents reflection of portions of the transmitted pulse from an interface. The darkness of each band is directly proportional to the extent by which the amplitude of the corresponding peak exceeds the print threshold in the recorder. (The triple band is the characteristic of the particular radar system with which the radar signals were obtained. This characteristic is, of course, undesirable since it limits the ability of this system to discriminate neighboring interfaces that are spaced closer than the typical width of a group of bands.)

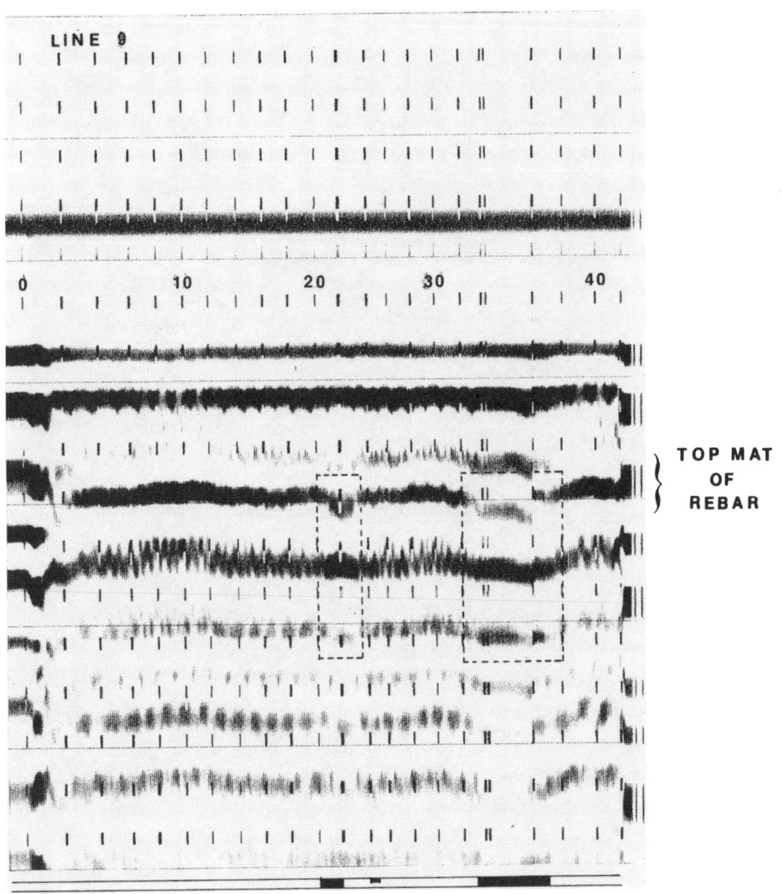

FIGURE 4. (A) A single waveform as seen by the receiver, and (B) a display of cascade of waveforms by a facsimile graphic recorder as antenna is scanned horizontally above a subsurface interface. (With permission from Geophysical Survey Systems, Inc.)

APPLICATIONS

DETECTION OF DELAMINATION IN CONCRETE

A major problem with the reinforced concrete bridge decks located in coastal areas or areas where de-icing salts are used on roadways during winter is the premature deterioration of the concrete. The intrusion of chloride from these salts into the concrete causes the embedded reinforcement to corrode, which eventually cause the concrete to crack or delaminate. Marine structures are also subjected to this type of deterioration.

The simplest method for detecting delamination in concrete involves sounding the concrete with a hammer or heavy chains, which produces a characteristic hollow sound when delamination is present. Although the method is effective, it is adversely affected by the presence of traffic noises. Since sounding is a contact method, its use requires closure of traffic lanes, which is often costly and undesirable. Because of the need for an alternative to the sounding method, the use of noncontact and nondestructive methods such as infrared thermography[12,13] and radar[14-18] have been studied.

Principle

When a beam of microwave energy is directed at a reinforced concrete slab (see Figure 5), a portion of the energy is reflected from the surface of the concrete, and the remaining

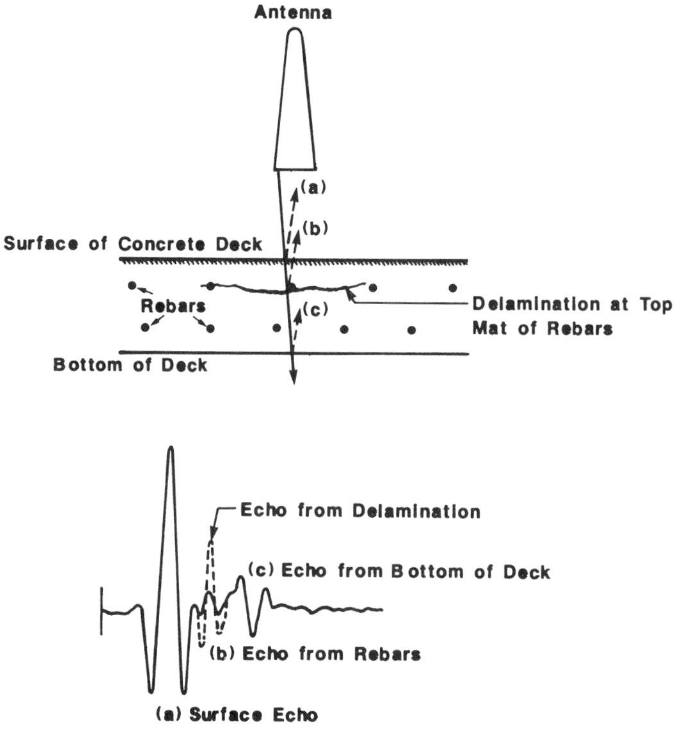

FIGURE 5. Radar echoes from the cross section of a reinforced concrete deck. The presence of a delamination causes additional reflection of the incident energy.

energy penetrates this interface. The surface reflection has a negative polarity since the dielectric constant of concrete, which has been reported to range from 6 when dry to about 12 when saturated,[10] is considerably higher than that of air, which is 1. (It must be noted that the actual *in situ* relative dielectric constant of concrete, and most materials, will vary because it is affected to varying degrees by not only its water content but also by its conductivity, mineral composition, etc.)

As the remaining microwave energy propagates into the concrete, a portion ot the beam will be completely reflected and scattered as it strikes the top mat of reinforcement. This reflection will also have a negative polarity, since the dielectric constant of metal is infinite compared with that of the surrounding concrete. The remaining energy will continue deeper into the concrete slab until a portion of it strikes the second mat of reinforcement and the same reflection and scattering processes occurs. Eventually, some portion of the original beam of microwave energy will reach the bottom of the concrete slab, and some of it will be reflected at the concrete/air interface to give a positive reflection signal. The remainder will penetrate through this interface and be lost from the receiving antenna.

When the concrete slab is delaminated, usually at the level of the top mat of reinforcement, there is an additional reflection from the deteriorated section. This additional reflection, usually of negative polarity, serves as an indicator of the presence of a delamination in the concrete slab.

Test Procedures

To inspect concrete members, including bridge decks, an antenna is placed with its transmitting face parallel to and at a distance from the surface of the concrete. For the inspection of bridge decks, the distance found to be most suitable is 8 to 12 in. (0.2 to 0.3 m). However, if the concrete member is relatively thick and the expected deterioration is deep or if the antenna does not have sufficient power or penetration, the antenna may be placed directly above the concrete. With the various adjustments in the control unit — such as range, gain, sensitivity, etc., — set at optimum levels, the resulting radar signal is recorded with a properly calibrated oscillographic recorder. This is often called the static mode of measurement, since the antenna is stationary with respect to the concrete being inspected.

In the inspection of bridge decks, in which a relatively large concrete area has to be inspected, the antenna is mounted on the front or the rear of an inspection vehicle, which is also instrumented with a horizontal distance-measuring device. If a single antenna radar system is used, the vehicle has to make several passes over each traffic lane from one end of a deck to the other at a selected speed (usually in the range of 5 to 10 mile/h, or 8 to 16 km/h). During each pass the antenna scans a different area in the lane. The stream of radar signals are recorded continuously with an instrumentation tape recorder. With a two-antenna or a multiple-antenna radar system, a single pass may be made over each lane. (It must be emphasized that when a lane is scanned with a two-antenna radar system in a single pass only, a significant portion of the lane will be missed.) This procedure creates continuous recordings of the stream of reflections from the entire depth of the deck along the paths of the antenna. These recordings are played back at a later time for signal interpretation.

Interpretation of Radar Signals

Perhaps the most difficult aspect of the inspection is the interpretation of the radar signals for locating delaminations or any other defects in the concrete. One procedure that has been used to analyze the tape-recorded streams of signals consisted of displaying the waveforms on an oscilloscope, photographing the display with a synchronized shutterless motion picture camera, and then projecting the image in the film on a viewing screen.[15] The image, which resembles ''cascading waveforms'', is basically an array of closely spaced waveforms with topographic features that differentiate delaminated concrete from sound concrete.

An easier procedure that achieved the same effect is to playback the stream of radar signals on a facsimile graphic recorder,[17] which provides a graphic chart similar to that shown in Figure 4B. For an example, Figure 6 shows a graphic profile of the radar signals obtained from scanning a 42-ft (14-m) section of a concrete bridge deck. This section of concrete contained two confirmed delaminations at the top mat of rebars, which were approximately 4 and 6 ft (1.2 to 1.8 m) long. These delaminations appeared, in this and other profiles, as recognizable depressions.

The signature for concrete delaminations often assumed the shape of a depression on the graphic profile (see Figure 7), because as the antenna moves over a delamination the major reflection at any instant would be that part of the microwave beam that is perpendicular to the nearest plane of delamination.[20] Although it has not been extensively confirmed by coring, it is believed that very severe delaminations and other concrete defects, such as honeycombing, may be distinguished by other signatures or as irregularities in the graphic profiles.

Although the use of the facsimile recorder is a simple and reasonably effective procedure, it is not completely objective. Furthermore, it can become cumbersome and time-consuming when large areas of bridge decks are involved. Consequently, computerized procedures based on simple physical models of concrete bridge decks have been developed. These models are based on predictions of how the radar waveform may be affected by such varying physical conditions as the thickness of the concrete over the top mat of rebars, salt and moisture content, spacing of the rebars, etc.[19,21,24]

Advantages and Limitations

Advantages

Radar has some advantages that are very desirable in the inspection of bridge decks and

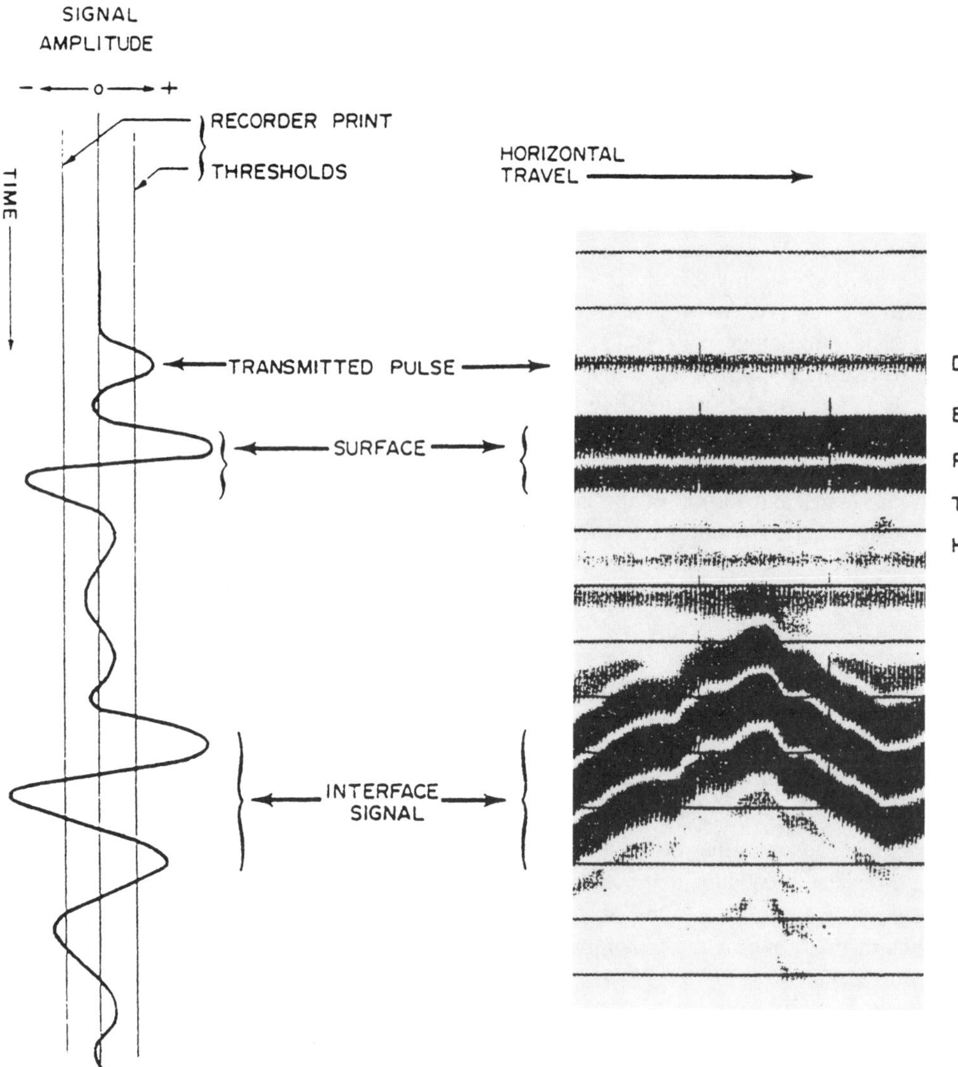

FIGURE 6. Radar scan of a 42-ft (14-meter) section of a concrete bridge deck as displayed on a facsimile graphic recorder.

other concrete structures. Unlike infrared thermography, which has been found to be relatively effective provided proper ambient conditions exist during inspection,[12,13] radar is free of such restriction.

Since radar yields information on the structural profile across the depth of the object being tested, it is at present the only commercially available nondestructive method for the inspection of concrete bridge decks that have asphalt overlays, which constitute a significant portion of bridges. Under favorable conditions, radar can also detect localized loss of bond between the overlay and the concrete, in addition to detecting delaminations in the concrete.

Since it is unnecessary for the antenna to be in contact with a bridge deck, the disruption to traffic during radar inspection is minimal.

Limitations

Some efforts have been made to estimate the accuracy of radar in locating delaminations in bridge decks. Cantor et al. reported that 90% of the area that was predicted by radar to

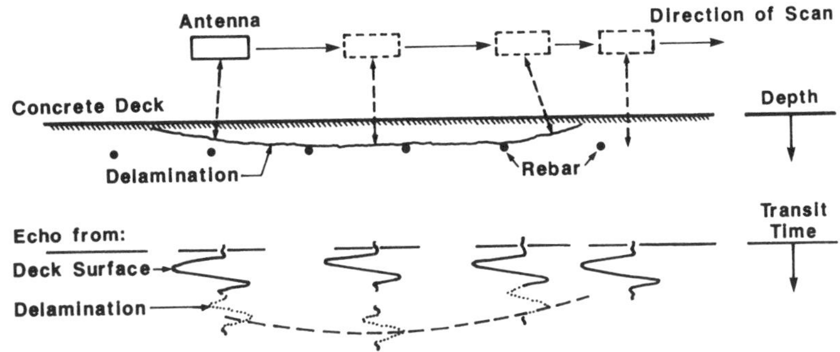

FIGURE 7. Development of an idealized radar signature (a depression) for a delamination in a concrete deck. (Echoes from the top mat of rebars are not shown.)

be distressed in a deck was confirmed as such by coring, and 91% of the area that was predicted to be sound concrete was confirmed.[16] In another study,[17] which involved comparison of the results of radar surveys on several bridge decks performed prior to the removal of existing overlays and the results of chain dragging conducted after the removal of the overlays, it was found that on the average radar detected only 80% of the existing concrete delamination, while falsely identifying 8% of the area tested as delaminated. The results from both studies indicated that radar occasionally gave an indication of delamination in what was actually sound concrete. Radar also failed to locate some of the delaminated areas, especially those that were only 1 ft (0.3 m) wide or less; partly because the strong reflections from the rebars tend to mask reflections from delaminations, which would be relatively weak when the concrete is dry.

This inaccuracy is partly due to the inherent limitation of current radar antennas to resolve consecutive reflections arriving at time intervals which are shorter than the characteristic pulse width of the antenna. Therefore, depending on the dielectric constant of the concrete, the reflection from a small delamination at the level of rebars that are relatively close to the surface may not be resolved from the reflection at the surface.

The presence of interfering signals, such as reverberations, in the radar signals from the "bowtie" antenna is a contributing factor too. Another cause of inaccuracy is the lack of a complete understanding of the relationship between different radar signatures and the various types of defects that are encountered in these concrete structures and how these signatures are affected by the condition in the structure, especially moisture content.

DETERMINATION OF THE DEGREE OF HYDRATION OF CEMENT

The durability of a concrete is greatly influenced by the degree of hydration attained by the cement. Therefore, a nondestructive method that can be used satisfactorily for determining the degree of hydration, especially for in-place concrete at an early age, can be a very useful tool. The use of microwave radar measurements for this application has not yet received wide interest, which is evidenced by the scarcity of literature on the topic.

In 1972 Rzepecka et al. reported that it was feasible to test concrete for strength during the curing process by observing changes in the dielectric property.[25] In 1982 Gorur et al. reported the use of laboratory microwave equipment and a relatively small wave-guide to monitor the curing of cement paste.[26] In 1983 the use of a portable short-pulse radar system to monitor the hydration of concrete blocks by using reflection measurements was reported.[27]

Principle

When portland cement is mixed with water, the various components in the cement undergo chemical reactions with the water (hydration) to form the tobermorite gel and other

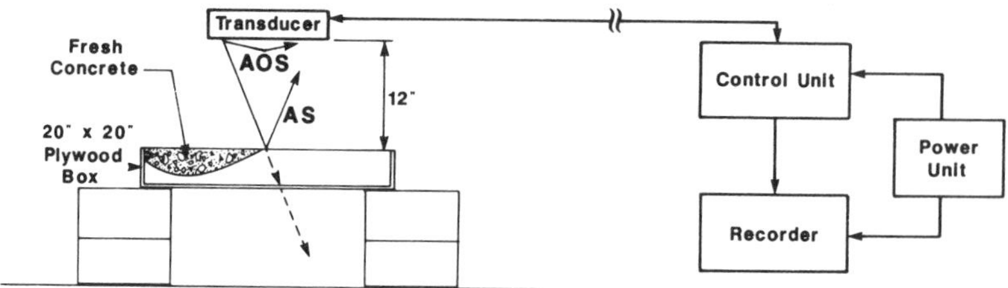

FIGURE 8. Measurement of the reflection of microwave pulses from the surface of fresh concrete. (The thickness of the specimen was 3.0 in. [7.5 cm].)

products that provide the desirable cementing action in a concrete. As these hydration processes progress, some of the mix water is consumed by the reactions. Since "free" water has a relatively high dielectric constant of 80 in contrast to approximately 5 for chemically bound water,[28] the dielectric properties of concrete at early ages are very much influenced by the degree of hydration. Therefore, the degree of hydration can be monitored by continuous measurement of the dielectric properties of the concrete as it cures.

Test Procedures

The dielectric property of concrete that is relatively convenient to deal with and simple to measure is its reflectivity or reflection coefficient. When a microwave beam is directed at the surface of a concrete specimen, the reflectivity of the concrete is the ratio of the energy reflected from its surface to the incident energy, in accordance with Equation 1.

A simple and effective setup for measuring these parameters is illustrated in Figure 8. After a fresh batch of concrete is mixed, it is placed into a plywood box that is at least 20-in (50-cm) wide by 20-in (50-cm) long by 3-in. (7.5-cm) deep. After the concrete has been consolidated and trowelled until its surface is reasonably flat, the antenna is centered 12 in. (30 cm) above the surface of the concrete. (An open-ended 12-in. (30-cm) high box made of 1-in. (2.5-cm) thick styrofoam is a convenient means of supporting the antenna above the concrete.) The amplitude of the reflection from the surface of the fresh concrete (AS) can then be measured with a calibrated oscilloscope or recorded on an oscillographic recorder. Figure 9A shows a typical waveform recorded from a fresh concrete specimen.

To measure the amplitude of the initial microwave pulse striking the surface of the concrete (AI), a piece of aluminum plate that measures at least 20 in. × 20 in. (50 cm × 50 cm) is placed on the surface of the concrete. (This causes the microwave pulse that would otherwise strike the surface of the concrete to be completely reflected back to the antenna.) The amplitude AI is similarly measured with an oscilloscope or oscillographic recorder. An example of this waveform is shown in Figure 9B. The reflection coefficient, or reflectivity, of the fresh concrete can be calculated from the expression

$$\rho = \frac{AS}{AI} \times \frac{AOI}{AOS} \tag{11}$$

where AOS and AOI are the radar signals traveling through air from the transmitting to the receiving antenna during the measurements of AS and AI, respectively. These parameters serve as convenient means to correct for electronic drifts in the radar system.

If desired, it is also possible to follow the hydration of the cement during curing by measuring the amplitude of the reflection from the bottom of the concrete. However, owing to the high reflectivity at the surface of concrete and their high attenuation during the first few days after their casting, this reflection would generally be too weak to observe during that period. This can be remedied by placing aluminum foil in the bottom of the wooden box before placing the concrete.

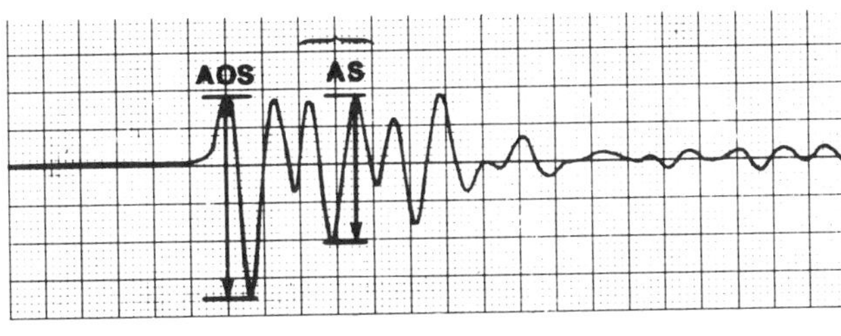

A

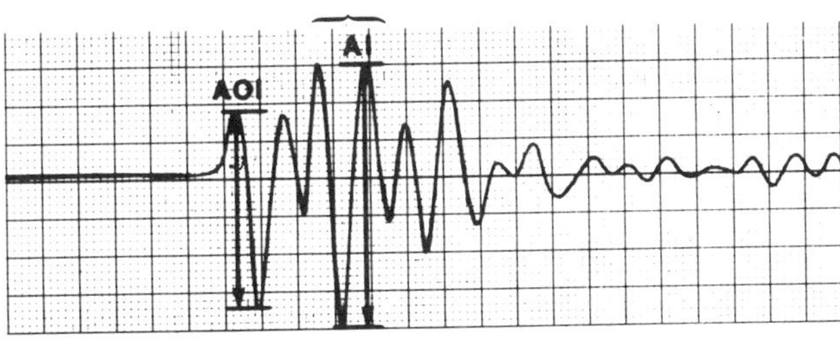

B

FIGURE 9. Waveforms of reflections from: (A) a concrete specimen (AS), and (B) an aluminum plate resting on top of the concrete (AI).

These measurements are repeated as often as necessary and for as long as the change in the dielectric properties of the concrete is significant and of interest. From the measured reflectivity of the concrete surface at any age, the relative dielectric constant of the concrete (ϵ_{r2}) can be calculated using equation 6, which becomes

$$\rho = \frac{1 - \sqrt{\epsilon_{r2}}}{1 + \sqrt{\epsilon_{r2}}} \tag{12}$$

since ϵ_{r1} equals 1.

If there is appreciable bleeding in the concrete, the measured reflectivity may represent the free water content in the surface layer of the fresh concrete more than the free water in the bulk of the concrete. In such cases, the dielectric constant of the entire thickness of the concrete specimen can be indirectly measured, as indicated in Equation 10, by using the two-way transit time of the reflection from the bottom of the specimen. Unfortunately, if the water-cement ratio of the concrete is high, or the concrete is too thick, or there is too much reinforcement, this reflection may be too weak to detect, especially during the first few days after the placing of the concrete. In such cases, aluminum foil or a metal plate can be placed at the bottom of the concrete to enhance the reflection, or the antenna can be placed directly on top of the concrete,[27] or an antenna with more penetration but less resolution can be used.

Example Test Data

Radar measurements can be utilized to compare concrete made with different types of cement. For example, Figure 10 shows the changes in the observed reflectivities and relative dielectric constants of two different concrete mixtures, starting 2 h after their mixing and continuing through the first 28 days.[29] These concrete mixtures were made with the same

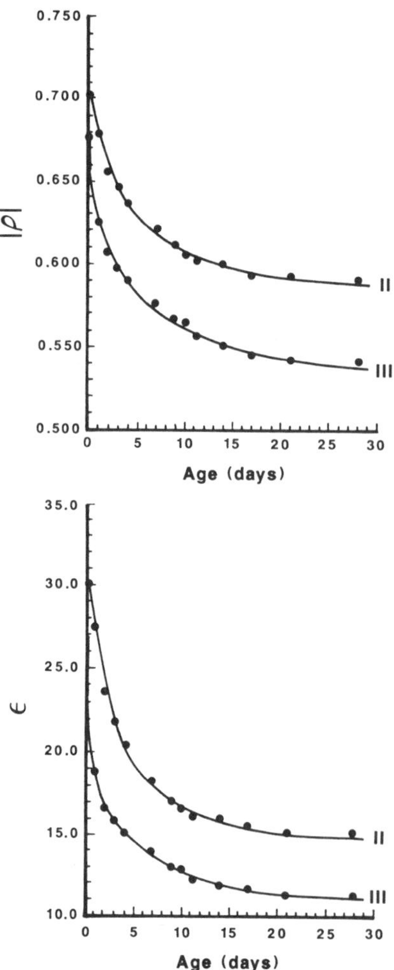

FIGURE 10. Influence of age on the reflectivity and the dielectric constant of concrete mixtures containing Type II and Type III cements.

aggregates and water-cement ratio (0.48) but with different types of cement (Type II and Type III). The concrete containing Type III cement exhibited a faster decrease in reflectivity and relative dielectric constant compared with the concrete containing Type II cement. This difference is a manifestation of the higher content of relatively fast reacting tricalcium silicate and tricalcium aluminate in Type III cement.

Since the dielectric properties and strength of a concrete are both affected by the same hydration processes, these properties can be related, if desired. Figure 11 shows a correlation of the compressive strength of each of the above two concrete mixtures, as measured by testing cylinders, which were prepared from these mixtures, at 1, 2, 4, 7, 10, 14, and 28 days after casting, with their corresponding reflectivities at the same ages. It can be seen that there was a very strong linear correlation between these two properties, with a correlation coefficient of 0.97 for all the data combined.

Advantages and Limitations

The standard deviation associated with the measurement of reflectivity, based on triplicate waveforms that were typically recorded during each measurement, was estimated to

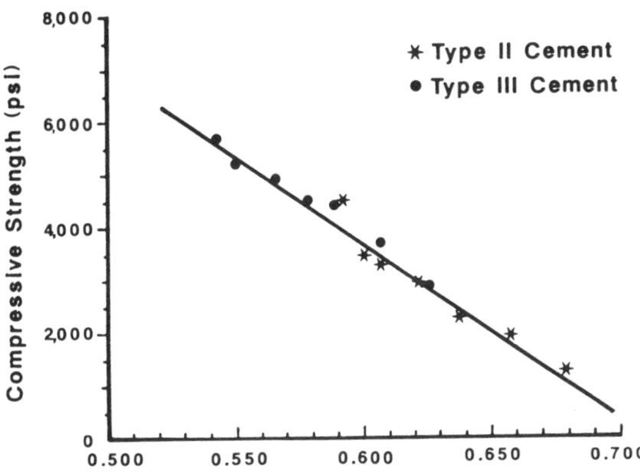

FIGURE 11. Correlation between the reflectivity and compressive strength of two concrete mixtures with the same W/C of 0.48 but different types of cement. (*Note:* 145 PSI = 1 MPa).

be approximately ±0.009. The repeatability of results is sensitive to fluctuation in the measurement setup, especially the position of the antenna with respect to the concrete surface.

There is indication that the effect of admixtures and pozzolans on the hydration rate of a concrete may also be manifested in the reflectivity of the concrete surface; therefore, it may be possible to use radar to supplement other techniques in the study of these materials.

DETERMINATION OF WATER CONTENT IN FRESH CONCRETE

Control of proper water and cement content during concrete batching operations is very important, since these factors have great influence on the durability of the concrete. Radar has recently been studied as a rapid and yet reasonably accurate method for determining the water content of fresh concrete mixtures.[30]

Principle

One of the bases for this potential application of radar is the very same principle involved in the monitoring of cement hydration in concrete. Figure 12 shows the observed amplitudes of the reflection from the bottom of three concrete specimens made with the same materials but of different water-cement ratios.[27,29] These data indicate that the dielectric property of a concrete is dependent on the amount of unreacted water in the concrete, which in turn is dependent on not only its age but also the amount of mixing water used.

The other basis for this potential application is the dependency of the attenuation of microwave pulses, during its travels through fresh concrete, on the relative dielectric constant of the concrete, as expressed in Equation 8. Since the dielectric constant and, therefore, the tendency of a concrete to attenuate microwave pulses is directly proportional to its water content, the ability of the concrete to transmit the microwave pulses must be inversely proportional to its water content.

Therefore, the water content of fresh concrete mixtures may be indirectly determined through measurement of either their microwave reflectivity or their transmission properties.

Test Procedures

To measure water content through measurement of reflectivity, the experimental setup discussed earlier and illustrated in Figure 8 can be used. For measurement of transmittance, however, a setup such as that illustrated in Figure 13, which utilized two identical antennas, has been used.[30] In this setup, one antenna functioned only as a transmitter while the other functioned only as a receiver; this arrangement was made possible through the use of a

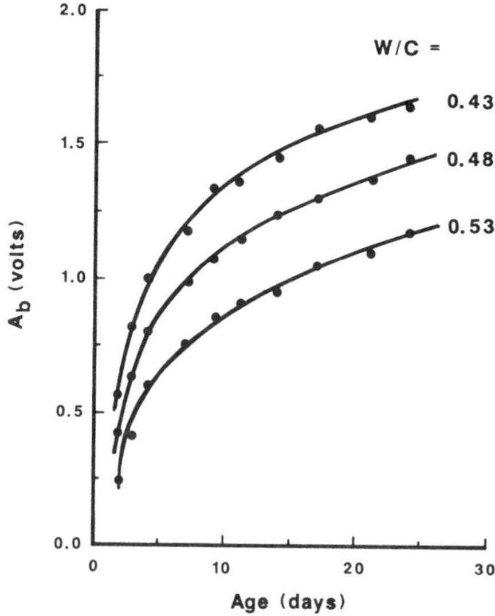

FIGURE 12. Influence of age and W/C on the intensity of reflection from the bottom of an 8-in. (20.3 cm) thick concrete specimen.

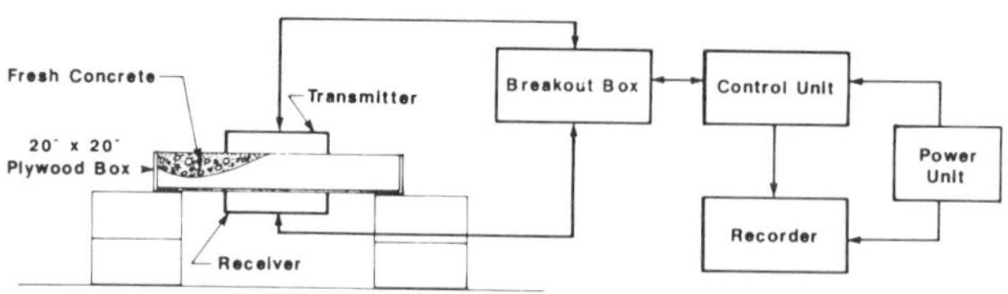

FIGURE 13. Measurement of transmission through a layer of freshly mixed concrete.

manufacturer-supplied electronic circuit, "breakout box", that interferes with the normal dual functions of each antenna.

Figure 14 shows a typical recorded waveform of the microwave energy (AT) that had reached the receiver after propagating through a layer of fresh concrete. The transmission property of the fresh concrete was then expressed in terms of a quantity (T), which is defined as the ratio of AT to the amplitude of the initial microwave pulses striking the surface of the concrete (AI), i.e.,

$$T = \frac{AT}{AI} \tag{13}$$

Immediately after AT was measured, the receiving antenna was disconnected to allow the transmitting antenna to resume its normal dual function and the measurement of AI according to the procedure that has been described earlier.

Since the potential benefit of this application is to allow for timely adjustment of the water content of any batch of fresh concrete mixture, if necessary, the measurement of either reflectivity or transmission should be performed as soon as the materials are uniformly mixed.

Example Test Data

Figure 15 shows a plot of measured reflectivities of a series of four air-entrained concrete

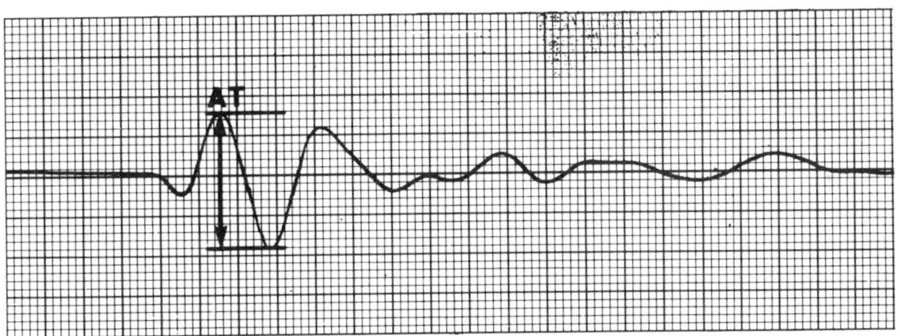

FIGURE 14. Waveform of transmission through a layer of fresh concrete.

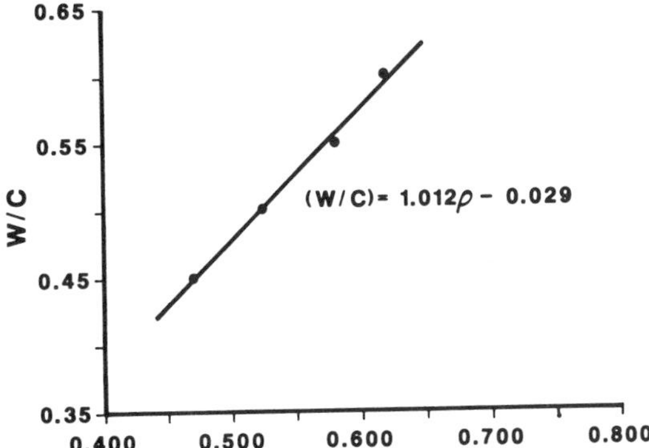

FIGURE 15. Relationship between reflectivity and W/C for concrete mixtures (containing Type II cement, granite aggregates, air-entraining agent, and water reducer) at about 3 min after mixing.

mixtures vs. their known water-cement ratios.[30] These mixtures contained the same amounts of fine and coarse aggregates, cement, and air-entraining agent; the only variable is the amount of mix water, or water-cement ratio, which ranged from 0.45 to 0.60. (Since the amount of cement used in these mixtures was constant, the water contents of these mixtures were expressed in terms of water-cement ratios.) It is evident that there is a reasonably good linear correlation between these parameters.

If instead of reflectivity, the measured transmission through these same concrete mixtures was correlated with their water-cement ratio (see Figure 16), a strong inverse linear relationship between these parameters was obtained.[30]

Advantages and Limitations

The use of radar to determine the water content of fresh concrete is a relatively new application and therefore has not been studied extensively yet. In perhaps the only study conducted to date, it was reported that, for the procedure used, the error associated with the measurement of water-cement ratio by the reflectivity method was estimated to be less than ± 0.038 (at a 95% confidence level); and the standard deviation was ± 0.011.[30] Although this degree of accuracy doesn't compare favorably to those reported for the modified Kelly-Vail method and the microwave-oven drying method,[31] which have received extensive study in the last several years, the method has merit and should be studied further to improve its reliability.

The error that was associated with the use of the transmission method to measure water-cement ratio was estimated to be ± 0.024, at a 95% confidence level; with a standard

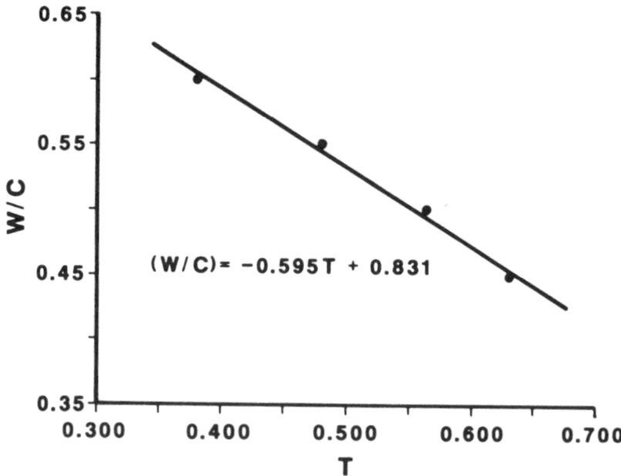

FIGURE 16. Relationship between transmission value and W/C for concrete mixtures (containing Type II cement, granite aggregates, air-entraining agent, and water reducer) at about 3 min after mixing.

deviation of ±0.005.[30] This accuracy is better than that corresponding to the reflectivity method and at least equal to that of the modified Kelly-Vail method.

A disadvantage with either of these radar methods is that a calibration line is probably necessary for every combination of materials used, because it appears that the type of cement, aggregate, and additive used can influence the parameters in the relationship between water content and reflectivity or transmission value.

The potential advantages of these radar methods are their speed and simplicity. The necessary handling of the fresh concrete sample prior to the use of either procedures is minimal, since it involves only transferring sufficient amount of the fresh concrete into a suitable sample container and consolidation of the concrete. After a calibration line appropriate to the selected concrete materials has been established, the actual measurement of either reflectivity or transmission usually requires less than 30 sec, which is faster than even the fastest of the two nonradar methods, which requires at least 15 min.

Another potential advantage is that these methods are amenable to automation. In a large capacity ready-mix plant, the procedure could conceivably be setup in such a manner that either the reflectivity or the transmission of fresh concrete could be measured automatically and continuously with the aid of microprocessors along a production line.

MEASUREMENT OF THICKNESS

In the construction of concrete pavements, strict compliance with the specified minimum slab thickness is important in ensuring long service life for the new pavements. To ensure compliance, cores are extracted at specified intervals in each newly constructed lane for direct measurement. In addition to being time consuming and costly, this destructive inspection method creates undesirable discontinuities in the pavements. Radar has been tested as a possible alternative for this application.[32]

Principle

According to Equation 10, assuming that the dielectric constant of a concrete slab is uniform and known, the two-way transit time of microwave pulses through the concrete is directly proportional to the thickness of slab.

Test Procedure

There are two ways by which this relationship may be applied in the nondestructive

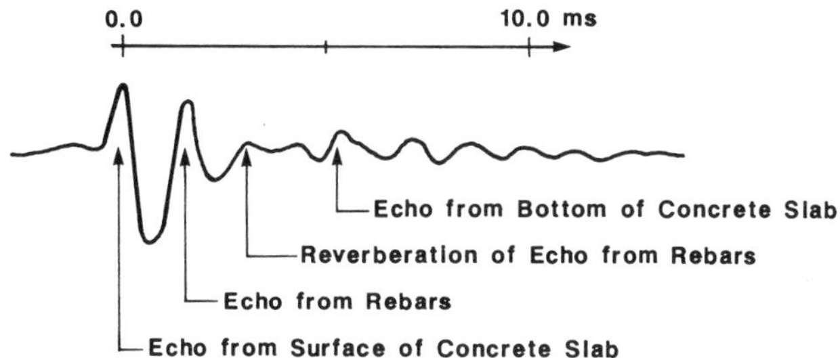

FIGURE 17. Radar echoes recorded at a test location on a continuously reinforced concrete pavement.

measurement of concrete slab thickness. First, the relative dielectric constant of the concrete at each test location can be calculated from the measured surface reflectivity. Then, the thickness of the slab at that location can be calculated from this relative dielectric constant and the measured two-way transit time.

The second approach involves the establishment of a suitable calibration line to relate slab thickness and the two-way transit time. From a review of the reflection waveforms recorded for all randomly selected test locations, several locations that appear to provide a suitable range of measured two-way transit times are selected for direct measurements of slab thickness by coring or any other preferred method. Then the measured slab thicknesses and transit times at these selected locations are correlated through a linear regression analysis to establish the calibration line from which the thickness of the slab at all other test locations may be estimated.

The first approach is susceptible to error, because the relative dielectric constant as determined from the surface reflectivity may not be representaive of the entire thickness of the concrete. The second approach is susceptible to error too, because it assumes that the relative dielectric constant of the concrete is uniform at all test locations.

Example Test Data

Figure 17 shows the reflection waveform recorded for one of 51 locations in a continuously reinforced concrete pavement where the use of radar to determine slab thickness was studied.[32] This waveform consists of the reflections from the surface of the pavement, from the rebars, and from the bottom of the slab, and sandwiched between the latter two reflections is the reverberation from the rebars.

In the investigation, a calibration line was established from the cores and radar data at seven locations selected from the 51 locations. Using the resulting calibration line, the thickness of the concrete slab at each of the remaining locations was estimated from its respective measured transit times. Figure 18 shows that these estimates compare favorably with the lengths of the cores, with absolute errors ranging from 0.0 to 0.9 in. (0.0 to 2.25 cm). This presence of a range of errors likely reflects the fallacy of the assumption, that the concrete at all locations have the same relative dielectric constant, which is inherent in this procedure. Further, some of these errors exceeded the ±0.25 in. (0.63 cm) that is considered acceptable for compliance testing.

Advantages and Limitations

The success of thickness measurement using radar depends on a reasonably detectable reflection from the backside (or the bottom) of the concrete member (or slab), because this allows for the precise identification of the reflection and, therefore, the accurate measurement of the transit time. Conditions that would prevent the reflection from being precisely detected include the presence of reinforcement, relatively high attenuation of the microwave pulses by the concrete, insufficient difference between the relative dielectric constants of the concrete and the subbase, and slabs that are too thick.

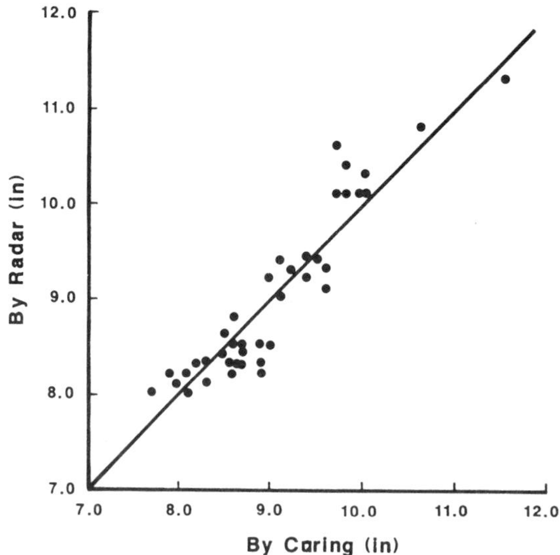

FIGURE 18. Comparison between concrete slab thickness as determined by coring and by radar using a calibration line. (*Note:* 1 in. = 2.54 cm.)

For some pavements, it is likely that there may be only small differences between the relative dielectric constants of the concrete and the subbase materials, so that this reflection would be very weak and difficult to identify, particularly when the concrete has cured long enough that its dielectric constant has approached that of the subbase. Consequently, prior to an actual inspection, it is difficult to predict how precise the radar will be in measuring the thickness of a particular pavement.

It is possible that the reflection may be more pronounced, if the measurement is made during the period between 3 to 10 days after the concrete is placed — when its relative dielectric constant is still relatively high (see Figure 10) and its ability to attenuate microwave pulses is probably low enough. This possibility, however, has not been studied yet.

CONCLUSIONS

Short-pulse radar is proving to be a powerful scientific tool with a wide range of applications, including the nondestructive detection of delaminations in bare or overlaid reinforced concrete bridge decks. Its accuracy and ability to detect other types of defects existing in concrete bridge decks and other structures need improvement, and it is believed that this can be achieved through improvements in the resolution of the antenna and by increasing the understanding of the various radar signatures encountered in these structures.

In addition to the detection of delaminations in concrete, radar shows potential for other applications such as the monitoring of cement hydration or strength development in concrete, the study of the effect of various admixtures and additives on concrete, the rapid determination of water content in fresh concrete, the measurement of the thickness of concrete members, and the locating of rebars. These potential applications have not yet received wide attention, as is evident in the scarcity of literature.

REFERENCES

1. **Bertram, C. L., Campbell, K. J., and Sandler, S. S.,** Locating large masses of ground ice with an impulse radar system, in *Proc. 8th Int. Symp. on Remote Sensing,* Willow Run Laboratory, University of Michigan, Ann Arbor, October 1972.
2. **Morey, R. M. and Harrington, W. S., Jr.,** Feasibility Study of Electromagnetic Subsurface Profiling, Rep. EPA-R2-72-082, U.S. Environmental Protection Agency, Washington, D.C., October 1972.
3. **Campbell, K. J. and Orange, A. S.,** A continuous profile of sea ice and freshwater ice thickness by impulse radar, *Polar Rec.,* 17, 31, 1974.
4. **Morey, R. M.,** Application of downward looking impulse radar, in *Proc. 13th Annu. Canadian Hydrographic Conf.,* Canada Centre for Inland Waters, Burlington, Ontario, Canada, March 1974.
5. **Porcello, L. J.,** The Apollo lunar sounder radar system, in *Proc. Inst. Electrical and Electronic Engineers,* 62, 769, 1974.
6. **Lord, A. E. and Koerner, R. M.,** Nondestructive Testing Techniques to Detect Contained Subsurface Hazardous Waste, Rep. No. EPA/600/2-87/078, Drexel University, Philadelphis, September 1987.
7. **Haeni, P. and Trent, R. E.,** Measuring scour with ground-penetrating radar, sonar, and seismic geographical methods, presented at the 67th Annu. Meet. of the Transportation Research Board, Washington, D.C., January 11 to 14, 1988.
8. **Bertram, C. L., Morey, R. M., and Sandler, S. S.,** Feasibility Study for Rapid Evaluation of Airfield Pavements, Rep. No. AFWL-TR-71-178, U.S. Air Force Weapons Laboratory, June 1974.
9. **Clemena, G. G. and McGhee, K. H.,** Applicability of radar subsurface profiling in estimating sidewalk undermining, *Transp. Res. Rec.,* 752, 21, 1980.
10. **Kovacs, A. and Morey, R. M.,** Detection of Cavities Under Concrete Pavement, CRREL Rep. 82-18, U.S. Army Cold Regions Research and Engineering Laboratory, Hanover, NH, July 1983.
11. **Clemena, G. G., Sprinkle, M. M., and Long, R. R., Jr.,** Use of ground-penetrating radar for detecting voids under a jointed concrete pavement, *Transp. Res. Rec.,* 1109, 1, 1987.
12. **Clemena, G. G. and Mckeel, W. T., Jr.,** Detection of delamination in bridge decks with infrared thermography, *Transp. Res. Rec.,* 664, 180, 1978.
13. **Holt, F. B. and Manning, D. G.,** Detecting delamination in concrete bridge decks, *Concr. Int.,* 2, 34, 1980.
14. **Alongi, A. V.,** Radar Examination of Condition of Two Interstate Highway Bridges, Penetradar Corp., Niagara Falls, NY, 1973.
15. **Cantor, T. and Kneeter, C.,** Radar and acoustic emission applied to the study of bridge decks, suspension cables and masonry tunnel, *Transp. Res. Rec.,* 676, 27, 1978.
16. **Cantor, T. and Kneeter, C.,** Radar as applied to evaluation of bridge decks, *Transp. Res. Rec.,* 852, 37, 1982.
17. **Clemena, G. G.,** Nondestructive inspection of overlaid bridge decks with ground-penetrating radar, *Transp. Res. Rec.,* 899, 21, 1983.
18. **Manning, D. and Holt, F.,** Detecting deterioration in asphalt-covered bridge decks, *Transp. Res. Rec.,* 899, 10, 1983.
19. **Alongi, A., Cantor, T., Kneeter, C., and Alongi, A., Jr.,** Concrete evaluation by radar — theoretical analysis, *Transp. Res. Rec.,* 853, 31, 1982.
20. **Clemena, G. G.,** Survey of Bridge Decks with Ground-Penetrating Radar — A Manual, Rep. VTRC 86-R3, Virginia Transportation Research Council, Charlottesville, VA, July 1985.
21. **Ulriksen, C. P. F.,** Application of Impulse Radar to Civil Engineering, Doctoral thesis, Lund University, Sweden, 1982.
22. **Chung, T., Carter, C. R., Manning, D. G., and Holt, F. B.,** Signature Analysis of Radar Waveforms Taken on Asphalt Covered Bridge Decks, Rep. ME-84-01, Ontario Ministry of Transportation and Communicaitons, Ontario, Canada, June, 1984.
23. **Maser, K. R.,** Detection of progressive deterioration in bridge decks using ground-penetrating radar, in *Proc. Experimental Assessment of Performance of Bridges,* ASCE Convention, Boston, MA, October 1986.
24. **Alongi, A., Jr.,** personal communication, 1987.
25. **Rzepecka, M. A., Hamid, A. K., and Soliman, A. H.,** Monitoring of concrete curing process by microwave terminal measurements, IEEE Trans. on Industrial Electronics and Control Instrumentation, 19, 120, 1972.
26. **Gorur, K., Smit, M. K., and Wittman, F. H.,** Microwave study of hydrating cement paste at early age, *Cement Concr. Res.,* 12, 447, 1982.
27. **Clemena, G. G.,** Microwave Reflection Measurements of the Dielectric Properties of Concrete, Rep. VTRC-84-R10, Virginia Transportation Research Council, Charlottesville, VA, 1983.
28. **Hasted, J. B.,** The dielectric properties of water, *Prog. Dielectrics,* 3, 101, 1961.
29. **Clemena, G. G.,** unpublished data, 1987.

30. **Clemena, G. G.,** Determining Water Content of Fresh Concrete by Microwave Reflection or Transmission Measurement, Rep. VTRC-88-R3, Virginia Transportation Research Council, Charlottesville, VA, 1987.
31. **Head, W. J., Phillippi, P. A., Howdyshell, P. A., and Lawrence, D.,** Evaluation of selected procedures for the rapid analysis of fresh concrete, *Cement Concr. Aggregates,* 5, 88, 1983.
32. **Clemena, G. G. and Steele, R. E.,** Inspection of the Thickness of In-Place Concrete with Microwave Reflection Measurements, Rep. VTRC-88-R16, Virginia Transportation Research Council, Charlottesville, VA, 1988.

Chapter 12

STRESS WAVE PROPAGATION METHODS

Mary Sansalone and Nicholas J. Carino

ABSTRACT

This chapter presents a review of nondestructive testing methods based on the use of stress waves. The pulse-echo, impact-echo, impulse-response, and spectral analysis of surface waves techniques for evaluation of concrete are discussed. The principles, test procedures, signal processing, and representative applications of each method are presented. It is shown that the common feature of the methods is that inferences about internal conditions of concrete structures are made based on the effect that the structure has on the propagation of stress waves. The methods differ in the source of the stress waves, the testing configuration, instrumentation, the characteristics of the measured response, and the signal processing techniques that are used. These differences make each method particularly suitable for specific applications.

INTRODUCTION

Except for visual inspection, the use of acoustic methods is the oldest form of nondestructive testing. Striking an object with a hammer and listening to the "ringing" sound is a common way of detecting the presence of internal voids, cracks, or delaminations. However, the technique is qualitative, and its success depends upon the experience of the user.

In 1929, Solokov in the USSR first suggested the use of ultrasonic waves to find defects in metal objects.[1] However, it was not until World War II spurred the development of sophisticated electronic instrumentation in the 1940s that significant progress was made. Ultrasonic* pulse-echo flaw detectors were first introduced in 1942 by Firestone of the University of Michigan and, independently, by Sproule of England. Since that time, ultrasonic pulse-echo testing of metals, plastics, and other homogeneous materials has developed into an efficient, reliable, and versatile nondestructive test method.

The development of test techniques and equipment for ultrasonic evaluation of less ideal materials, such as concrete, has been hindered by the difficulties inherent in obtaining and interpreting a signal record from a heterogeneous material. Because high frequency (1 MHz or greater) stress pulses cannot penetrate into concrete, none of the commercially available transducers are satisfactory for pulse-echo testing of concrete. An alternative approach is to generate low frequency stress pulses using mechanical impact. Several techniques are being developed for testing concrete. Some of these techniques are being used in field inspections, others are still in the developmental stages. However, at present, there is no standard or routine method based on the use of stress wave propagation for finding flaws in concrete structures.

This chapter begins with a review of the basic principles of stress wave propagation in solids. Next, the methods that are being used and being developed for finding flaws in concrete are discussed. Test procedures, signal processing, and representative applications of each method for evaluation of concrete are presented.

* Ultrasonic refers to sound waves with frequencies above the audible range which is generally taken to be about 20 kHz. Most ultrasonic pulse-echo devices operate at 1 MHz or greater.

BASIC PRINCIPLES

The purpose of this section is to provide background information on elastic wave propagation in solids.

WAVE TYPES

When a stress is applied suddenly to a solid, the disturbance which is generated travels through the solid as stress waves. There are three primary modes of stress wave propagation through isotropic, elastic media: dilatational, distortional, and Rayleigh waves. Dilatational and distortional waves, commonly referred to as compression and shear waves, or P- and S-waves, are characterized by the direction of particle motion with respect to the direction the wavefront is propagating. In a P-wave, motion is parallel to the direction of propagation; in the S-wave, motion is perpendicular to the direction of propagation. P-waves can propagate in all types of media; S-waves can propagate only in media with shear stiffness, i.e., in solids. Rayleigh waves, or R-waves, are waves which propagate along the surface of a solid. The particular motion in an R-wave near the surface is retrograde elliptical.

The wavefront defines the leading edge of a stress wave as it propagates through a medium. The shapes of the P-, S-, and R-wavefronts depend upon the characteristics of the source used to generate the waves. There are three idealized types of wavefronts: planar, spherical, and cylindrical. For example, when the stress waves are generated by impact at a point on the surface of a solid, the resulting P- and S-wavefronts are spherical and the R-wavefront is cylinderical.

WAVE SPEED

In most applications of stress wave propagation, the input is a pulse of finite duration and the resulting disturbance propagates through the solid as transient waves. The propagation of transient stress waves through a heterogeneous bounded solid, such as a structural concrete member, is a complex phenomenon. However, a basic understanding of the relationship between the physical properties of a material and the wave speed can be acquired from the theory of wave propagation in isotropic elastic media.[2]

In infinite elastic solids, the P-wave speed, C_p, is a function of Young's modulus of elasticity, E, the mass density, ρ, and Poisson's ratio, ν:

$$C_p = \sqrt{\frac{E\,(1-\nu)}{\rho\,(1+\nu)(1-2\nu)}} \tag{1}$$

In bounded solids, such as thin plates or long rods, P-wave speed can vary depending on the dimensions of the solid relative to the component wavelength(s) of the propagating wave. For rod-like structures, such as piles, P-wave speed is independent of Poisson's ratio if the rod diameter is much less than the wavelength(s) of the propagating wave.[3] In this case, C_p is given by the following equation:

$$C_p = \sqrt{\frac{E}{\rho}} \tag{2}$$

For a Poisson's ratio of 0.2, a typical value in concrete, the P-wave speed is five percent higher in an infinite solid than in a long thin rod.

The S-wave speed, C_s, in an infinite solid is given by the following equation:

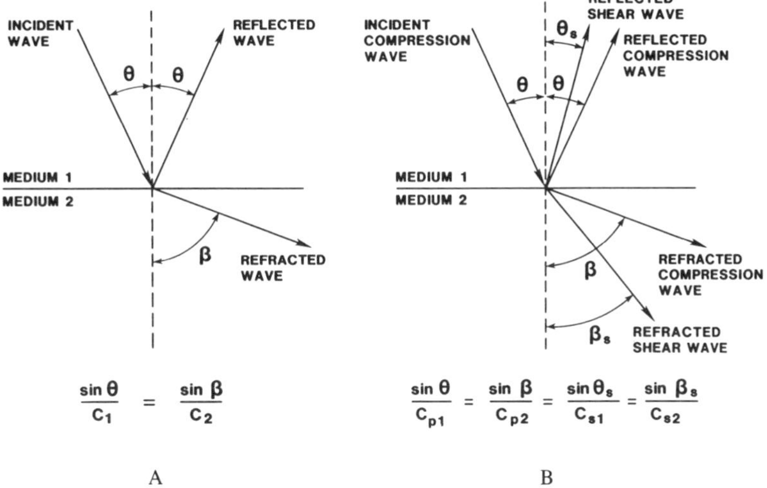

FIGURE 1. The behavior of a P-wave incident upon an interface between two dissimilar media; (A) reflection and refraction; (B) mode conversion.

$$C_s = \sqrt{\frac{G}{\rho}} \tag{3}$$

where G = shear modulus of elasticity = E/2(1-ν).
A useful parameter is the ratio, α, of the S- to P-wave speeds:

$$\alpha = \frac{C_s}{C_p} = \sqrt{\frac{(1-2\nu)}{2(1-\nu)}} \tag{4}$$

For a Poisson's ratio of 0.2, α is 0.61.

R-waves propagate at a speed, C_R, which can be determined from the following approximate formula:[3]

$$C_R = \frac{0.87+1.12\nu}{1+\nu} C_s \tag{5}$$

For Poisson's ratio of 0.2, the R-wave speed is 92% of the S-wave speed, or 56% of the P-wave speed.

REFLECTION AND REFRACTION

When a P- or S-wavefront is incident upon an interface between dissimilar media, "specular" reflection occurs. (The term specular reflection is used because the reflection of stress waves is similar to the reflection of light by a mirror.) As shown in Figure 1A, stress waves can be visualized as propagating along ray paths. The geometry of ray reflection is analogous to that of light rays, that is, the angle of reflection of any ray is equal to the angle of incidence, θ, for that ray.

At a boundary between two different media only a portion of a stress wave is reflected. The remainder penetrates into the underlying medium (wave refraction). The angle of refraction, β, is a function of the angle of incidence, θ, and the ratio of wave speeds, C_2/C_1, in the different media, and is given by Snell's law:

TABLE 1
Specific Acoustic Impedances

Material	Density (kg/m³)	P-wave velocity (m/s)	Specific acoustic impedance (kg/(m² − s))
Air	1.205	343	0.413
Concrete[a]	2300	3000—4500	$6.9—10.4 \times 10^6$
Granite	2750	5500—6100	$15.1—16.8 \times 10^6$
Limestone	2690	2800—7000	$7.5—18.8 \times 10^6$
Marble	2650	3700—6900	$9.8—18.3 \times 10^6$
Quartzite	2620	5600—6100	$14.7—16.0 \times 10^6$
Soils	1400—2150	200—2000	$0.28—4.3 \times 10^6$
Steel	7850	5940	46.6×10^6
Water	1000	1480	1.48×10^6

[a] The mass density of concrete depends on the mixture proportions and the specific gravities of the mix ingredients. The given density is for an average, normal weight concrete.

$$\sin \beta = \frac{C_2}{C_1} \sin \theta \tag{6}$$

Unlike light waves, stress waves can change their mode of propagation when striking the surface of a solid at an oblique angle. Depending on the angle of incidence, P-waves can be partially reflected as both P- and S-waves and can be refracted as both P- and S-waves. Since S-waves propagate at a lower velocity than P-waves, they will reflect and refract at angles (determined using Snell's law), θ_s and β_s, that are less than the angles of reflection and refraction for P-waves, as shown in Figure 1B.

The relative amplitudes of reflected waves depend upon the mismatch in specific acoustic impedances at an interface, the angle of incidence, the distance of an interface from the pulse source, and the attenuation along the wave path. The influence of each of these factors is considered in the following discussion.

The portion of an incident ray of a P-wave that is reflected at an interface between two media depends on the specific acoustic impedances of each medium. The specific acoustic impedance, Z, of a medium is

$$Z = \rho C_p \tag{7}$$

Specific acoustic impedance values for P-waves in selected materials are given in Table 1.[5] Equation 7 is also valid for S-waves if the S-wave velocity is used to calculate specific acoustic impedance.

The amplitude in a reflected ray is maximum when the angle of incidence of the ray is normal to the interface. The amplitude of the reflected ray relative to the amplitude of the incident ray can be determined using the following equation:

$$R_n = \frac{Z_2 - Z_1}{Z_2 + Z_1} \tag{8}$$

where R_n is the reflection coefficient for normal incidence and Z_1 and Z_2 are the specific acoustic impedances of media 1 and 2, respectively. If Z_1 is greater than Z_2, then R_n is negative, indicating that the reflected wave will have the opposite sign, that is, a phase change occurs. For example, if the stress in an incident P-wave is compressive, then the stress in the reflected P-wave is tensile. If Z_2 is greater than Z_1, no phase change occurs.

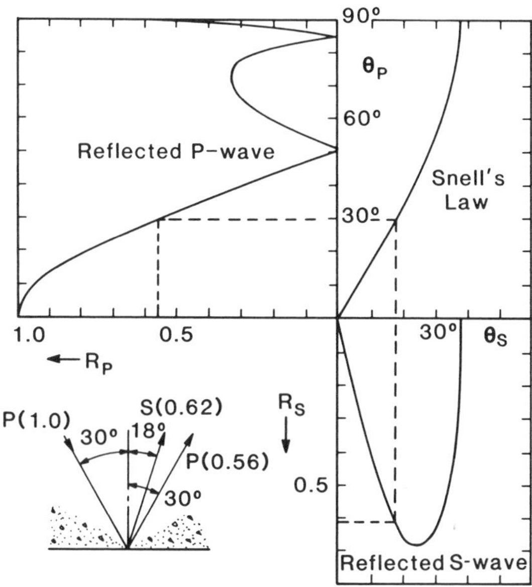

FIGURE 2. Reflection coefficients at a concrete/air interface for an incident P-wave as a function of the incidence angle (Poisson's ratio = 0.2).[5]

For incident angles other than normal to an interface, the reflection coefficients are dependent on the angle of incidence, and they can be determined using the formulas in Krautkramer and Krautkramer,[6] which are applicable for plane waves incident upon plane boundaries. These formulas were used to calculate the reflection coefficients for a concrete/air interface. Figure 2 shows reflection coefficients for an incident P-wave. It is assumed that the incident wave has an amplitude equal to unity. A similar figure can be constructed for an incident S-wave.[7] The figure is composed of three graphs. The graph in the upper left gives the reflection coefficients, R_p, for the reflected P-wave. The graph in the lower right gives the reflection coefficients, R_s, for the mode-converted S-wave. The graph in the upper right gives the angular relationship between the incident wave and the mode-coverted wave, which is determined by Snell's law. The drawing in the lower left gives an illustrative example. Note that for P-waves with a low angle of incidence, R_p is approximately equal to one and R_s is small.

In the previous discussion it was assumed that reflection and refraction of wavefronts occurred at planar interfaces between two dissimilar media. This type of analysis is also applicable to flaws or discontinuities within a medium. The ability (sensitivity) of stress wave propagation methods to detect flaws or discontinuities depends on the component frequencies (or wavelengths) in the propagating wave and on the size of the flaw or discontinuity. A general rule is that the size of the flaw must be approximately equal to or larger than the wavelengths in the propagating wave. Wave velocity, C, frequency, f, and wavelength, λ, are related by the following equation:

$$C = f\,\lambda \tag{9}$$

For example, to detect a flaw with a diameter of about 0.1 m, it is necessary to introduce into the concrete (P-wave velocity of about 4000 m/s) a stress pulse that contains frequencies greater than approximately 40 kHz (wavelengths less than approximately 0.1 m).

As a wave propagates through a solid its amplitude decreases with path length due to

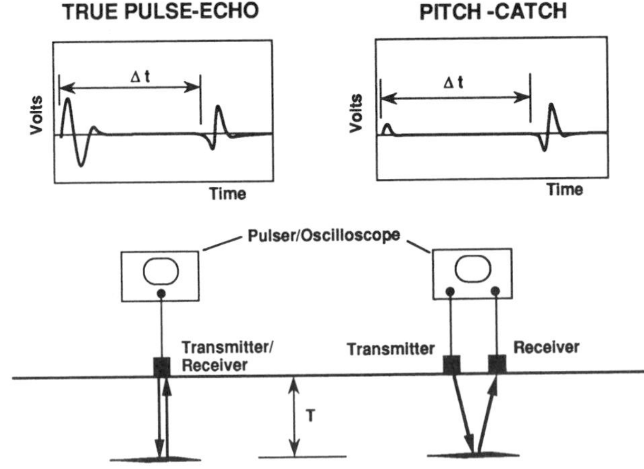

FIGURE 3. Schematic of pulse-echo and pitch-catch techniques.

attenuation (scattering and absorption) and divergence (beam spreading). Divergence causes the amplitude of spherical waves to decrease in proportion to the inverse of the distance from the source. In evaluation of concrete, low frequency (long wavelength) waves must be used to reduce the attenuation of wave energy due to scattering (reflection and refraction from mortar-aggregate interfaces). If the wavelength of the propagating wave is less than the size of the aggregate, the mismatch in acoustic impedances between the mortar and the aggregate causes scattering of incident waves at each mortar-aggregate interface. For example, if the maximum size aggregate is 25 mm in a concrete with a P-wave speed of 4000 m/s, frequencies lower than 4000/0.025 = 160 kHz should be used to reduce scattering. The concrete will appear homogeneous to these lower frequency waves. However, use of low frequency waves reduces the sensitivity of the propagating waves to small flaws. Thus there is an inherent limitation in the flaw size that can be detected within concrete using stress wave propagation methods.

TEST METHODS

Several test methods based on stress wave propagation are being used for nondestructuve testing of concrete structures. The echo methods (Impact-Echo and Pulse-Echo) are being used for thickness measurements, flaw detection and integrity testing of piles. The Impulse-Response (or Shock) method is also being used to test piles. The Spectral Analysis of Surface Waves (SASW) method is being used to determine the thickness of pavements and elastic moduli of layered pavement systems. The following sections describe the principles of these methods, the required test equipment, and the signal processing techniques. The current uses for each method and examples of laboratory and field studies are presented.

PULSE-ECHO

Principle

In the pulse-echo method a stress pulse is introduced into an object at an accessible surface by a transmitter. The pulse propagates into the test object and is reflected by flaws or interfaces. The surface response caused by the arrival of reflected waves, or echoes, is monitored by either the transmitter acting as a receiver (true pulse-echo) or by a second transducer located near the pulse source (pitch-catch). Figure 3 illustrates the principle of these echo methods. The receiver output is displayed on an oscilloscope, and the display is

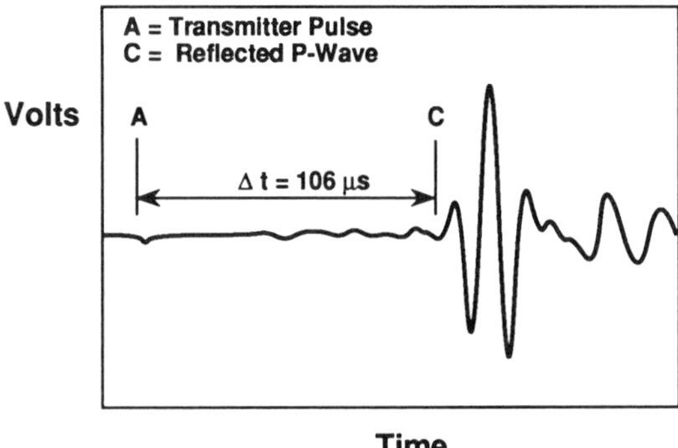

FIGURE 4. Waveform obtained from pitch-catch test on concrete slab. (Adapted from Reference 8.)

called a time-domain waveform. Using the time base of the display, the round-trip travel time of the pulse is determined. If the wave speed in the material is known, this travel time can be used to determine the depth of the reflecting interface using the following equation:

$$T = \frac{1}{2}(\Delta t)C_P \tag{10}$$

where Δt is the round-trip travel time, T is the depth, and C_p is the P-wave speed. The factor of one-half is used because the actual depth is one-half the travel path of the wave. This equation is approximate for a pitch-catch system and is applicable only if the separation between the sending and receiving transducers is small.

Signal Processing

Time-domain analysis has been used exclusively in all applications where the pulse-echo or pitch-catch methods have been used to test concrete structures. Figure 4 shows a waveform obtained from a 0.235-m thick slab using a pitch-catch test system.[8] The start of the transmitted pulse and the arrival of the P-wave echo reflected from the bottom of the slab are indicated on the waveform by the letters A and C, respectively. Using the time base of the oscilloscope, the elapsed time, Δt, between points A and C was determined to be 106 μs. Therefore, solving for C_p in Equation 10, the P-wave speed is 4430 m/s.

Instrumentation

The key components of a pulse-echo or pitch-catch test system are the transmitting and receiving transducer(s) and the oscilloscope which is used to record waveforms.

The majority of modern transducers utilize piezoelectric materials in the generation and reception of stress waves. Generally these materials are manufactured ceramics such as lead zirconate titanate (PZT) and lead metaniobate. A piezoelectric material subjected to an electrical field will change dimension suddenly, and then gradually "ring down" to its initial state. Conversely, a piezoelectric material subjected to deformation generates an electrical charge which is directly proportional to the applied strain. Thus a single transducer can be used for both generation and reception of stress waves. A fluid couplant, such as oil or grease, is needed between the transducer and the test object in order to transmit and receive the low amplitude stress waves used in pulse-echo testing.

When a transmitting transducer is excited by a high voltage pulse, its vibration time-history is approximately that of a damped sinusoidal curve. The vibration has a characteristic (or resonant) frequency. When a single transducer is used as the transmitter and receiver,

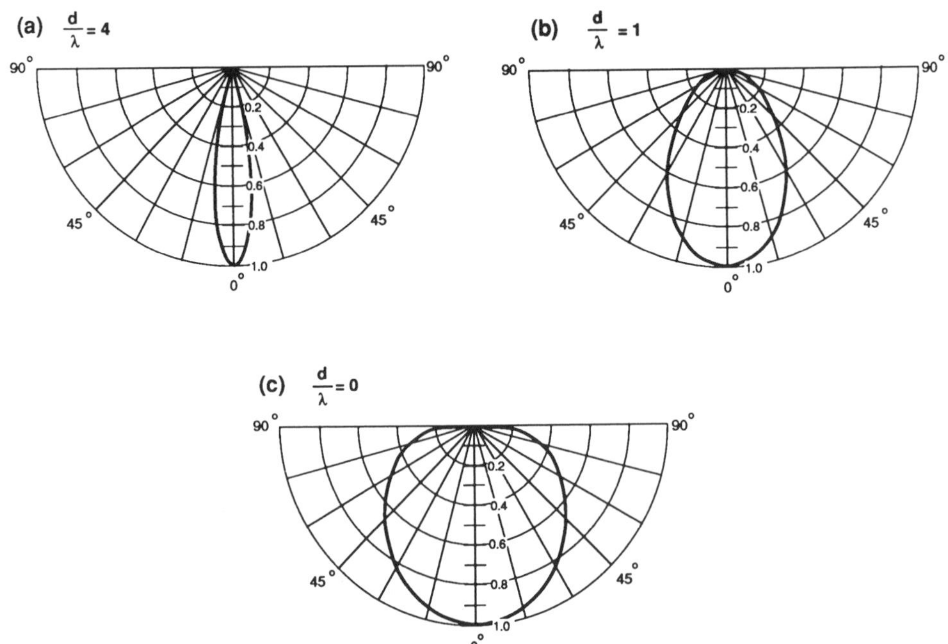

FIGURE 5. Polar diagrams of radiation patterns for transducers with various ratios of diameter to wavelength (Poisson's ratio = 0.2).

the transmitted pulse must be sufficiently short so as not to overlap with the arrival of the echoes. The transducer must be damped or the "ringing" (vibrations) of the piezoelectric element will render it unresponsive to echoes. However, the more heavily a transducer is damped, the less sensitive it is to the displacements caused by the arrival of echoes. Thus an acceptable balance must be achieved in the design and construction of a piezoelectric pulse-echo transducer. The development of a suitable pulse-echo transducer for testing concrete is a difficult undertaking. Although some success has been reported,[7] most researchers have resorted to pitch-catch systems. In this case a heavily-damped transducer is used to send the pulse, and a lightly damped transducer is used to receive the echoes.

Most of the energy transmitted into the test object is contained within a cone-shaped region which has its apex at the transducer. The ratio of transducer diameter, d, to the wavelength of the transmitted waves, λ, determines the radiation pattern of the stress pulse. Figure 5 shows the directional characteristics in concrete for $d/\lambda = 4$, $d/\lambda = 1$, and $d << \lambda$ (a point source). These figures represent the variation of the wave amplitude with the direction of radiation. The radial distances from the origin to the curves represent the relative amplitude with the highest amplitude occurring in the direction of the transducer axis. As d/λ decreases, the apex angle of the cone increases, that is the directionality of the pulse decreases. With decreasing directionality, a greater volume of material is probed by the pulse, and it becomes more difficult to identify the boundaries of internal defects as the transducer is scanned across the test object. It is generally much simpler to interpret test results using a transducer that emits a focused beam ($d/\lambda = 4$). If a 100-kHz focused transducer ($d/\lambda = 4$) were to be used to evaluate concrete with a P-wave speed of 4000 m/s, the transducer must have a diameter of 0.16 m.

In summary, low frequency transducers are needed for testing concrete. However, it is difficult to construct low frequency transducers that generate short duration, focused stress pulses. Often the dimensions of the transducer become very large, which makes the transducer cumbersome and also creates difficulties in coupling the transducer to the surface of concrete.

Because of these difficulties, there are currently no commercially available transducers for pulse-echo testing of concrete.

Applications

Since the early 1960s, experimental pulse-echo and pitch-catch systems have been developed for concrete. Successful applications have been limited to measuring the thickness of thin slabs, pavements, and walls, measuring the length of piles, and in one instance, locating surface-opening cracks in submerged structures. Each of these applications is reviewed in the following paragraphs. For specific details on the construction and characteristics of the various transducers, the reader is referred to the cited references.

In 1964, Bradfield and Gatfield[9] of England reported the development of an echo technique for measuring the thickness of concrete pavements. Using two 100-kHz resonant transducers (0.16 m tall, 0.10 m wide, and 0.25 m long) in a pitch-catch arrangement, they were able to measure the 0.3-m thickness of a concrete specimen with an accuracy of 2%. However, this system could not be field tested due to the impracticality of the test configuration.[9] Besides being bulky, the transducers were coupled to the concrete by a large plastic block which required a smooth flat concrete surface for good coupling. Difficulties were also reported in obtaining reflections from rough textured bottom surfaces.

In 1968, Howkins et al.,[10] at IIT Research Institute, independently investigated available echo techniques in an attempt to identify a feasible method for pavement thickness measurements. Using a pitch-catch technique similar to that developed by Bradfield and Gatfield, the IIT researchers were able to measure the thickness of 0.18-m and 0.25-m thick, portland cement concrete slab specimens and 0.13-m thick asphalt concrete slab specimens with an accuracy of 2%. However, it was concluded that the transducer arrangement was not practical for field use.

A pitch-catch system was developed at Ohio State University in the late 1960s to measure pavement thickness.[11] A large transmitter was needed to produce a focused pulse with a low resonant frequency. The transmitter was a hollow cylinder, with a 0.46-m outer diameter, a 0.15-m inner diameter, and a 200 kHz resonant frequency. The receiving transducer was placed at the center of the transmitter. Thicknesses were measured with accuracies of $\pm 3\%$ at more than 90% the pavement test locations. The accuracy and good performance of the Ohio State thickness gauge was confirmed in independent field tests conducted in 1976 by Weber et al.[12] However, they concluded that the Ohio State instrument needed to be redesigned to better withstand the rigors of field use before it could be considered as practical nondestructive testing equipment.

Claytor and Ellingson, at Argonne National Laboratory, attempted to use a pulse-echo system to measure the thickness of 0.305-m thick refractory concrete specimens.[13] It was found that for frequencies below 100 kHz, the use of a single transducer was impractical because the ringing after transducer excitation obscured the echo signal. Tests were also carried out using two transducers in a pitch-catch arrangement; however, the transmitting transducer generated strong R-waves which interfered with the reception of the echo signal by the receiving transducer. To reduce R-wave interference, large diameter (0.18 m) transducers were constructed. As the response of a transducer is an averaged phenomenon over the contact area, the sensitivity of a larger diameter transducer to localized surface disturbances (R-waves) was reduced and the thickness of the concrete specimens could be determined.

In 1977, Forrest, at the Naval Civil Engineering Laboratory, reported the use of a pulse-echo system for measuring the length of concrete piles.[14] A large transducer with a resonant frequency of 12 kHz was used to measure the length of piles up to 24-m. A pulse-echo system works for long piles because there is sufficient time for transducer ringing to damp out before a reflection from the bottom of the pile arrives.

In 1984, Smith[15] demonstrated that R-waves can be used to detect surface opening cracks in submerged concrete structures, such as concrete tanks and offshore structures. Two 500 kHz, 0.025-m diameter, P-wave transducers were used as transmitter and receiver. When a transmitted P-wave strikes the surface of a submerged solid at a critical angle (defined by Snell's law, Equation 6), mode conversion occurs producing an R-wave which propagates along the solid-liquid interface. As the R-wave propagates, mode conversion also occurs, producing a P-wave which radiates into the liquid at the same critical angle. This P-wave is picked up by the receiving transducer. The distance between the two transducers can be adjusted to optimize the amplitude of the received signal. If the path of the propagating R-wave is crossed by a crack, reflection occurs and no signal will be picked up by the receiving transducer. If a crack is favorably oriented (a crack at 90° to the propagating wave is the best orientation), the P-waves produced by mode conversion of the reflected R-wave will be picked up by the transmitting transducer. From an analysis of the received signals obtained from a complete scan, i.e., from moving the transducers parallel to and over the surface of the test object in a prearranged pattern, surface opening cracks were located.

Since 1983, Thorton and Alexander, at the U.S. Army Engineer Waterways Experiment Station (WES), have been working to develop a pitch-catch system, concentrating their efforts mainly on transducer development.[8,16] Their system uses a transmitting transducer made of four plates of lead metaniobate forming a 120-mm square. The mosaic transducer has high damping and a resonant frequency of about 190 kHz. The receiver is made of PZT. The transducers are coupled to the concrete with castor oil, and a relatively flat surface is needed to achieve proper coupling of the transducer faces. Slab thickness measurements up to 0.25 m have been reported.[8,16] Figure 4, presented earlier, is an example of a measurement made using the WES pitch-catch system.

IMPACT-ECHO

The idea of using impact to generate a stress pulse is an old idea which has the great advantage of eliminating the need for a bulky transmitting transducer. However, the stress pulse generated by impact at a point does not have the directionality of a pulse from a large diameter transducer. The energy propagates into a test object in all directions, and reflections may arrive from many directions. For this reason, impact methods have been primarily used for testing piles. The pile boundary acts as a wave guide and confines most of the energy within the pile. Recently, the authors have developed an impact method for testing of thin concrete structures, and they coined the term "impact-echo" to describe their technique.

Principle

The principle of the impact-echo technique is illustrated in Figure 6. A transient stress pulse is introduced into a test object by mechanical impact on the surface. The stress pulse propagates into the object along spherical wavefronts as P- and S-waves. In addition, a surface wave (R-wave) travels along the surface away from the impact point. The P- and S- stress waves are reflected by internal interfaces or external boundaries. The arrival of these reflected waves, or echoes, at the surface where the impact was generated produces displacements which are measured by a receiving transducer and recorded on a digital oscilloscope. Because of the radiation patterns associated with P- and S-waves,[5,7] if the receiver is placed close to the impact point, the waveform is dominated by the displacements caused by P-wave arrivals. As will be discussed, the success of the method depends on using the correct impact.

Signal Analysis

Pulses generated by impact are composed of low frequency waves which have the ability to penetrate concrete. However, for a point source the pulse propagates in all directions

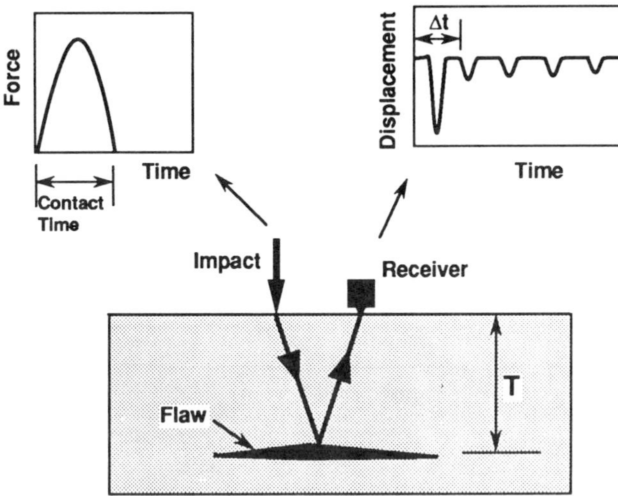

FIGURE 6. Principle of the impact-echo method.

(Figure 5c) rather than as a focused beam such as in pulse-echo systems. If a ray strikes a favorably oriented reflector within a test object, the ray is reflected back to the receiver and a surface displacement is recorded. The waveform can be simple or complex depending upon the test object. Interpretation of waveforms in the time domain using Equation 10 as described above has been successful for long slender structures, such as piles and drilled shafts. These types of structures are wave guides, that is, the geometry of the structure acts to focus the energy in the propagating wave within a narrowly defined boundary. Thus reflections recorded at the top surface are relatively easy to interpret. In addition, there is sufficient time between the generation of the stress pulse and the reception of the wave reflected from the bottom surface or from an inclusion or other flaw so that the arrival time of the reflected wave is generally easy to determine even if long duration impacts produced by hammers are used. An example of a waveform obtained from an impact test of a pile is shown in Figure 7.[17] In this example, an accelerometer was used as the receiving transducer. Time zero in the waveform is the start of the impact. The R-wave produced by the impact caused the first set of peaks, and the echo from the bottom of the 15.3-m pile gives rise to the second set of peaks. The time from the start of the impact to the arrival of the echo is about 7.5 ms.

For relatively thin structures such as slabs and walls, time-domain analysis is feasible if short duration impacts are used, but it is time consuming and can be difficult depending on the geometry of the structure.[5,7,18] An alternative approach, which is much quicker and simpler, is frequency analysis of the displacement waveforms.[19] The principle of frequency analysis is as follows. In Figure 6, the stress pulse generated by the impact propagates back and forth between the flaw and the top surface of the plate. Each time the pulse arrives at the top surface it produces a characteristic displacement. Thus the waveform is periodic, and the period is equal to the travel path, 2T, divided by the P-wave speed. Since frequency is the inverse of the period, the frequency, f, of the characteristic displacement pattern is:

$$f = \frac{C_P}{2T} \tag{11}$$

where C_p is the P-wave speed determined from an impact-echo test on a part of the structure

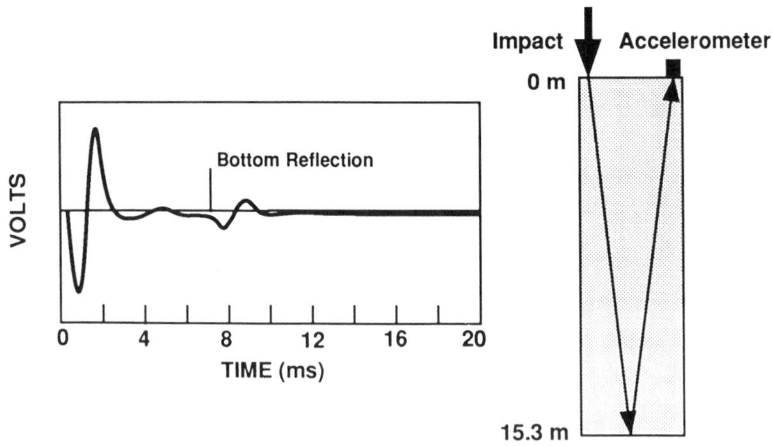

FIGURE 7. Waveform from impact-echo test on a pile. (Adapted from Reference 17.)

of known thickness. Thus, if the frequency content of an waveform can be determined, the thickness of the plate (or distance to a reflecting interface) can be calculated:

$$T = \frac{C_P}{2f} \tag{12}$$

In practice, the frequency content of the digitally recorded waveforms is obtained using the fast Fourier transform (FFT) technique.[20] This technique is based on the principle of the Fourier transform which states that any waveform can be represented as a sum of sine curves, each with a particular amplitude, frequency, and phase shift. The FFT is used to calculate the amplitude spectrum of the waveform, which gives the relative amplitude of the component frequencies in the waveform.

Figure 8A shows an amplitude spectrum obtained from an impact-echo test performed over a solid portion of a 0.5-m thick concrete slab.[21] In the amplitude spectrum there is a frequency peak at 3.42 kHz, which corresponds to multiple reflections between the bottom and top surfaces of the slab. Using Equation 11 and solving for C_p, the P-wave speed is calculated to be 3420 m/s. Figure 8B shows the amplitude spectrum obtained from a test over a portion of the slab containing a disk-shaped void.[21] The peak at 7.32 kHz results from multiple reflections between the top of the plate and the void. Using Equation 12, the calculated depth of the void is 3420/(2 * 7320) = 0.23 m, which compares favorably with the known distance of 0.25 m.

The resolution in the amplitude spectrum, that is, the frequency difference between adjacent discrete points, is equal to the sampling frequency used to capture the waveform divided by the number of points in the waveform record. This imposes a limit on the resolution of the depth calculated according to Equation 12. Because depth and frequency are inversely related, it can be shown that for a fixed resolution in the frequency domain, the resolution of the calculated depth improves as the frequency increases, that is, as depth decreases.

In using the impact-echo method to determine the locations of flaws within an object, tests are performed at regularly spaced points along "scan" lines marked on the surface. Examination of the amplitude spectra from these scans reveals the depth and approximate size of defects which may be present. One technique for visualization of the location of defects along a given scan is to create a "waterfall plot" of the individual spectra. In a waterfall plot, the spectra are plotted consecutively in the order corresponding to the test location along the scan line. Figure 9 is an example of a waterfall plot which was obtained for a scan across a section of a 0.5-m thick, reinforced concrete slab containing two disk-

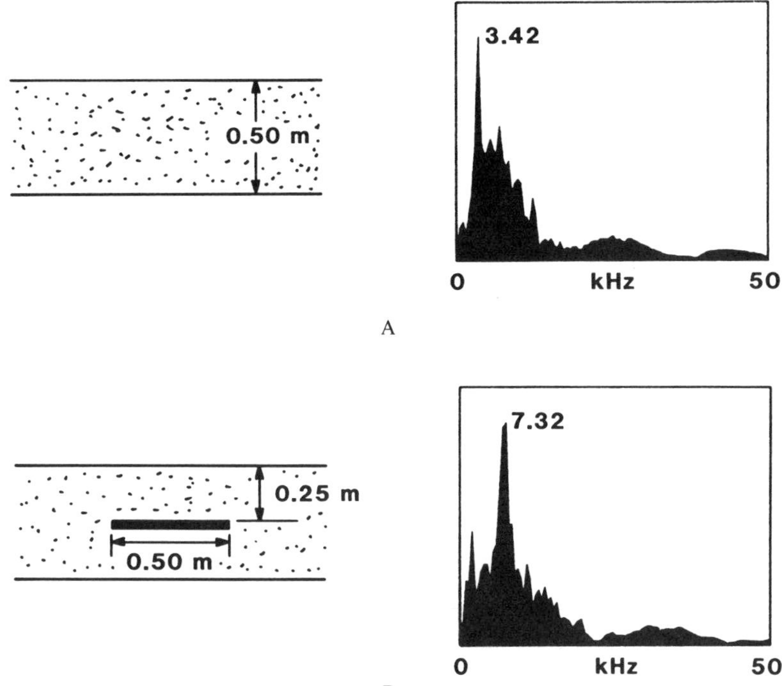

FIGURE 8. Examples of amplitude spectra: (A) test over a solid portion of concrete slab; (B) test over a disk-shaped void embedded in same slab.[21]

shaped voids.[21] The waveforms at each test point are shown on the left and the corresponding spectra are on the right. Spectrum No. 1 with a single large amplitude peak at 3.91 kHz is typical of spectra obtained over the solid portion of the slab. Using Equation 11, the P-wave speed in the slab is 3910 m/s. Spectra Nos. 2, 3, and 6 reveal the presence of voids. Equation 12 can be used to calculate the depth of each void. The peak at 7.81 kHz in spectrum No. 3 corresponds to reflections from a void at 3910/2(7810) = 0.25 m, and the peak at 15.9 kHz in spectrum No. 6 corresponds to reflections from a void at 3910/2(15900) = 0.12 m. These calculated depths agree with the known depths of the artificial voids.

The information contained in a waterfall plot can be further processed to produce a cross-sectional view of the test object along the scan line. The technique that has been developed is called ''spectral peak plotting''.[22] The series of operations which are performed for each spectrum in the scan are illustrated in Figure 10 and outlined as follows:

1. Locate the peak in the spectrum which has the highest amplitude.
2. Normalize the amplitude values by dividing by the highest amplitude.
3. Select a threshold value for the relative amplitude.
4. Locate all peaks in the spectrum whose relative amplitudes exceed the threshold value.
5. For each peak selected in step 4, compute the depth corresponding to the frequency values using the P-wave speed and Equation 12.
6. Using the depth calculated in step 5, plot the depths at the test point along the scan line.

The above procedure has been incorporated into a computer program which uses the individual spectra as input and generates a picture of the cross section along the scan line.[22] Spectral peak plotting permits a rapid assessment of impact-echo results by allowing the user to ''see'' inside the test object.

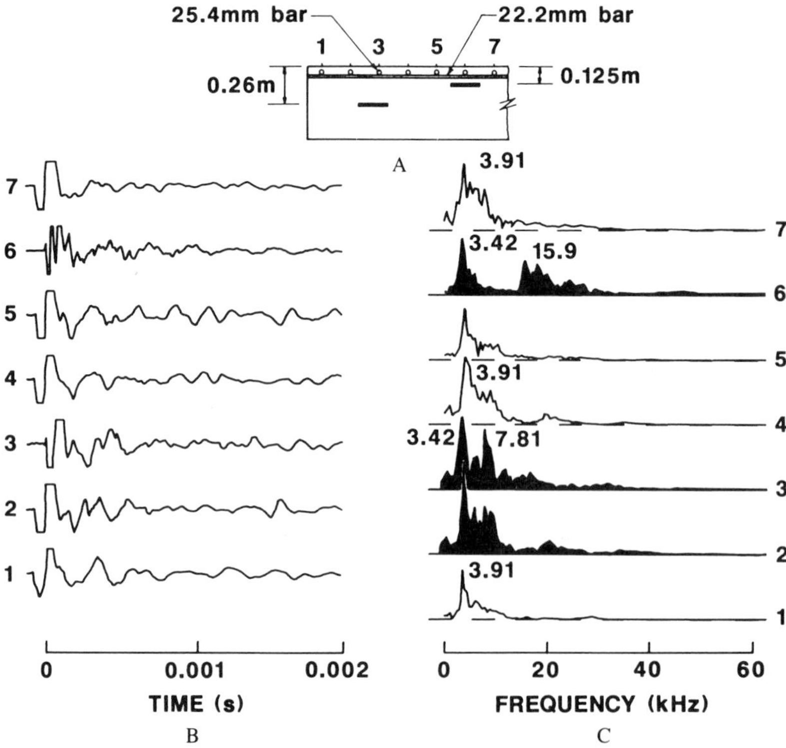

FIGURE 9. Impact-echo results across section of 0.5-m thick slab containing 0.2-m diameter disk-shaped voids: (A) slab cross-section; (B) waveforms; and, (C) amplitude spectra.[21]

The spectra in Figure 9 were processed using the spectral peak plotting computer program, and the resulting computer generated cross-section is shown in Figure 11. For this case, the maximum depth for the plot was chosen as 0.5 m therefore, each tick mark along the depth axis represents 0.5 m. The tick marks along the top surface represent the individual test locations. The depth calculated at each test point is plotted as a "dash" between the midpoints of adjacent test points. The presence of the two voids in the slab can be clearly seen.

Instrumentation

An impact-echo test system is composed of three components: an impact source; a receiving transducer; and a digital processing oscilloscope or waveform analyzer which is used to capture the transient output of the transducer, store the digitized waveforms, and perform signal analysis. A waveform analyzer should have a sampling frequency of at least 500 kHz. Presently, impact-echo test systems are not commercially available, however, a suitable system can be readily assembled from commercially available components.

The selection of the impact source is a critical aspect of a successful impact-echo test system. The force-time history of an impact may be approximated as a half-sine curve, and the duration of the impact is the "contact time." The contact time determines the frequency content of the stress pulse generated by the impact.[19] The shorter the contact time, the higher the range of frequencies contained in the pulse. Thus, the contact time determines the size of the defect which can be detected by impact-echo testing. As the contact time decreases and the pulse contains higher frequency (shorter wavelength) components, smaller defects can be detected. In addition, short duration impacts are needed to accurately locate shallow defects. The stress pulse must have frequency components greater than the frequency corresponding to the flaw depth (Equation 11). As an approximation, the highest frequency

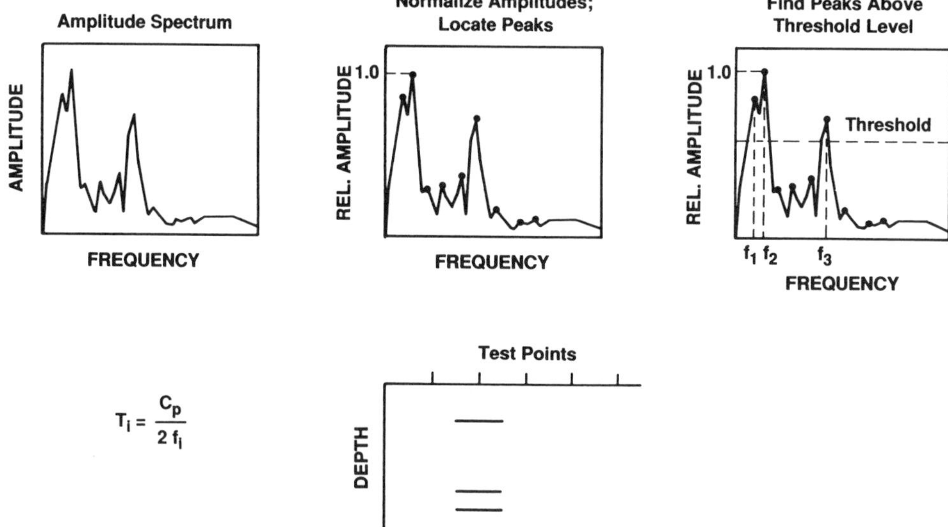

FIGURE 10. Spectral peak plotting procedure for constructing cross-section of test object from amplitude spectra.

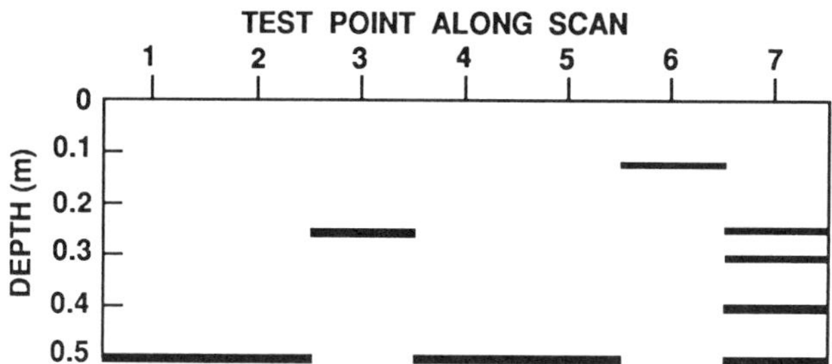

FIGURE 11. Spectral peak plot of the amplitude spectra shown in Figure 9C.

component of significant amplitude in a pulse equals the inverse of the contact time. For example, for a contact time of 100 μs, the maximum frequency having significant amplitude is about 10 kHz. For a P-wave speed of 4000 m/s, a pulse with a contact time greater than 100 μs could not be used to determine the depth of defects shallower than about 0.2 m. Thus, for impact-echo testing of slab-like structures, short duration impacts are preferable. However, experience has shown that as the contact time decreases, the amplitude spectra become more complex.

Many impact sources have been tried. In evaluation of piles, hammers are used.[23-27] Hammers produce energetic impacts with long contact times (greater than 1 ms) which are acceptable for testing long, slender structures but are not suitable for detecting flaws within thin structures such as slabs or walls. Impact sources with shorter duration impacts (20 to 60 μs), such as small steel spheres and spring-loaded spherically-tipped impactors, have been used for detecting flaws within slab and wall structures ranging from 0.15 to 1 m thick.[5,7,18,28,29] Steel spheres are convenient impact sources because the contact time is proportional to the diameter of the sphere.

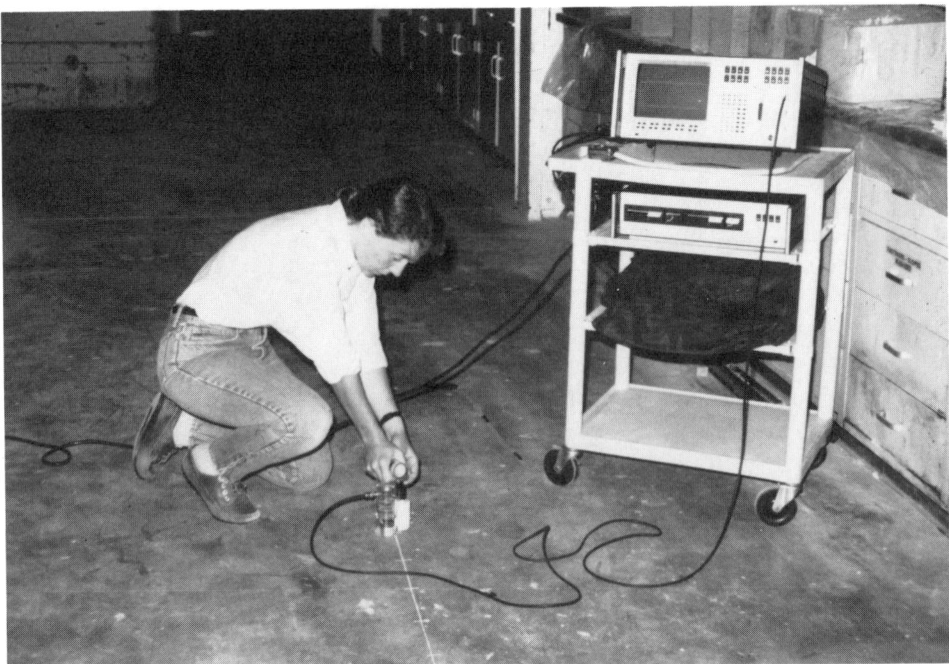

FIGURE 12. Prototype impact-echo test system being used to determine slab thickness.

In evaluation of piles, geophones (velocity transducers) or accelerometers have been used as the receiving transducer.[23-27] For impact-echo testing of slab and wall structures, Sansalone and Carino[5] have used a conical-tipped piezoelectric transducer as the receiver. This transducer responds to surface displacement over a broad frequency range. It was developed at the National Bureau of Standards as a secondary reference standard for calibrating acoustic emission transducers.[30] A thin lead strip is used to provide acoustic coupling between the conical transducer and the test surface. No liquid couplants are required. Figure 12 shows the impact-echo system developed by the authors being used for a scan across a slab-on-grade. The instrument includes the receiver and impact source and is connected to the waveform analyzer shown on the right.

Applications

Since the early 1970s, the impact-echo method has been widely used for evaluation of concrete piles.[23-27] Steinbeck and Vey[24] were among the first to apply the technique to piles. A pulse was introduced at the top surface by impact and the returning echoes were monitored by an accelerometer mounted on the same surface. The time-domain signal record was used to detect partial or complete discontinuities, such as voids, abrupt changes in cross-section, very weak concrete, and soil intrusions, as well as the approximate location where such irregularities existed. In the absence of major imperfections, the location of the bottom of a sound pile could be determined. The success of the method depends upon the pile length and the characteristics of the surrounding soil; echoes from the bottom of a long pile in a stiff, dense soil with an acoustic impedance similar to that of concrete may be too weak to be detected.[23]

Carino and Sansalone initiated experimental and theoretical studies at the National Bureau of Standards in 1983 to develop an impact-echo method for testing structures other than piles. Details of these studies may be found in the cited references.[5,7,18,19,21,28,29,31-37] In the laboratory they have used the method to detect various types of interfaces and defects in concrete slab and wall structures, including cracks and voids in plain and reinforced concrete, the depth of surface-opening cracks, voids in prestressing tendon ducts, honeycombed concrete, the thickness of slabs and overlays and detecting delaminations in slabs with and without asphalt concrete overlays. Most of their experimental studies have been controlled-flaw studies were specimens containing flaws at known locations have been constructed and tested. However, two of their studies were carried out on specimens constructed by other researchers for studies to evaluate nondestructive test methods based on stress wave propagation. These studies involved detecting delaminations in a reinforced concrete slab[28] and locating voids in tendon ducts in a 1-m thick, reinforced wall which was built to simulate a wall for an arctic offshore structure. In both of these studies, the locations of the flaws were not known to the authors prior to testing; the impact-echo method successfully detected the flaws in both studies.

In conjunction with their experimental studies, Sansalone and Carino have performed a series of numerical studies using the finite element method. There are no theoretical solutions available for calculating the transient impact response of a bounded solid containing flaws. Thus the finite element method was used to study how transient waves propagate in bounded solids and how they interact with flaws. The results of these studies form the theoretical basis for the impact-echo method. Variations in the parameters important in impact-echo testing were easily studied using finite element models. Conclusions were drawn about the effect of contact time of the impact, flaw size and depth, and specimen geometry on waveforms and spectra.[5,33,35]

The impact-echo method for structures other than piles has been used in the field.[21,27] As an example, the method was used successfully to detect pockets of unconsolidated concrete below cooling pipes in a 0.15-m thick ice-skating rink slab.[21] Briefly, portions of the slab were known to contain pockets of unconsolidated concrete about 50 to 100 mm in length below the small diameter, closely spaced, cooling pipes. Regions of the skating rink slab were selected for testing. The thickness of the slab was measured at existing core-holes in the slab. A P-wave speed of 4000 m/s was obtained from impact-echo tests performed adjacent to these holes. To be able to detect the small voids, an impact with a contact time of about 20 μs was used.

Figure 13A shows the cross section obtained using spectral peak plotting of the amplitude spectra along one scan line. The results indicate that this scan was carried out over generally sound concrete because, except for test point 10, the cross section shows only the bottom of the slab. In contrast, Figure 13B shows the cross section obtained along a scan line in an area of the slab believed to contain voids. The presence of voids is indicated by lines within the slab and by the multiple lines associated with the bottom of the slab. Note that this portion of the slab is considerably thinner than that shown in Figure 13A. Thus it was shown that, by using an appropriate short duration impact, the impact-echo method could identify the presence of small voids in a thin slab.

IMPULSE-RESPONSE METHOD

Principle

The principle of the impulse-response method[23,38,39] is similar to the impact-echo method. A stress pulse is generated by mechanical impact on the surface of an object. The force-time function of the impact is monitored by using an instrumented hammer or by using a hammer to strike a load cell. A transducer located near the impact monitors the velocity of the surface as it vibrates in response to the arrival of reflected echoes. The waveforms of the force and velocity transducers are recorded and processed on a dynamic signal analyzer. The analysis reveals information about the condition of the structure.

Signal Processing

To calculate the dynamic response of a structure to a given input, the force-time function

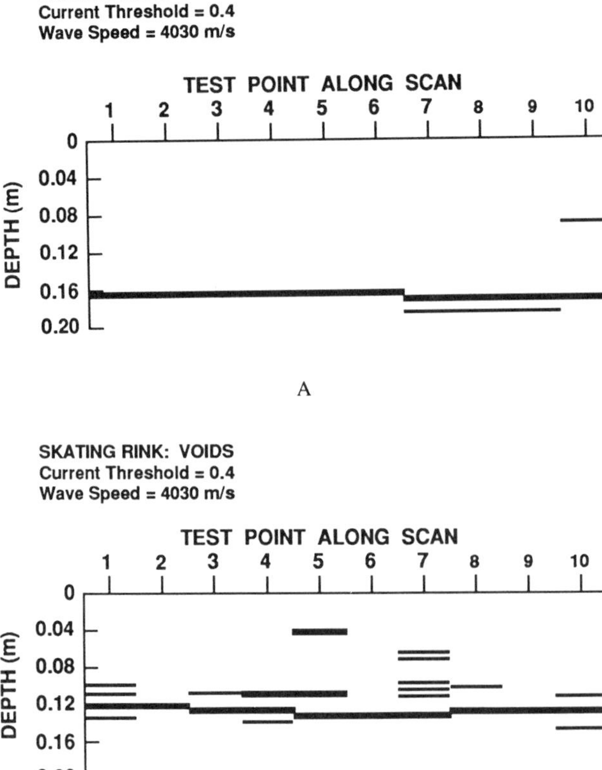

FIGURE 13. Spectral peak plots of impact-echo results for ice-skating rink slab: (A) cross-section in portion of slab containing sound concrete; (B) cross-section in portion of slab containing voids.[22]

of the input is convolved* with the impulse response function of the structure. The impulse response is the structure's response to an input having a force-time function that is a single spike at time zero (impulse). The impulse response function is a characteristic of a structure, and it changes depending on geometry, support conditions, and the existence of flaws or cracks. Alternatively, the impact response can be calculated in the frequency domain by multiplying the Fourier transform of the force input with the Fourier transform of the impulse response function.[41]

In the impulse-response method, the time history of the impact force and the time history of the structure's response are recorded, and the impulse response is calculated. This can be accomplished by deconvolution or, in the frequency domain, by dividing the Fourier transform of the response waveform by the Fourier transform of the impact force-time function. In the frequency domain, the resultant response spectrum indicates structural response as a function of the frequency components of the input. Digital signal processing techniques are used to obtain the impulse response function, often referred to as the transfer function. A procedure for computing the transfer function is outlined in Higgs[38] and involves the following steps:

* The convolution of two functions f(x) and g(x) is: $h(x) = \int f(u)g(x-u)du$. See Reference 40 for a more detailed explanation of the convolution integral.

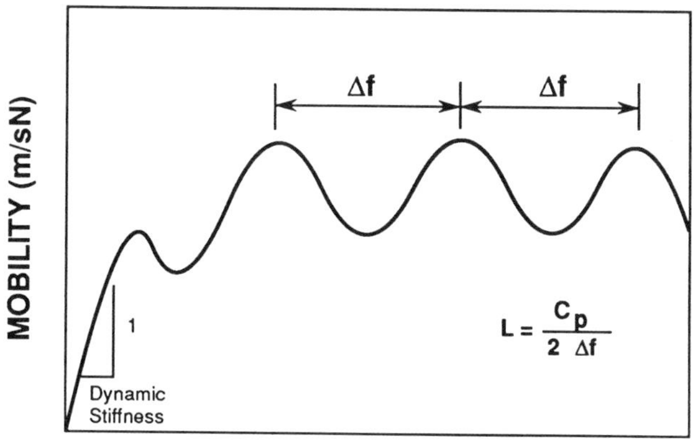

FIGURE 14. Idealized mobility plot for a pile. (Adapted from Reference 29.)

1. Calculate the Fourier transforms of the measured force-time function, f(t), and the measured response, v(t). These will be denoted as F(ω) and V(ω).
2. Using the complex conjugate** of the Fourier transform of the force-time function, F* (ω), compute the cross-power spectrum, V(ω)·F*(ω).
3. Compute the power spectrum of the force-time function, F(ω)·F*(ω).
4. Divide the cross-power spectrum by the power spectrum to obtain the transfer function:

$$H(\omega) = \frac{V(\omega)\cdot F^*(\omega)}{F(\omega)\cdot F^*(\omega)} \tag{13}$$

To improve the results, the test can be repeated, and average power spectra can be used to compute the transfer function. These calculations can be carried out automatically with a dynamic signal analyzer.

Depending on the measured physical quantity of the structural response, the response spectrum obtained by the division in Equation 13 has different meanings. Typically, velocity is measured and the resulting impulse-response spectrum has units of velocity/force which is referred to as "mobility", and the spectrum is often called a "mobility plot". At frequency values corresponding to resonant frequencies of the structure, mobility values are maximum. Figure 14 shows an idealized mobility plot for a pile. The series of peaks correspond to the fundamental and higher modes of vibration; the difference between any two adjacent peaks, Δf, is equal to the fundamental longitudinal frequency.[23,38,39]

The length, L, of the pile can be calculated using the following equation:

$$L = \frac{C_P}{2(\Delta f)} \tag{14}$$

where C_p is the P-wave speed. Note the similarity between this equation and Equation 12.

In addition to length, other information can be obtained from impulse-response tests on pile structures. At low frequencies, the pile and soil vibrate together, and the mobility plot provides information on the dynamic stiffness of the soil-pile structure.[23,39] In this low

** The complex conjugate of a vector F = x + iy is F* = x − iy.

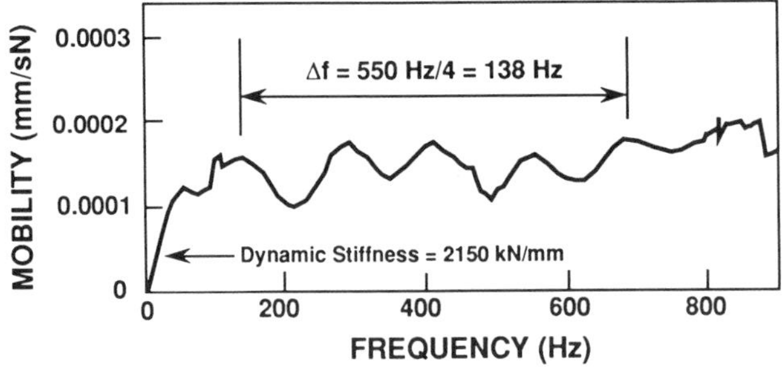

A

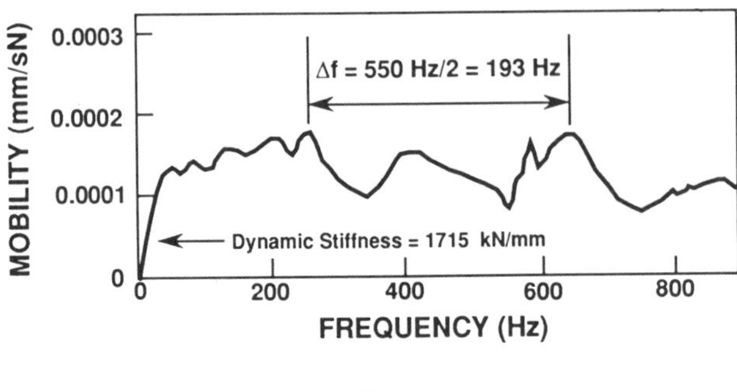

B

FIGURE 15. Mobility plots: (A) sound pile; (B) pile containing a defect across its full width. (Adapted from Reference 17.)

frequency range, the mobility plot is approximately a straight line and the slope of the straight line represents the dynamic flexibility of the pile head. The dynamic stiffness is the inverse of the dynamic flexibility. Thus mobility plots with steeper initial slopes correspond to a lower dynamic stiffness of the pile head. The pile head stiffness is a function of the dynamic stiffness of the pile and the dynamic stiffness of the surrounding soil.

Instrumentation

The components of a typical impulse-response test system are the impact source which can be an instrumented hammer or a hammer striking a load cell which is located on the surface of the test object; a geophone (low frequency velocity transducer); and a two-channel dynamic signal analyzer.

Applications

The impulse-response method evolved from a forced vibration method first used in France by Paquet and Briard in 1968.[42] In forced vibration testing, an electrodynamic vibrator was attached to the top of the pile, and the pile response was measured. To obtain the response spectrum, the response was measured for different applied frequencies. Thus considerably more effort was required compared with the impulse-response technique. Many examples of integrity testing of piles using forced vibration and impulse-response have been reported.[17,23,38,39]

To illustrate how the method works, impulse-response spectra obtained from two test piles having the same dimensions are shown in Figure 15.[17] Figure 15A is the response spectrum obtained from a sound pile. The P-wave speed in this pile was 4140 m/s. The initial straight-line part of the curve was used to determine the dynamic stiffness which was reported to be 2150 kN/mm. The fundamental longitudinal frequency of 138 kHz was calculated by determining the average frequency difference between four successive peaks. Using Equation 13, the length of the pile was calculated to be 15.0 m. The known length of the pile was 15.2 m.

For comparison, Figure 15B shows the response spectrum obtained from the second test pile which contained a defect across the full width of the pile at a depth of 9.8 m. The P-wave speed in this pile was 4200 m/s. The measured dynamic stiffness of 1715 kN/mm was much lower than that for the sound pile. The peaks in the response curve are much less regular than for the sound pile. The fundamental frequency was estimated to be 193 Hz. This was obtained by averaging the difference between three successive frequency peaks as shown on Figure 15B. Using Equation 13, the depth to the reflecting interface was calculated to be about 11 m. Thus the presence of the defect was indicated by a reduction in the dynamic stiffness and an increase in the fundamental resonant frequency.

The results obtained from actual piles can be complicated by a number of factors which make interpretation of response spectra more difficult than the above examples. Davis and Dunn[39] list the following complicating factors:

1. Variations in the diameter of a pile
2. Variations in the quality of concrete within a pile
3. Variations in the stiffness and damping characteristics of the soil through which the pile passes
4. The top part of the pile may be exposed above the ground surface

In addition, the length to diameter ratio (L/D) for a pile relative to the damping characteristics of the soil must be considered. Higgs[38] states that for L/D greater than 20, test results are not likely to be definitive unless the pile passes through a very soft soil deposit onto a rigid stratum.

A simple guide taken from Higgs[38] is shown in Table 2. This table offers insight into how results obtained from impulse-response tests on a series of piles might be interpreted. It also demonstrates how the variables calculated from response spectrum (pile length, dynamic stiffness, and the geometric mean of the mobility, N) are interrelated.

SPECTRAL ANALYSIS OF SURFACE WAVES (SASW) METHOD

Principle

The SASW method is a technique for determining stiffness profiles and layer thicknesses in layered systems, such as pavement systems. The method developed out of past attempts to use R-waves to determine the properties of soil and pavement sites.[43,44]

In the method, a transient stress pulse is generated by impact on the surface of the test site. Two receivers, located as shown in Figure 16, monitor the movement of the surface as the waves produced by the impact propagate past the receivers. Because the amplitude of particle motion in the R-wave is very large at the surface relative to the amplitude of motion in the P- and S-waves, surface movement caused by the R-wave dominates the measured response. The waveform measured by the two receivers contains information that is used to construct the stiffness profile of the underlying materials.

The R-wave produced by impact contains a range of frequencies or components of different wavelengths. This range depends on the contact time of the impact; the shorter the contact time the broader the range of frequencies or wavelengths. The amplitude of particle motion in each component of the R-wave decays exponentially with depth. At a depth below the surface of about 1.5 wavelengths, the amplitude in each component wavelength is 1/10 the amplitude at the surface. Thus longer wavelength components penetrate more deeply, and this is the key to gaining information about the properties of the underlying layers.

TABLE 2
Interpretation of Impulse-Response Tests on Piles

Stiffness	Length	N Value[a]	Conclusion
As expected	As built	As expected	Regular pile
Low	As built	High	Possible reduction in pile section or lower grade concrete in pile
Very low	Short	Low	Fault at depth indicated
High	Near or as built	Low	General oversized pile section
Very high	Short	Low	Bulb at depth indicated
High	Multiple length	Variable/low	Irregular pile section in pile shaft (enlargements)
Low	Multiple length	Variable/high	Irregular pile section in pile shaft (constrictions), or changeable quality of concrete
High	As built	As expected	Regular pile with strong anchorage; low settlement expected
Low	As built	As expected	Regular pile with weak anchorage; high settlement expected

[a] The N-value is the geometric mean of the mobility values in the resonance portion of the response spectrum.

From Higgs, J., *Concrete*, October 31, 1979. With permission.

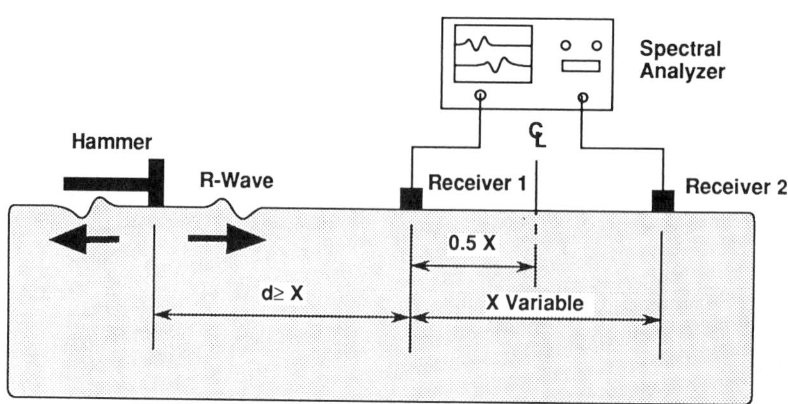

FIGURE 16. Schematic of the SASW test method.

In the SASW method, the impact is chosen so that there are high frequency (short wavelength) components in the R-wave which will propagate entirely within the top layer of the layered system. These components propagate with a speed determined by the S-wave speed (dependent on shear modulus of elasticity and density) and Poisson's ratio of the top layer (see Equations 3 and 5). Lower frequency components penetrate into the underlying layer or layers; therefore their speed of propagation will be affected by the properties of these layers. Thus, a layered system is a dispersive medium for R-waves, which means that different frequency components in the R-wave will propagate with different speeds. The speeds of the individual frequency components are called "phase velocities".

Phase velocities are calculated by determining the time it takes for each component frequency to travel between the two receivers. These travel times are determined from the phase difference of the frequency components when they arrive at the receivers. The meaning of "phase difference" is illustrated in Figure 17, which shows two points, A and B, on a sine curve with a characteristic period equal to t^*. The frequency, f, of the sine curve is the inverse of the period. The phase difference between points A and B is defined as:

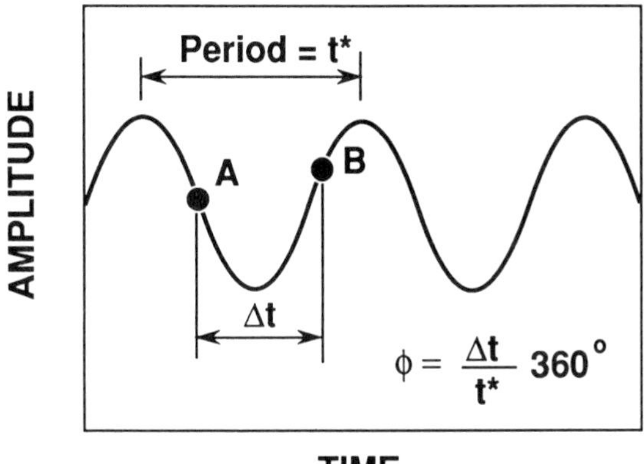

FIGURE 17. Definition of phase difference between two points on a periodic waveform.

$$\phi = \frac{\Delta t}{t^*} 360^\circ = \Delta t \, f \, 360^\circ \tag{15}$$

where Δt is the time difference between points A and B. In the SASW method, the waveforms from the two receivers are processed, as explained in the next section, to obtain the phase differences of the component frequencies. Thus, for each frequency, the travel time between the receivers, Δt_f, can be calculated:

$$\Delta t_f = \frac{\phi}{360^\circ \, f} \tag{16}$$

From the travel time, the velocity of a component frequency, $C_{R(f)}$, can be determined since the distance, X, between the two receivers is known:

$$C_{R(f)} = \frac{X}{\Delta t_f} = x \frac{360}{\phi} f \tag{17}$$

The wavelength, λ_f, corresponding to a component frequency is calculated using the following equation:

$$\lambda f = \frac{C_{R(f)}}{f} = x \frac{360}{\phi} \tag{18}$$

By repeating these calculations for each component frequency, a plot of phase velocity vs. wavelength is obtained. Such a plot is called a ''dispersion curve'' and is used to obtain the stiffness profile.

A process called ''inversion'' is used to obtain the approximate stiffness profile at the test site from the experimental dispersion curve. Nazarin and Stokoe at The University of Texas at Austin developed an interactive computer program which uses an iterative procedure to perform the inversion.[43-48] First, the test site is modeled using layers of varying thickness. The number and thickness of the layers depends on expected stiffness gradients in the underlying materials. For widely varying properties, thinner layers are needed to accurately define the stiffness profile. Each layer is assigned a density, Poisson's ratio, and an S-wave speed. Using this information, the solution for surface wave propagation in a layered system

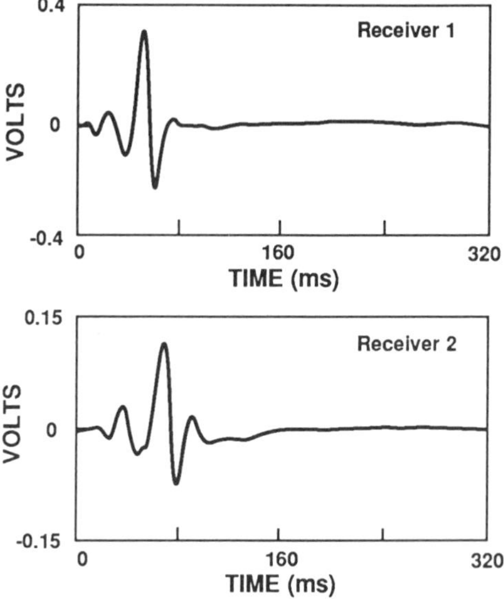

FIGURE 18. Waveforms obtained from an SASW test on a pavement. (Adapted from Reference 47.)

is determined and a theoretical dispersion curve is calculated for the assumed layered system.[47] The theoretical curve is compared with the experimental dispersion curve. If the curves match, the problem is solved and the assumed stiffness profile is correct. If there are significant discrepancies, the assumed layered system is changed or refined and a new theoretical curve is calculated. This process continues until there is good agreement between the theoretical and experimental curves.

Signal Processing

The phase information needed to construct the experimental dispersion curve can be obtained from the experimental waveforms by digital signal processing. A brief explanation of the analysis procedure is given here, and the reader is referred to the work of Nazarin and Stokoe for further details.

Figure 18 is an example of the waveforms recorded by the two receivers. (The receiver closest to the impact point is connected to channel 1 of the signal analyzer.) The first step is to calculate the cross power spectrum, G_{yx}, as follows:

$$G_{yx} = (S_y)(S^*_x) \tag{19}$$

where S_y is the Fourier transform of the waveform recorded on channel 2, and S^*_x is the complex conjugate of the Fourier transform of the waveform recorded on channel 1. To improve the quality of the data, typically, a test is repeated about three to five times, and the average cross power spectrum is used in subsequent steps.[45]

The cross power spectrum can be represented by its amplitude and phase spectra. An example of a phase spectrum is shown in Figure 19. The phase spectrum gives the phase difference between the two receivers for each component frequency. Applying Equation 17, the phase difference for each frequency component is used to compute the phase velocity of that component. Finally, by using Equation 18, the wavelength of each component is calculated, and the experimental phase velocity vs. wavelength curve is established.

The coherence function is also generally calculated as a means of assessing the quality of observed signals.[47] The coherence function spectrum is obtained from power spectra and is defined as:

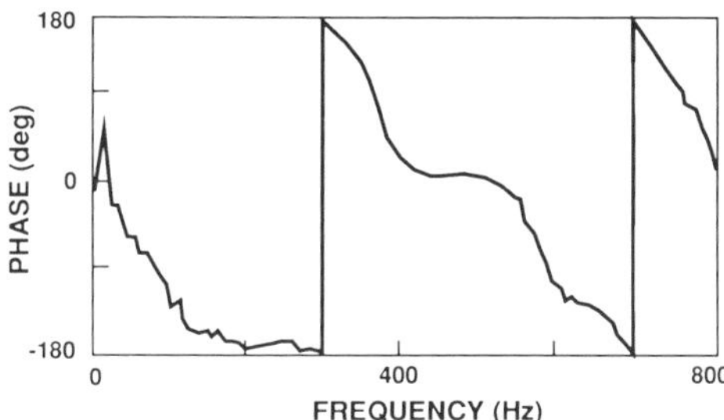

FIGURE 19. Phase spectrum of averaged cross power spectrum determined from five waveforms such as those shown in Figure 18. (Adapted from Reference 47.)

$$\gamma^2 = \frac{G_{yx}\,(G^*_{yx})}{G_{xx}\,(G^*_{yy})} \tag{20}$$

where G^*_{yx} is the complex conjugate of the average cross power spectrum; G_{xx} is the averaged auto power spectrum (S_x) (S^*_x) of the channel 1 waveform; and G_{yy} is the average auto power spectrum (S_y) (S^*_y) of the channel 2 waveform. A coherence value close to one at a given frequency indicates good correlation between that frequency component in the input signal and in the measured waveform. A low coherence can indicate the presence of noise or other problems in the measured signals.[47] In constructing the experimental dispersion curves, frequencies (wavelengths) with low coherence values are not considered.

Instrumentation

There are three components to a SASW test system: the impact source which is usually a hammer; two receivers which are geophones (velocity transducers) or accelerometers; and, a two channel spectral analyzer for recording and processing the waveforms.

The general test configuration was shown in Figure 16. It has been found that because of experimental limitations, reliable phase velocities are calculated only for components with wavelengths greater than one-half the receiver spacing and less than three times the spacing.[46] Thus, to construct a reliable dispersion curve over a wide range of wavelengths, tests are repeated with different receiver spacings. The test arrangement which gives best results is illustrated in Figure 20, and it is known as the common receivers midpoint (CRMP) geometry.[46] In this arrangement, the receivers are always located equidistant from a chosen centerline. The receivers are first located close together, and for subsequent tests the receiver spacing is increased by a factor of two. The source is moved so that the distance between source and nearest receiver is equal to or greater than the distance between the two receivers. As a check on the measured phase information, for each receiver spacing, a second series of tests is carried out by reversing the position of the source. Typically, five receiver spacings are used at each test site. For tests of concrete pavements, the closest spacing is usually about 0.15 m.[47]

The required characteristics of the impact source depend on the stiffnesses of the layers, the distances between the two receivers, and the depth to be investigated.[46] When investigating concrete pavements, the receivers are located relatively close together. In this case, a small hammer is required so that a short duration pulse is produced with sufficient energy at frequencies up to about 10 to 20 kHz. As an approximation, the highest frequency component with significant energy can be taken as the inverse of the pulse duration (contact time). As the depth to be investigated increases, the distance between receivers also increases, and an impact that generates a pulse with greater energy at lower frequencies is required. Thus heavier hammers, such as a sledge hammer, are used.

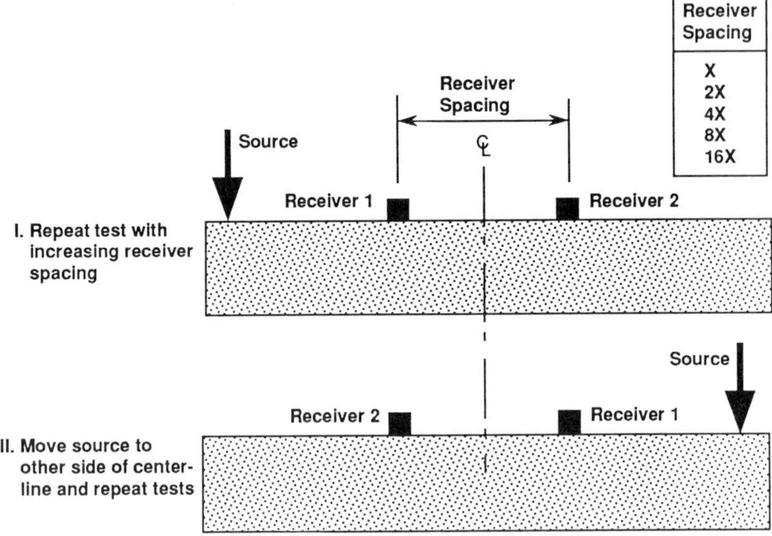

FIGURE 20. Common receivers midpoint geometry used in repeated SASW tests to improve reliability.

The two receiving transducers measure vertical surface velocity or acceleration. The selection of transducer type depends in part on the test site.[46] For tests at soil sites where deep layers are to be investigated and larger receiver spacings are used, geophones are generally used because of their superior low-frequency sensitivity. For tests of concrete pavements where shallower depths are investigated and closer spacings are used, the receivers must provide accurate measurements at higher frequencies. Thus, for pavements a combination of geophones and accelerometers is often used.[46]

Applications

The SASW method has been used to determine the S-wave velocity profiles of soil sites and stiffness profiles of flexible and rigid pavement systems. Many case studies are presented in the References by Nazarin and Stokoe.[46-48] Only the use of the SASW method for testing reinforced concrete pavement systems will be discussed here.

Concrete pavement systems present the most difficult challenge for the SASW method because the large contrast in stiffness between the concrete pavement and the underlying subgrade makes the inversion procedure numerically more difficult.[46] However, Nazarin and Stokoe have developed a technique which has been shown to be successful in a number of studies on concrete pavements. At many of these sites they have been able to verify their results by comparison with pavement and soil profiles obtained from boreholes. Following are some typical results.

At one test site,[46] accelerometers were used so that frequencies up to 12.5 kHz could be measured. Figure 21 shows the experimental dispersion curve obtained from this series of tests. The S-wave velocity profile obtained after inversion is shown in Figure 22A. This profile shows that the S-wave velocity in the concrete is 2950 m/s. This velocity was found to be representative of the good quality concrete used in construction of the pavement. For comparison, Figure 22B shows the profile obtained from borings. Borings showed that the pavement system consisted of approximately 255 mm of reinforced concrete, 100 mm of asphaltic-concrete base, 150 mm of lime-treated subbase, and subgrade. The shape of the S-wave velocity profile obtained from the SASW testing is in excellent agreement with the actual depths of the subsurface layers. If desired, the S-wave velocity profile can be converted to a profile of Young's modulus of elasticity. This requires knowledge of the density and

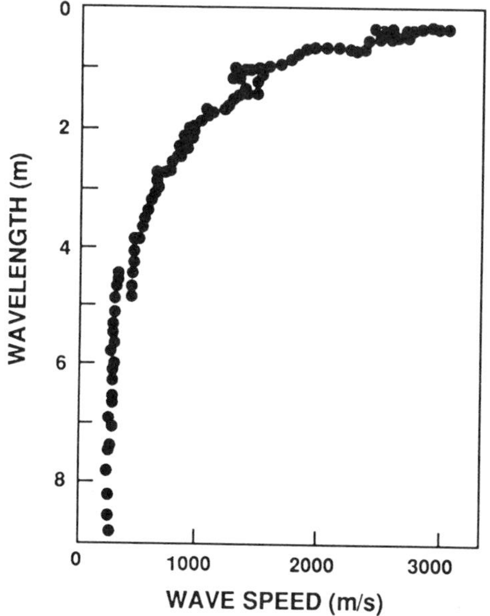

FIGURE 21. Experimental dispersion curve from SASW tests on concrete pavement. (Adapted from Reference 46.)

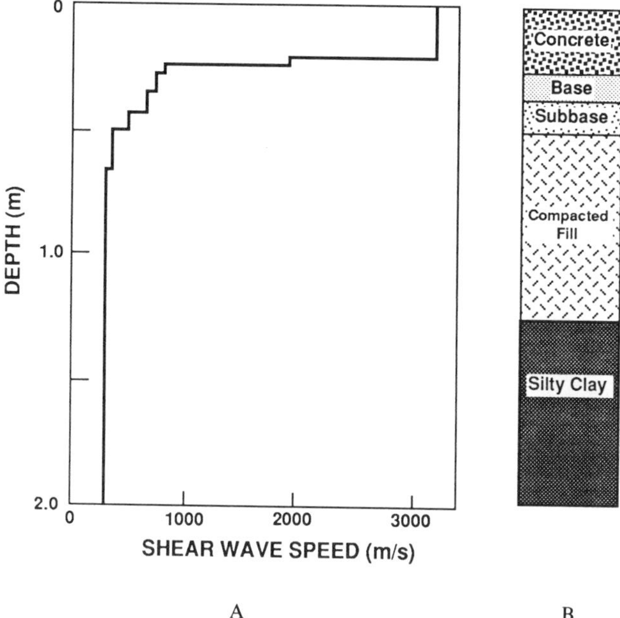

FIGURE 22. (A) S-wave speed profile obtained from inversion of dispersion curve shown in Figure 21; (B) soil profile obtained from borehole. (Adapted from Reference 46.)

Poisson's ratio for each of the layers. However, it must be realized that the computed stiffnesses are representative of behavior only at the low strain levels associated with the SASW test.

SUMMARY

This chapter has presented a review of the test methods which are based on stress wave propagation. The principles, instrumentation, signal processing techniques, and applications of each test method were discussed. The common feature of the various methods is that inferences about internal conditions of concrete structures are made based on the effect that the structure has on the propagation of stress waves. In all cases, stress waves are introduced into the test object and the surface response is monitored. Access to only one surface is required. Depending on the details of the testing configuration and the measured response, different information is gained about the structure.

Conceptually, the pulse-echo method is the simplest technique. The method involves measuring the travel time from the generation of the stress pulse to the arrival of the reflected echo. Knowing the wave speed, the depth of the reflector is calculated. A reflector is any interface where there is a change in the specific acoustic impedance, such as that occurring at an internal defect or the boundaries of the structure. Development of practical pulse-echo test systems for concrete has been hindered by the difficulties inherent in developing an adequate low frequency transducer which can emit a short duration pulse. Some success has been achieved in the development of dual transducer, pitch-catch systems. However maximum penetration is limited, and a practical field system is yet to be developed.

The impact-echo method uses mechanical impact to generate a high energy stress pulse. Surface displacements are measured as the stress pulse undergoes multiple round trips between the test surface and the reflecting interface. This permits frequency analysis of the recorded surface displacement waveforms. The dominant frequency in the amplitude spectrum is used to quickly determine the depth of the reflecting interface from the known wave speed. The amplitude spectra along a scan line are used to construct a cross-section of the structure which displays the location of the reflecting interfaces. The theoretical basis of the impact-echo method has been established using numerical simulations of transient wave propagation in bounded solids. The method's capability to detect a variety of defects has been demonstrated. Although a commercial impact-echo test is not yet available, one can be assembled from readily available components.

The impulse-response method is similar to the impact-echo method, except that the force-time history of the impact is recorded and more complex signal processing is used. The frequency content of the received waveform is correlated with the frequency content of the force-time function, from which inferences can be made about structural conditions. The method is widely used for testing of piles. Information about pile length, presence of defects, and overall pile head stiffness is obtained from the test results. A test system includes a receiver (velocity transducer or accelerometer), a hammer instrumented with a force load cell, and a two-channel dynamic signal analyzer.

The spectral analysis of surface waves method (SASW) uses signal processing techniques similar to those used in the impulse-response method, but information about the structure is extracted from the surface wave created by impact. It is the most complex of the impact methods covered in this review. The SASW method is based on the principle that the various wavelength components in the impact-generated surface wave penetrate to different depths in the test structure. By monitoring the surface motion at two points a known distance apart, information is extracted about the velocity of the various wavelength components which can then be used to infer the elastic properties of the underlying materials. It has been used successfully to construct the stiffness profiles of pavement systems.

The three impact techniques appear similar in terms of the physical test procedure. However, by using different sensors and signal processing methods, the user can obtain different information about the test object. Each method is best for particular kinds of applications. Persons interested in using NDT methods based on stress wave propagation will probably find it advantageous to develop the capability to use all the methods so that the most appropriate one can be used for a particular situation.

Impact techniques indicate a complete departure from the high frequency, pulse-echo technology, which works so well for metals, to test methods where low frequency stress waves are generated by mechanical impact. This appears to be the key for overcoming many of the difficulties involved with testing heterogeneous materials with stress waves. As research continues, the capabilities of these methods will continue to grow and eventually low-cost test systems will become available.

REFERENCES

1. **Slizard, J., Ed.,** *Ultrasonic Testing,* John Wiley & Sons, New York, 1982.
2. **Timoshenko, S. P. and Goodier, J. N.,** *Theory of Elasticity,* 3rd ed., McGraw-Hill, New York, 1970.
3. **Banks, B., Oldfield, G., and Rawding, H.,** *Ultrasonic Flaw Detection in Metals,* Prentice-Hall, New Jersey, 1962.
4. **Viktorov, I.,** *Rayleigh and Lamb Waves,* translated by W. P. Mason, Plenum Press, New York, 1967.
5. **Sansalone, M. and Carino, N. J.,** Impact-Echo: a method for flaw detection in concrete using transient stress waves, NBSIR 86-3452, National Bureau of Standards, September 1986. (NTIS PB #87-104444/AS)
6. **Krautkramer, J. and Krautkramer, H.,** *Ultrasonic Testing Fundamentals,* Springer-Verlag, New York, 1969.
7. **Carino, N. J.,** Laboratory study of flaw detection in concrete by the pulse-echo method, in *In Situ/Nondestructive Testing of Concrete,* Malhotra, V. M., Ed., ACI SP-82, American Concrete Institute, 1984, 557.
8. **Thorton, H. T. and Alexander, A. M.,** Development of nondestructive testing systems for in-situ evaluation of concrete structures, Tech. Rep: REMR-CS-10, Waterways Experiment Station, U.S. Army Corps of Engineers, December 1987.
9. **Bradfield, G. and Gatfield, E.,** Determining the thickness of concrete pavements by mechanical waves: directed beam method, *Mag. Concr. Res.,* 16, (46), 49, 1964.
10. **Howkins, S. et al.,** Measurement of pavement thickness by rapid and nondestructuve methods, NCHRP Rep. 52, 1968.
11. **Mailer, H.,** Pavement thickness measurement using ultrasonic techniques, *Highw. Res. Rec.,* 378, 20, 1972.
12. **Weber, W., Jr., Grey, R., and Cady, P.,** Rapid measurement of concrete pavement thickness and reinforcement location — field evaluation of nondestructive systems, NCHRP Rep. 168, 1976.
13. **Claytor, T. and Ellingson, W.,** Development of ultrasonic methods for the nondestructive inspection of concrete, Ultrasonics Symp. Proc., Halifax, Nova Scotia, July 1983.
14. **Forrest, J.,** In-situ measuring techniques for pile length, TN-1475, Civil Engineering Laboratory, Naval Construction Battalian Center, Port Hueneme, CA, March 1977.
15. **Smith, R.,** The use of surface scanning waves to detect surface opening cracks in concrete, presentation at Int. Conf. on *In-Situ Nondestructive Testing of Concrete,* Ottowa, 1984, 557.
16. **Alexander, A. M. and Thorton, H. T., Jr.,** Ultrasonic pitch-catch and pulse-echo measurements in concrete, in *Nondestructive Testing of Concrete,* Lew, H. S., Ed., ACI SP-112, March, 1989, 21.
17. **Olson, L. and Church, E.,** Survey of nondestructive wave propagation testing methods for the construction industry, Proc. 37th Annual Highway Geology Symp., Helena, MT, August 1986.
18. **Carino, N. J., Sansalone, M., and Hsu, N. N.,** A point source-point receiver technique for flaw detection in concrete, *J. Am. Conc. Inst.,* 83(2), 199, 1986.

19. **Carino, N. J., Sansalone, M., and Hsu, N. N.,** Flaw detection in concrete by frequency spectrum analysis of impact-echo waveforms, *International Advances in Nondestructive Testing,* 12th ed., McGonnagle, W. J., Ed., Gordon & Breach, New York, 1986, 117.
20. **Stearns, S.,** *Digital Signal Analysis,* Hayden Book Company, Rochelle Park, NJ, 1975.
21. **Sansalone, M. and Carino, N. J.,** Laboratory and field study of the impact-echo method for flaw detection in concrete, in *Nondestructive Testing of Concrete,* Lew, H. S., Ed., SP-112, American Concrete Institute, March 1989, 1.
22. **Carino, N. J. and Sansalone, M.,** Impact-echo: a new method for inspecting construction materials, to be published in *Proc. Conf. NDT&E for Manufacturing and Construction,* dos Reis, H. L. M., Hemisphere, 1990, 209.
23. **Stain, R.,** Integrity testing, *Civil Eng. (London),* April 1982, 53.
24. **Steinbach, J. and Vey, E.,** Caisson evaluaiton by stress wave propagation method, *J. Geotech. Eng. Div. ASCE,* 101, No. GT4, April 1975, 361.
25. **Vankuten, H. and Middendorp, P.,** Testing of foundation piles, *Heron,* Vol. 26, No. 4, 1982, 3.
26. **Brendenberg, H., Ed.,** Proc. Int. Semin. on the Application of Stress-Wave Theory on Piles, Stockholm, June 1980.
27. **Olson, L. D. and Wright, C. C.,** Seismic, sonic, and vibration methods for quality assurance and forensic investigation of geotechnical, pavement, and structural systems, *Proc. Conf. on Nondestructive Testing and Evaluation for Manufacturing and Construction,* dos Reis, H. L. M., Ed., Hemisphere, 1990, 263.
28. **Sansalone, M. and Carino, N. J.,** Detecting delaminations in concrete slabs with and without overlays using the impact-echo method, *J. Am. Concr. Inst.,* 86(2), 175, 1989.
29. **Sansalone, M. and Carino, N. J.,** Impact-echo method: detecting honeycombing, the depth of surface-opening cracks, and ungrouted ducts, *Concr. Int.,* 10(4), 38, 1988.
30. **Proctor, T. M., Jr.,** Some details on the NBS conical transducer, *J. Acoustic Emission,* 1(3), 173.
31. **Sansalone, M. and Carino, N. J.,** Nondestructive assessment of concrete structures using the impact-echo method, *Proc. Int. Symp.* on Re-Evaluation of Concrete Structures, June 13—15, 1988, The Technical University of Denmark, Lyngby.
32. **Sansalone, M. and Carino, N. J.,** Transient impact response of thick circular plates, *J. Res. Natl. Bur. Stand.,* 92(6), 355, 1987.
33. **Sansalone, M. and Carino, N. J.,** Transient impact response of plates containing flaws, *J. Res. Natl. Bur. Stand.,* 92(6), 369, 1987.
34. **Sansalone, M., Carino, N. J., and Hsu, N. N.,** A finite element study of transient wave propagation in plates, *J. Res. Natl. Bur. Stand.,* 92(4), 267, 1987.
35. **Sansalone, M., Carino, N. J., and Hsu, N. N.,** A finite element study of the interaction of transient stress waves with planar flaws, *J. Res. Natl. Bur. Stand.,* 92(4), 279, 1987.
36. **Sansalone, M., Carino, N. J., and Hsu, N. N.,** Finite element studies of transient wave propagation, *Review in Progress in Quantitative Nondestructive Evaluation,* Vol 6A, Thompson, D. O. and Chimenti, D. E., Eds., Plenum Press, New York, 1987, 125.
37. **Sansalone, M., Carino, N. J., and Hsu, N. N.,** Flaw detection in concrete and heterogeneous materials using transient stress waves, *J. Acoustic Emission,* 5(3), S24, 1986.
38. **Higgs, J.,** Integrity testing of piles by the shock method, *Concrete,* October 1979, 31.
39. **Davis, A. and Dunn, C.,** From theory to field experience with the nondestructive vibration testing of piles, *Proc. Inst. of Civil Engineers,* Vol. 57, Part 2, December 1974, 571.
40. **Bracewell, R.,** *The Fourier Transform and its Applications,* 2nd ed., McGraw-Hill, New York, 1978.
41. **Papoulis, A.,** *Signal Processing,* McGraw-Hill, 1977.
42. **Paquet, J. and Briard, M.,** Controle non destrucht des pieux en beton, *Ann. de l'Institut Technique du Batiment et des Travaux Publics,* No. 337, March 1976, 49.
43. **Heisey, J. S., Stokoe, K. H., II, and Meyer, A. H.,** Moduli of pavement systems from spectral analysis of surface waves, *Transp. Res. Rec.* 853, 22, 1982.
44. **Nazarin, S., Stokoe, K. H., II, and Hudson, W. R.,** Use of spectral-analysis-of-surface-waves method for determination of moduli and thicknesses of pavement systems, *Transp. Res. Rec.,* 930, 38, 1983.
45. **Nazarin, S. and Stokoe, K. H., II,** Nondestructive testing of pavements using surface waves, *Transp. Res. Rec.,* 930, 1984.
46. **Nazarin, S. and Stokoe, K. H., II,** In-situ determination of elastic moduli of pavement systems by spectral-analysis-of-surface-waves method (practical aspects), Res. Rep. 368-1F, Center for Transportation Research, The University of Texas at Austin, May 1986.
47. **Nazarin, S. and Stokoe, K. H., II,** In-situ determination of elastic moduli of pavement systems by spectral-analysis-of-surface-waves method (theoretical aspects), Res. Rep. 437-2, Center for Transportation Research, The University of Texas at Austin, August 1986.
48. **Nazarin, S. and Stokoe, K. H., II,** Use of surface waves in pavement evaluation, *Transp. Res. Rec.,* 1070, 1986.

Chapter 13

INFRARED THERMOGRAPHIC TECHNIQUES

Gary J. Weil

ABSTRACT

Concrete is one of our world's most useful building materials. Most concrete has a design life of 25 years and, when it begins to fail, it does so slowly at first and builds to a catastrophic ending. Infared thermography, a non-destructive, remote sensing technique, has been proven to be an effective, convienent and economical method of testing concrete. It can detect and show, both large and small, internal voids, delaminations and cracks in concrete structures such as bridges, highway pavements, garage decks, and buildings.

INTRODUCTION

Concrete is one of our world's most useful building materials. It is used in almost every phase of society's infrastructure: from the buildings that house our people to the roads and bridges that allow us to travel from place to place; from the dams that help control nature's forces to the launch pads that will help us explore the heavens. This building material has strength and rigidity along with versatility, but it does have its limits. Most concrete structures have a design life of 20 to 25 years, and when they begin to deteriorate they do so slowly at first and then gradually progress to failure. This failure can be expensive in terms of both dollars and lives, but this scenario can be avoided. Planned restoration can extend the life of our concrete structures almost indefinitely, and testing of concrete structures to establish the existing conditions is the basis of economically viable restoration. In order for any testing technique to be widespread, it must have the following qualities:

1. It must be accurate.
2. It must be nondestructive.
3. It must be able to inspect large areas as well as localized areas.
4. It must be efficient in terms of both labor and equipment.
5. It must be economical.
6. It must not be obtrusive to the surrounding environment.
7. It must not inconvenience the structure's users.

One technique for testing in-place concrete has emerged during the last 10 years that fulfills these requirements to a considerable degree. That technique is called infrared thermographic testing. During its gestation period, it has been used to test concrete on bridge decks, highways, dams, garages, airport taxiways, and around sewer lines. It has shown itself to be both accurate and efficient in locating subsurface voids, delaminations, poor binding, and other anomalies in concrete structures.

HISTORICAL BACKGROUND

Infrared thermography is based on the principle that subsurface anomalies in a material affect heat flow through that material. These changes in heat flow cause localized differences in surface temperature. Thus, by measuring surface temperature under conditions of heat flow, one can determine the location of the subsurface anomalies.

The first documented experimental paper on using infrared thermography to detect concrete subsurface delaminations was published by the Ontario Ministry of Transportation and Communication in 1973*. It illustrated effective methods, although they depended on relatively crude, inefficient techniques.[1] Using these basic techniques, additional research was performed.[2] These later studies were performed on concrete bridge decks, again located in Canada. They were based upon the use of a basic portable infrared scanner, to measure surface temperature, without the use of computer enhancements. They were carried out using a variety of techniques, i.e., both day- and night-time scanning, and both ground and helicopter based data gathering. They proved that infrared thermographic techniques could be used to detect concrete subsurface delaminations on bridge decks.

During the next 10 years, the Ontario Ministry of Transportation and Communications was a strong advocate of research on these infrared thermographic techniques.[2] At the same time, research was progressing in the U.S.,[3,4] and is still being continued.[5-10] An early study was performed for the Wisconsin DOT along a four lane, 16 mile (27 km) portion of Interstate 90/94. In this study a videotape was used to record both visible and infrared images of the highway. These tests used manual methods to transfer the delamination data to scaled plan drawings.

In 1983, a major concrete bridge deck delamination analysis was performed on the Dan Ryan Expressway located in Chicago. This test was significant because it showed that infrared thermography could be used efficiently on congested highways. The fieldwork was performed from a mobile van with traffic control provided by two sign-board vehicles behind the data collection van. Permanent lane closure was not required, thereby reducing costs and inconvenience, particularly for the motorists using the expressway. Field data on the 11 mile (17.6 km), eight-lane expressway was collected in 14 h on 5 separate days, significantly less time than would have been needed for other inspection techniques.

In 1985, delamination inspections were performed on the entrance and exit ramps and bridge decks spanning the Mississippi River at St. Louis, Missouri for the Illinois Department of Transportation. The bridges are a major part of the highway system on Interstate 55-70 and include approximately 40 lane miles (65 km) of bridge deck roadways. These were crucial structures because over 90% of the traffic between Missouri and Illinois, near St. Louis, crossed over one of these bridges. Traffic stoppages had to be kept to a minimum. Emergency asphalt patching repairs were also being performed in close proximity. Five techniques were evaluated: (1) visual inspections; (2) infrared thermography; (3) ground penetrating radar; (4) corings; and (5) chloride measurements. The various tests were performed by separate firms, and the results were analyzed by an independent engineering firm. All data were recorded on a scaled CAD system to allow overlaying of the data and comparisons of the results of the various techniques at individual locations as well as overall statistics. Infrared thermography proved to be the most accurate nondestructive method as well as the most efficient to perform.

Probably the largest individual infrared thermographic inspection to date occurred in 1987 at the Lambert St. Louis International Airport. This involved testing concrete taxiways. The concrete slabs ranged from 14 to 18 in. (26 to 252 mm) in thickness. The rules set up by the airport engineering department dictated that the testing must be performed during low air traffic periods (11:00 p.m. to 5:00 a.m.) and no loading gates could be blocked. The field inspection was completed in 5 working nights. Approximately 2,000,000 square feet (222,200 m^2) of concrete was inspected with production rates approaching 1000 square feet (112 m^2) per minute. In addition to determining individual slab conditions, the use of infrared thermography with computer enhancements allowed the determination of damage caused by traffic patterns and underground erosion caused by soil migration and subsurface moisture problems.

* Personal communication from Tony Masliwec.

THEORETICAL CONSIDERATIONS

An infrared thermographic scanning system measures surface temperatures only, but the surface temperatures of a concrete mass are dependent upon three subsystems: (1) the subsurface configuration; (2) the surface conditions; and (3) the environment.

The subsurface configuration effects are based upon the principle that heat cannot be stopped from flowing from warmer to cooler areas, it can only be slowed down by the insulating effects of the material through which it is flowing. Various types of construction materials have different insulating abilities or thermal conductivities. In addition, differing types of concrete defects have different thermal conductivity values. For example, an air void has a lower thermal conductivity compared with the surrounding concrete.

There are three ways of transferring thermal energy from a warmer to a cooler region: (1) conduction; (2) convection; and (3) radiation. Sound concrete should have the least resistance to conduction of heat, and the convection effects should be negligible. However, the various types of anomalies associated with poor concrete, namely, voids and low density, decrease the thermal conductivity of the concrete by reducing the energy conduction properties, without substantially increasing the convection effects because we are talking about dead air spaces which do not allow the formation of convection currents.

In order to have heat energy flow, there must be a heat source. Since concrete testing can involve large areas, the heat source should be both low cost and able to give the concrete surface an even distribution of heat. The sun fulfills both these requirements. Allowing the sun to warm the surface of the concrete areas under test will normally supply the required energy. During nighttime hours, the process may be reversed with the warm ground acting as the heat source.

For concrete areas not accessible to sunlight, an alternative is to use the heat storage ability of the earth to draw heat from the concrete under test. The important point is that in order to use infrared thermography, heat must be flowing through the concrete. It doesn't matter in which direction it flows.

The second important factor to consider when using infrared thermography to measure temperature differentials due to anomalies is the surface condition of the test area.

It was mentioned that there are three ways to transfer energy. Radiation is the process that has the most profound effect upon the ability of the surface to transfer energy. The ability of a material to radiate energy is measured by the emissivity of the material. This is defined as the ability of the material to radiate energy compared with a perfect blackbody* radiator. This is strictly a surface property. The emissivity value is higher for rough surfaces and lower for smooth surfaces. For example, rough concrete may have an emissivity of 0.95 while shiny metal may have an emissivity of only 0.05. In practical terms, this means that when using thermographic methods to scan large areas of concrete, the engineer must be aware of differing surface textures caused by such things as broom textured spots, rubber tire tracks, oil spots, or loose sand and dirt on the surface.

The final factor that affects the temperature measurement of a concrete surface is the environmental system which surrounds that surface. Some of the various parameters that affect the surface temperature measurements are

1. Solar Radiation: Testing should be performed during times of the day or night when the solar radiation or lack of solar radiation would produce the most rapid heating or cooling of the concrete surface.
2. Cloud Cover: Clouds will reflect infrared radiation, thereby slowing the heat transfer

* A blackbody is a hypothetical radiation source which radiates the maximum energy theoretically possible at a given temperature. The emissivity of a blackbody equals 1.0.

process to the sky. Therefore, night-time testing should be performed during times of little or no cloud cover in order to allow the most efficient transfer of energy out of the concrete.

3. Ambient Temperature: This should have a negligible effect on the accuracy of the testing since the important consideration is the rapid heating or cooling of the concrete surface. This parameter will affect the length of time (i.e., the window) during which high contrast temperature measurements can be made. It is also important to consider if water is present. Testing while ground temperatures are less than 32°F should be avoided since ice can form, thereby filling subsurface voids.
4. Wind Speed: High gusts of wind have a definite cooling effect and reduce surface temperatures. Measurements should be taken at wind speeds of less than 15 mph (25 kw/h).
5. Surface Moisture: Moisture tends to disperse the surface heat and mask the temperature differences and thus the subsurface anomalies. Tests should not be performed while the concrete surface is covered with standing water or snow.

Once the proper conditions are established for scanning, a relatively large area should be selected for calibration purposes. This should encompass both good and bad concrete areas (i.e., areas with voids, delaminations, cracks, or powdery concrete). Each type of anomaly will display a unique temperature pattern depending on the conditions present. If, for example, the scanning is performed at night, most anomalies will be between 0.1° and 5°C cooler than the surrounding solid concrete depending on configuration (Plate 1A and B).* A daylight survey will show reversed results, i.e., damaged areas will be warmer than the surrounding sound concrete.

TESTING EQUIPMENT

In principle, in order to test concrete for subsurface anomalies, all that is really needed is a sensitive contact thermometer. However, even for a small test area, thousands of readings would have to be made simultaneously in order to outline the anomaly precisely. Since this is not practical, high resolution infrared thermographic scanners are used (Figures 1 and 2) to inspect large areas of concrete efficiently and quickly. This type of equipment allows large areas to be scanned, and the resulting data can be displayed as pictures with areas of differing temperatures designated by differing gray tones in a black and white image or by various colors on a color image. A wide variety of auxiliary equipment can be used to facilitate data recording and interpretation.

A complete thermographic scanning and analysis system can be divided into four main subsystems. The first is the infrared scanner head and detector which normally can be used with interchangeable lenses. It is similar in appearance to a portable video camera. The scanner's optical system, however, is transparent only to shortwave infrared radiation with wavelengths in the range of 3 to 5.6 μm, or to medium wave infrared radiation with wavelengths in the range of 8 to 12 μm. Normally the infrared scanner's highly sensitive detector is cooled by liquid nitrogen to a temperature of −196°C, and it can detect temperature variations as small as 0.1°C. Alternate methods of cooling the infrared detectors are available which use either compressed gases or electric cooling. These last two cooling methods may not give the same resolution, since they cannot bring the detector temperatures as low as liquid nitrogen. In addition, compressed gas cylinders may present safety problems during storage or handling.

The second major component of the infrared scanning system is a real-time micropro-

* Plate 1 appears after page 318.

FIGURE 1. High resolution infrared thermographic scanner with body harness as used to inspect short bridge decks.

cessor coupled to a black and white display monitor. With this component, cooler items being scanned are normally represented by darker gray tones, while warmer areas are represented by lighter gray tones. In order to make the images easier to understand for those unfamiliar with interpreting gray-tone images, a color monitor may also be installed. The microprocessor will quantize the continuous gray-tone energy images into 2, 3, or more "buckets" of energy levels and assign them contrasting visual colors representing relative temperatures. Thus, the color monitor displays the different temperature levels as different colors.

The third major component of the infrared scanning system is the data acquisition and analysis equipment. It is composed of an analog to digital converter, a computer with a high resolution color monitor, and data storage and analysis software. The computer allows the transfer of instrumentation videotape or live images of infrared scenes to single frame computer images. The images can then be stored individually and later retrieved for enhancement and individual analysis. The use of the computer allows the engineer in charge of testing to set specific analysis standards based upon destructive sample tests, such as corings, and apply them uniformly to the entire pavement. Standard, off-the-shelf image analysis programs may be used or custom written software may be developed.

A

FIGURE 2. (A) High resolution infrared thermographic scanner mounted on a custom-designed mobile van; (B) electronic and (C) computer enhancement system in van.

The fourth major component consists of various types of image recording and retrieving devices. These are used to record both visual and thermal images. They may be composed of instrumentation video tape recorders, still frame film cameras with both instant and 35 mm or larger formats, or computer printed images.

All of the above equipment may be carried into the field or parts of it may be left in the laboratory for additional use. If all of the equipment is transported to the field to allow simultaneous data acquisition and analysis, it is prudent to use an automotive van to set up and transport the equipment. This van should include power supplies for the equipment, either batteries and inverter, or a small gasoline driven generator. The van should also include a method to elevate the scanner head and accompanying video camera to allow scanning of the widest area possible depending upon the system optics used.

Several manufacturers produce infrared thermographic equipment. Each manufacturer's equipment has its own strengths and weaknesses. These variations are in a constant state of change as each manufacturer alters and improves his equipment. Therefore, equipment comparisons should be made before purchase.

FIGURE 2B

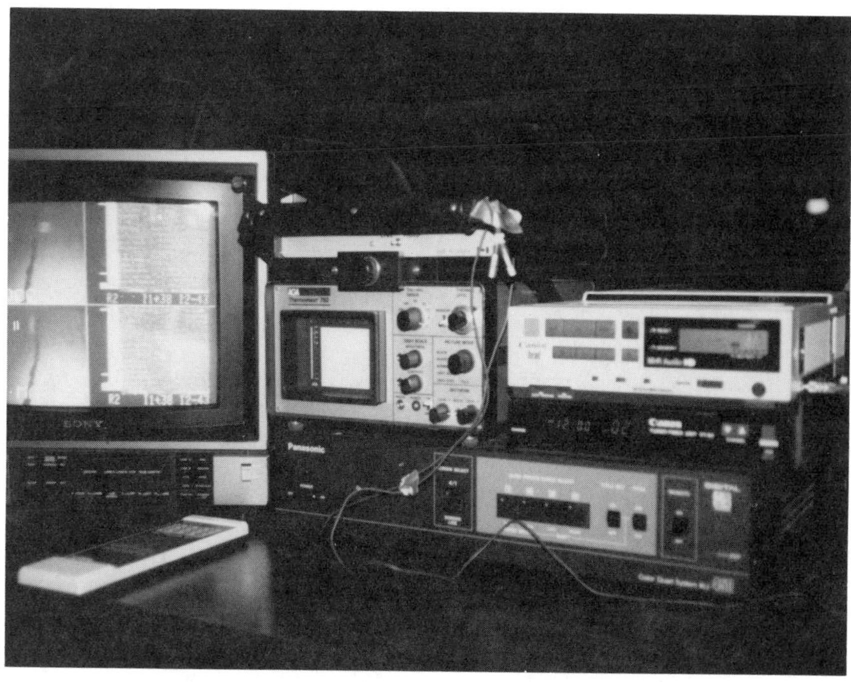

FIGURE 2C

TESTING PROCEDURES

In order to perform an infrared thermographic inspection, a movement of heat must be established in the structure. Our first example deals with the simplest and most widespread situation. Assume that we desire to test an open concrete bridge deck surface. The day preceding the inspection should be dry with plenty of sunshine. The inspection may begin either 2 to 3 h after sunrise or 2 to 3 h after sunset, both are times of rapid heat transfer. The deck should be cleaned of all debris. Traffic control should be established to prevent accidents and to prevent traffic vehicles from stopping or standing on the pavement to be tested. The infrared scanner will be assumed to be mounted in a mobile van along with other peripheral equipment such as recorders for data storage and a computer for assistance in data analysis. The scanner head and either a regular film-type camera or a standard video camera should be aligned to view the same sections to be tested.

The next step is to locate a section of concrete deck and, by coring, establish that it is sound concrete. Scan the reference area and set the equipment controls so that an adequate temperature image is viewed and recorded.

Next, locate a section of concrete deck known to be defective by containing a void, delamination, or powdery material. Scan this reference area and again make sure that the equipment settings allow the viewing of both the sound and defective reference areas in the same image with the widest contrast possible. These settings will normally produce a sensitivity scale such that full scale represents no more than 5°.

If a black and white monitor is used, better contrast images will normally be produced when the following convention is used; black is defective concrete and white is sound material. If a color monitor or computer enhanced screen is used, three colors are normally used to designate definite sound areas, definite defective areas, and indeterminate areas.

As has been mentioned, when tests are performed during daylight hours, the defective concrete areas will appear warmer, while during tests performed after dark, defective areas will appear cooler.

Once the controls are set and traffic control is in place, the van may move forward as rapidly as images can be collected, normally 1 to 10 miles (1.6 to 16 km)/h. If it is desired to mark the pavement, white or metallic paint may be used to outline the defective deck areas. At other times, videotape may be used to document the defective areas, or a scale drawing may be drawn with reference to bridge deck reference points. Production rates of up to 1,200,000 ft^2/day (133,333/m^2/day) have been attained.

During long testing sessions, reinspection of the reference areas should be performed approximately every 2 h, with more calibration retests scheduled during the early and later periods of the session when the teating ''window'' may be opening or closing.

For inside areas where the sun cannot be used for its heating effect, it may be possible to use the same techniques except for using the ground as a heat sink. The equipment should be set up in a similar fashion to that designed above, except that the infrared scanner's sensitivity will have to be increased. This may be accomplished by setting the full scale so it represents 2°C and/or using computer enhancement techniques to bring out detail and to improve image contrast.

Once the data are collected and analyzed, the results should be plotted on scale drawings of the area inspected. Defective areas should be clearly marked so that trends can be observed. Computer enhancements can have varying effects on the accuracy and efficiency of the inspection systems. Image contrast enhancements can improve the accuracy of the analysis by bringing out fine details, while automatic plotting and area analysis software can improve the efficiency in preparing the finished report.

A word of caution. When inspecting areas that contain shadow-causing areas, such as bridges with superstructures or pavements near buildings, it is preferable to perform the

FIGURE 3. Engineer, driver, and mobile infrared scanning equipment used to transverse Dr. Martin Luther King Bridge at St. Louis, MO.

inspection after sundown (Plate 2).* Since the shadows will constantly move, their resulting temperature variations will average out to a uniform level.

CASE HISTORIES

In order to illustrate some diverse applications for infrared thermographic testing, three case histories are reviewed:

1. Bridge deck concrete
2. Airport taxiway concrete
3. Garage deck concrete

Each of these inspections highlight a different important feature of this nondestructive, remote sensing evaluation technique.

The first case history reviews the inspection of a concrete deck on the Dr. Martin Luther King Bridge spanning the Mississippi River at St. Louis, Missouri. In St. Louis, the weather conditions can quickly change, going from clear sky to fog and rain in a matter of minutes. Therefore, it was decided to perform the field inspection within one 8-h period or less. Because the deck and its associated ramps were four lanes wide, almost a mile long, and included various sections with and without an overhead superstructure, it was decided to use a mobile lift platform capable of constant movement at up to $^3/_4$ mile (1.2 km)/h. The platform lifted the infrared equipment, and operator/engineer, and a vehicle driver to heights sufficient to view all four lanes in a single pass (Figure 3). The data, both infrared and visual, were recorded on both instrumentation videotape and 35 mm film formats. Due to

* Plate 2 appears after page 318.

FIGURE 4. Quality control check being performed by personnel on Dr. Martin Luther King Bridge at St. Louis, MO.

the traffic restrictions the bridge could not be fully closed during the inspection. Therefore, the survey was performed on a weekend night between 10:00 p.m. and 2:00 a.m.

Before, during, and after the inspection, reference areas were scanned to determine equipment settings which would give the greatest contrast on the infrared imager. The main sensitivity of the equipment was set at 5° for full scale (Plate 1B).

After the inspection, a simple technique was used to confirm the infrared data and interpretations to the supervising engineering company. Three separate areas were chosen from the void, delamination, and anomaly drawings developed from the infrared data. Then an 8 in. (200-mm) nail was driven into the pavement and its penetration was determined under a standard blow delivered by the supervising engineer. The locations where the nail penetrated deeply correlated exactly with the test inspection party's determinations for the locations of anomalies (Figure 4).

A second case study involves the inspection of over 3,125 slabs of reinforced concrete on the taxiway of one of the busiest airports in America, Lambert St. Louis International Airport. This inspection was performed during August and the field inspection took a total of 7 nights, during 2 of which no surveying could be done because of rain.

Because air traffic flow could not be interrupted on the taxiways, the inspection was performed from 11:00 p.m. to 5:00 a.m. when traffic was slowest. In order to move the infrared equipment about rapidly and to move out of traffic flow quickly, a van was used to carry all the infrared testing equipment along with associated surveying tools, such as power supplies, drawing equipment, and various recording devices (Figure 2). The van was custom designed to allow the scanner head and visual cameras to be raised to a 14 ft (4.8 m) height during scanning runs to allow the surveying of a 25 ft (8.0 m) wide by 25 ft (8.0 m) long slab in a single view. Production rates, which included the scanning operation, storage of images on computer discs and videotape, occasional 35-mm photographs, and all analysis allowed the inspection of up to 500,000 ft^2 (55,556 m^2)/night.

Prior to beginning the inspection, reference and calibration areas were determined for sound concrete and for concrete with subsurface voids and delaminations. These areas were rescanned at regular intervals during the inspection to ensure that equipment settings allowed for accurate data collection. This information was fed continuously into a computer and a color monitor was used to assist in location of anomalies. To speed up data interpretation, the thermal data presented on the monitor were divided into three categories represented by three colors: green for solid concrete; yellow for concrete areas with minor temperature deviations most likely caused by minor surface deterioration; and red for concrete areas with serious subsurface cracks/voids. The computer was also used to determine the area on each slab that appeared in the above colors. These data were used to designate each individual slabs for no corrective action, spot repairs, or major replacement.

The final case history involves the inspection in 1986 of garage concrete and adjacent roadway concrete at the same facility, Lambert St. Louis International Airport. The same techniques as described above were used, but particular attention was paid to the expansion joint areas between concrete slabs. Plate 3 shows one of the computer enhanced thermograms and a visual picture of an expansion joint in good condition. Plate 4 shows a nearby joint in a deteriorated subsurface condition.* The surface visual photograph shows no visual surface deterioration.

The deteriorated areas were confirmed and rehabilitated the following year.

ADVANTAGES AND LIMITATIONS

Infrared thermographic testing techniques for determining concrete subsurface voids, delaminations, and other anomalies have advantages over destructive tests such as coring and other NDT techniques such as radioactive/nuclear, electrical/magnetic, acoustic, and ground probing radar.

The obvious advantages of infrared thermographic analysis over the destructive testing methods is that major concrete areas need not be destroyed during the testing. Only small calibration corings are used. This results in major savings in time, labor, equipment, traffic control, and scheduling problems. In addition, when esthetics are important, no disfiguring occurs on the concrete to be tested. Rapid set up and take down are also advantages when vandalism is possible. Finally, no concrete dust and debris is generated which could cause environmental problems.

There are other advantages of infrared thermographic methods over other NDT methods. Infrared thermographic equipment is safe as it emits no radiation. It only records thermal radiation which is naturally emitted from the concrete, as well as from all other objects. It is similar in function to an ordinary thermometer, only much more efficient.

The final and main advantage of infrared thermography is that it is an area testing technique, while the other NDT methods are either point or line testing methods. Thus, infrared thermography is capable of forming a two dimensional image of the test surface showing the extent of subsurface anomalies.

The other methods including radioactive/nuclear, electrical/magnetic, acoustic, and ground probing radar are all point tests. They depend upon a signal propagating downward through the concrete at a discrete point. This gives an indication of the concrete condition at that point. If an area is to be tested, then multiple readings must be taken.

Ground probing radar has the advantage over the other point testing techniques in that the sensor may be mounted on a vehicle and moved in a straight line over the test area. This improves efficiency somewhat, but if an area is wide, many line passes would have to be made.

There is one major disadvantage to infrared thermographic testing. At this stage of

* Plates 3 and 4 appear after pages 318.

development, the depth or thickness of a void cannot be determined, although its outer dimensions are evident. It cannot be determined if a subsurface void is near the surface or farther down at the level of the reinforcing bars. Techniques such as ground penetrating radar or stress wave propagation methods can determine the depth of the void, but again these methods cannot determine the other dimensions in a single measurement.

In most testing instances, the thickness of the anomaly is not nearly as important as its other dimensions. But in those instances where information on a specific anomaly thickness or depths are needed, it is recommended that infrared thermography be used to survey the large areas for problems. Once specific problem locations are established, ground penetrating radar can be used to spot check the anomaly for its depth and thickness. This combined technique would give the best combination of accuracy, efficiency, economy, and safety.

CONCLUSIONS

1. Infrared thermographic testing techniques are based upon the principle that various subsurface defects change the rate at which heat flows through a structure.
2. Infrared thermographic testing may be performed during both day- and night-time hours depending upon environmental conditions.
3. After proper calibration using test borings, infrared thermographic techniques can distinguish various types of anomalies.
4. Infrared thermographic scanning techniques are more efficient than other destructive and nondestructive manual and electronic methods when testing large concrete areas.
5. Computer analysis of thermal images greatly improves the accuracy and speed of test interpretation.
6. At this stage of development, infrared thermography techniques cannot determine the depth or thickness of a void, but it can determine location and horizontal dimensions.

REFERENCES

1. **Moore, W. M., Swift, G., and Milberger, L. J.** An instrument for detecting delamination of concrete bridge decks, *Highw. Res. Rec.* 451, 44, 1973.
2. **Holt, F. B. and Manning, D. G.,** Infrared thermography for the Detection of Delaminations in Concrete Bridge Decks, Proceedings of the Fourth Biennial Infrared Information Exchange, pp. A61-A71; 1978.
3. **Weil, G. J.,** Infrared Thermal Sensing of Sewer Voids, *Proc. Thermosense VI,* 446, 116, 1983.
4. **Weil, G. J.,** Computer-aided infrared analysis of bridge deck delaminations, Proc. 5th Infrared Information Exchange, A85-A93, 1985.
5. **Weil, G. J.,** Infrared Thermal Sensing of Sewer Voids — 4-Year Update, *Proc. Thermosense X,* 934, 155, 1988.
6. ASTM D4788-8, Test method for detecting deluminations in bridge decks using infrared thermography.
7. **Weil, G. J.,** Toward an integrated non-destructive pavement testing management information system using infrared thermography, *Proc. U.S. Trans. Res. Board,* June 22, 1989, Washington, D.C.
8. **Weil, G. J.,** Remote sensing of land based voids using computer enhanced infrared thermography, *Proc. Int. Congr. Optical Sci. and Eng.,* April 14, 1989, Paris.
9. **Weil, G. J.,** Detecting the defects, *Civil Eng. Mag.,* 59(9), 74, 1989.
10. **Weil, G. J.,** Non-destructive remote sensing of subsurface utility distribution pipe problems using infrared thermography, *Proc. 2nd Int. Conf. on Pipeline Constr. Cong.,* Centrum Hamburg, October 26, 1989.

Chapter 14

ACOUSTIC EMISSION METHODS

Sidney Mindess

ABSTRACT

Acoustic emission refers to the sounds, both audible and sub-audible, that are generated when a material undergoes irreversible changes, such as those due to cracking. Acoustic emissions from concrete have been studied for the past 30 years, and can provide useful information on concrete properties. This review deals with the parameters affecting acoustic emissions from concrete, including discussions of the Kaiser effect, specimen geometry, and concrete properties. There follows an extensive discussion of the use of AE to monitor cracking in concrete, whether due to externally applied loads, drying shrinkage, or thermal stresses. AE studies on reinforced concrete are also described. It is concluded that while AE is a very useful laboratory technique for the study of concrete properties, its use in the field remains problematic.

INTRODUCTION

It is common experience that the failure of a concrete specimen under load is accompanied by a considerable amount of audible noise. In certain circumstances, some audible noise is generated even before ultimate failure occurs. With very simple equipment — a microphone placed against the specimen, an amplifier, and an oscillograph — subaudible sounds can be detected at stress levels of perhaps 50% of the ultimate strength; with the sophisticated equipment available today, sound can be detected at much lower loads, in some cases below 10% of the ultimate strength. These sounds, both audible and subaudible, are referred to as *acoustic emission.*

In general, acoustic emissions are defined as "the class of phenomena whereby transient elastic waves are generated by the rapid release of energy from localized sources within a material".[1] These waves propagate through the material, and their arrival at the surfaces can be detected by piezoelectric transducers. Acoustic emissions, which occur in most materials, are caused by irreversible changes, such as dislocation movement, twinning, phase transformations, crack initiation, and propagation, debonding between continuous and dispersed phases in composite materials, and so on. In concrete, since the first three of these mechanisms do not occur, acoustic emission is due primarily to:

1. Cracking processes
2. Slip between concrete and steel reinforcement
3. Fracture or debonding of fibers in fiber reinforced concrete

HISTORICAL BACKGROUND

The initial published studies of acoustic emission phenomena, in the early 1940s, dealt with the problem of predicting rockbursts in mines; this technique is still very widely used in the field of rock mechanics, in both field and laboratory studies. The first significant investigation of acoustic emission from metals (steel, zinc, aluminum, copper, and lead) was carried out by Kaiser.[2] Among many other things, he observed what has since become known as the Kaiser effect: "the absence of detectable acoustic emission at a fixed sensitivity level, until preveiously applied stress levels are exceeded".[1] While this effect is not present in all materials, it is a very important observation, and will be referred to again later in this review.

The first study of acoustic emission from concrete specimens under stress appears to have been carried out by Rüsch,[3] who noted that during cycles of loading and unloading below about 70 to 85% of the ultimate failure load, acoustic emissions were produced only when the previous maximum load was reached (the Kaiser effect). At about the same time, but independently, L'Hermite[4,5] also measured acoustic emission from concrete, finding that a sharp increase in acoustic emission coincided with the point at which Poisson's ratio also began to increase (i.e., at the onset of significant matrix cracking in the concrete). In 1965, however, Robinson[6] used more sensitive equipment to show that acoustic emission occurred at much lower load levels than had been reported earlier, and hence, could be used to monitor earlier microcracking (such as that involved in the growth of bond cracks in the interfacial region between cement and aggregate). Wells[7], in 1970, built a still more sensitive apparatus, with which he could monitor acoustic emissions in the frequency range from about 2 to 20 kHz. However, he was unable to obtain truly reproducible records for the various specimen types that he tested, probably due to the difficulties in eliminating external noise from the testing machine. Also in 1970, Green[8] reported a much more extensive series of tests, recording acoustic emission frequencies up to 100 kHz. Green was the first to show clearly that acoustic emissions from concrete are related to failure processes within the material; using source location techniques, he was also able to determine the locations of defects. It was this work that indicated that acoustic emissions could be used as an early warning of failure. Green also noted the Kaiser effect, which suggested to him that acoustic emission techniques could be used to indicate the previous maximum stress to which the concrete had been subjected. As we will see below, however, a true Kaiser effect appears not to exist for concrete.

Nevertheless, even after this pioneering work, progress in applying acoustic emission techniques remains slow. An extensive review by Diederichs et al.[10] covers the literature on acoustic emissions from concrete up to 1983. However, as late as 1976, Malhotra[9] noted that there was little published data in this area, and that "acoustic emission methods are in their infancy". Even in January, 1988, a thorough computer-aided search of the literature found only some 90 papers dealing with acoustic emissions from concrete over about the previous 10 years; while this is almost certainly not a complete list, it does indicate that there is much work to be carried out before acoustic emission monitoring becomes a common technique for testing concrete. Indeed, there are still no standard test methods which have even been suggested for this purpose.

THEORETICAL CONSIDERATIONS

When an acoustic emission event occurs at a source with the material, due to inelastic deformation or to cracking, the stress waves travel directly from the source to the receiver as body waves. Surface waves may then arise from mode conversion. When the stress waves arrive at the receiver, the transducer responds to the surface motions that occur. It should be noted that the signal captured by the recording device may be affected by the nature of the stress pulse generated by the source, the geometry of the test specimen, and the characteristics of the receiver, making it difficult to interpret the recorded waveforms. Two basic types of acoustic emission signals can be generated (Figure 1):

> *Continuous emission* is "a qualitative description of the sustained signal level produced by rapidly occurring acoustic emission events."[1] These are generated by events such as plastic deformations in metals, which occur in a reasonably continuous manner.

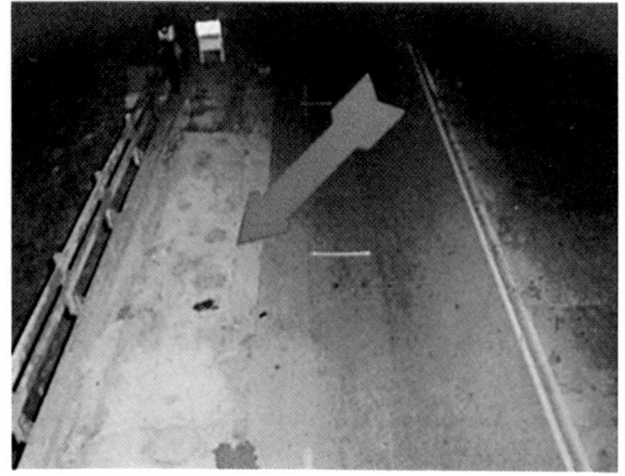

A

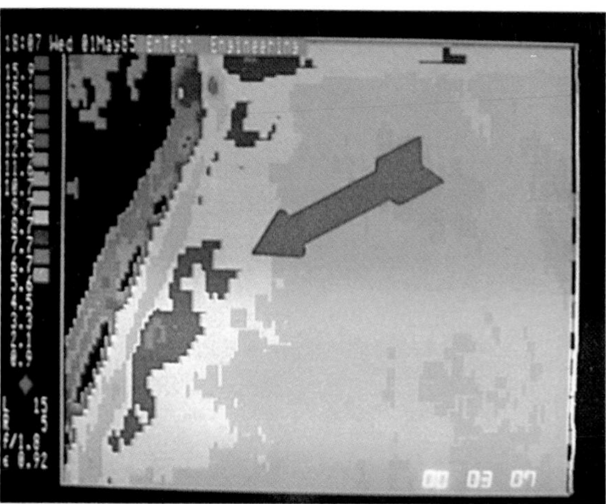

B

PLATE 1. (A) Visual image, (B) thermogram of powdery concrete on the Martin Luther King Bridge in St. Louis. Red areas represent powdery concrete.

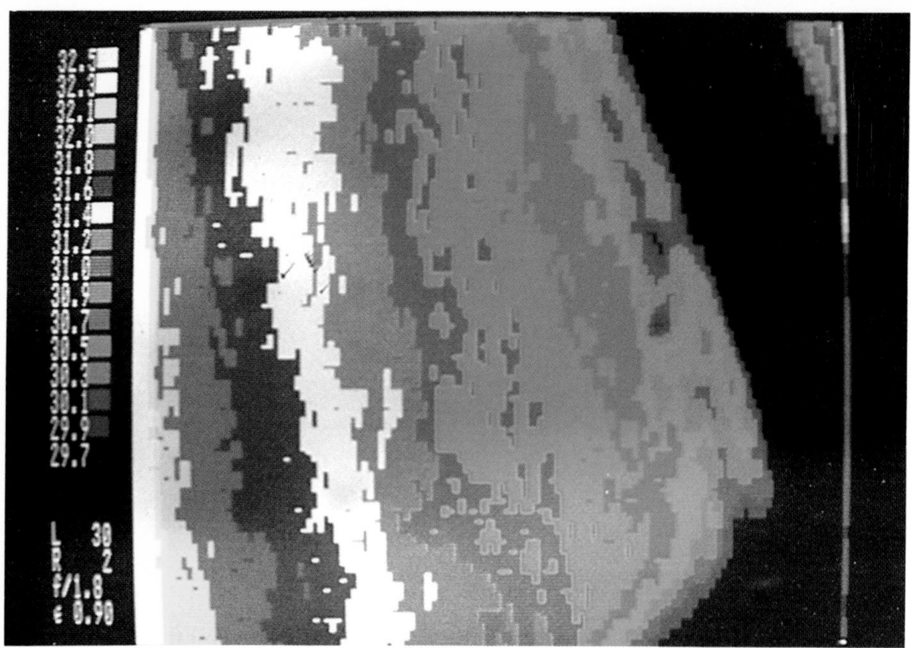

PLATE 2. Computer enhanced thermographic image of concrete pavement next to three-story building. Bands of temperatures are caused by shadows restricting solar energy being absorbed by pavement as shadows move.

A

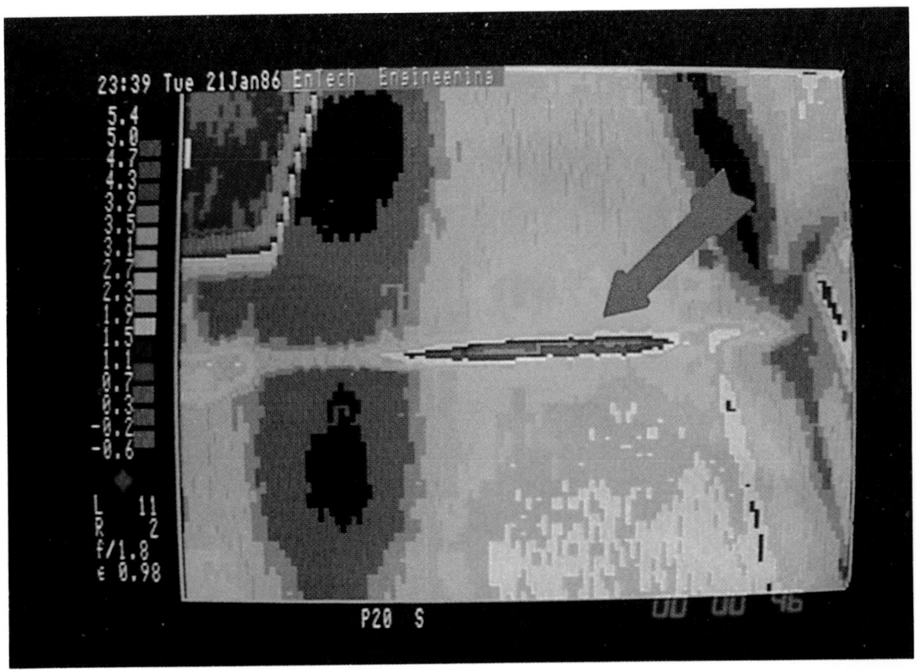

B

PLATE 3. (A) Visual and (B) thermal images showing a good roadway expansion joint located at Lambert St. Louis International Airport.

A

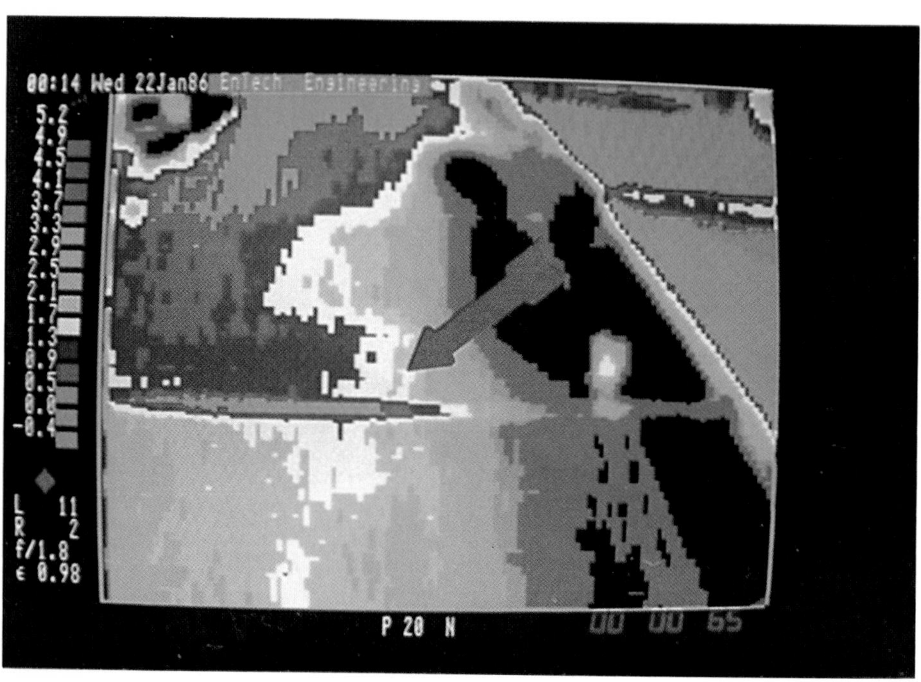

B

PLATE 4. (A) Visual and (B) thermal images showing a deteriorated expansion joint containing both voids and water located at St. Louis International Airport.

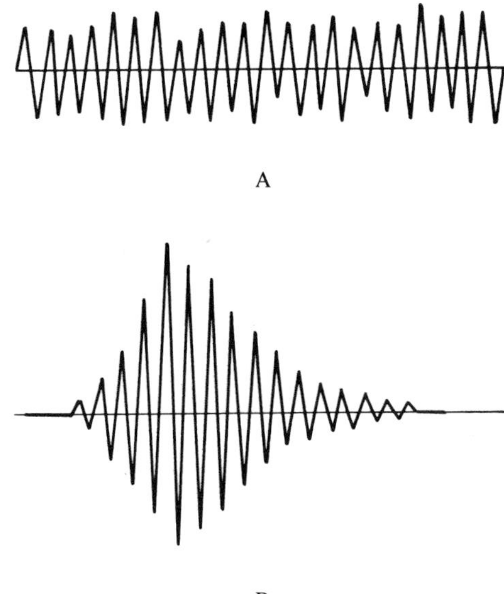

FIGURE 1. The two basic types of acoustic emission signals. (A) Continuous emission. (B) Burst emission.

Burst emission is "a qualitative description of the discrete signal related to an individual emission event occurring within the material",[1] such as that which may occur during crack growth or fracture in brittle materials. These burst signals are characteristic of the acoustic emission events resulting from the loading of cementitious materials.

Acoustic emissions from concrete occur over a very wide range of frequencies. The earliest work concentrated on rather low frequencies. Robinson[6] recorded acoustic emissions mainly at two frequencies: 2 kHz and 13 to 14 kHz; Wells[7] worked in the frequency range of 2 to 20 kHz; and Green[8] recorded emissions only up to 100 kHz. More modern instrumentation, however, can record much higher frequencies, typically in the range of 50 kHz to about 2 MHz. At lower frequencies, extraneous background noises from the test equipment or the laboratory environment become a problem; this was the difficulty faced in the earlier investigations referred to above. On the other hand, at very high frequencies, the attenuation of the signals is too severe, and thus, the distance from the piezoelectric transducer to the acoustic emission source must be reduced. The precise frequency range which is monitored does not appear to be very important for concrete. Detailed studies by Tanigawa et al.[11] in the frequency range up to 400 kHz showed that at low stresses, emissions tended to be in the frequency range below 150 kHz; at higher stresses, the higher frequency components become more significant. However, the relative shapes of the acoustic emission output vs. load curves were much the same for all of the frequency ranges studied.

EVALUATION OF ACOUSTIC EMISSION SIGNALS

A typical acoustic emission signal from concrete is shown in Figure 2.[12] However, when such acoustic events are examined in much greater detail, as shown in Figure 3,[13] the complexity of the signal becomes even more apparent; the scatter in noise shown in Figure 3, makes it difficult to determine exactly the time of arrival of the signal; this means that very sophisticated equipment must be used to get the most information out of the acoustic emission signals. In addition, to obtain reasonable sensitivity, the acoustic emission signals must be amplified. In concrete, typically, system gains in the range of 80 to 100 decibels (dB) are used.

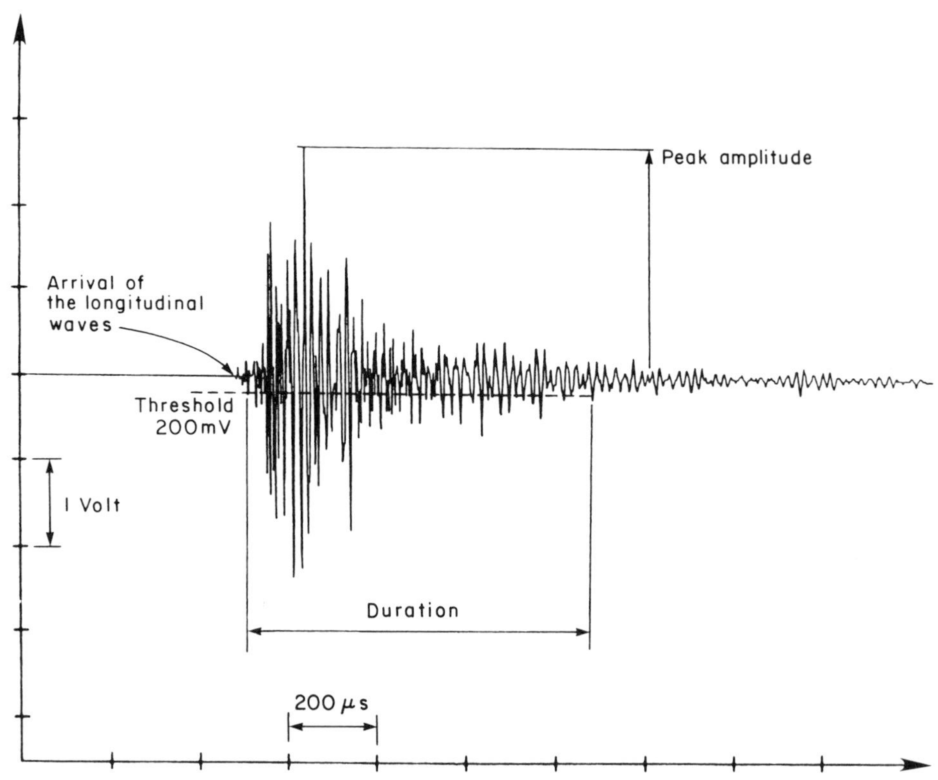

FIGURE 2. A typical acoustic emission signal from concrete. (From Berthelot, J. M. et al., private communication, 1987. With permission.)

There are a number of different ways in which acoustic emission signals may be evaluated.

Acoustic Emission Counting (ring-down counting) — This is the simplest way in which an acoustic emission event may be characterized. It is "the number of times the acoustic emission signal exceeds a preset threshold during any selected portion of a test",[1] and is illustrated in Figure 4. A monitoring system may record:

1. The *total number of counts* (e.g., 13 counts in Figure 4). Since the shape of a burst emission is generally a damped sinusoid, pulses of higher amplitude will generate more counts.
2. The *count rate*. This is the number of counts per unit of time; it is particularly useful when very large numbers of counts are recorded.
3. The *mean pulse amplitude*. This may be determined by using a root-mean square meter, and is an indication of the amount of energy being dissipated.

Clearly, the information obtained using this method of analysis depends upon both the gain and the threshold setting. Ring-down counting is affected greatly by the characteristics of the transducer, and the geometry of the test specimen (which may cause internal reflections) and may not be indicative of the nature of the acoustic emission event. In addition, there is no obvious way of determining the amount of energy released by a single event, or the total number of separate acoustic events giving rise to the counts.

Event counting — Circuitry is available which counts each acoustic emission event only once, by recognizing the end of each burst emission in terms of a predetermined length of

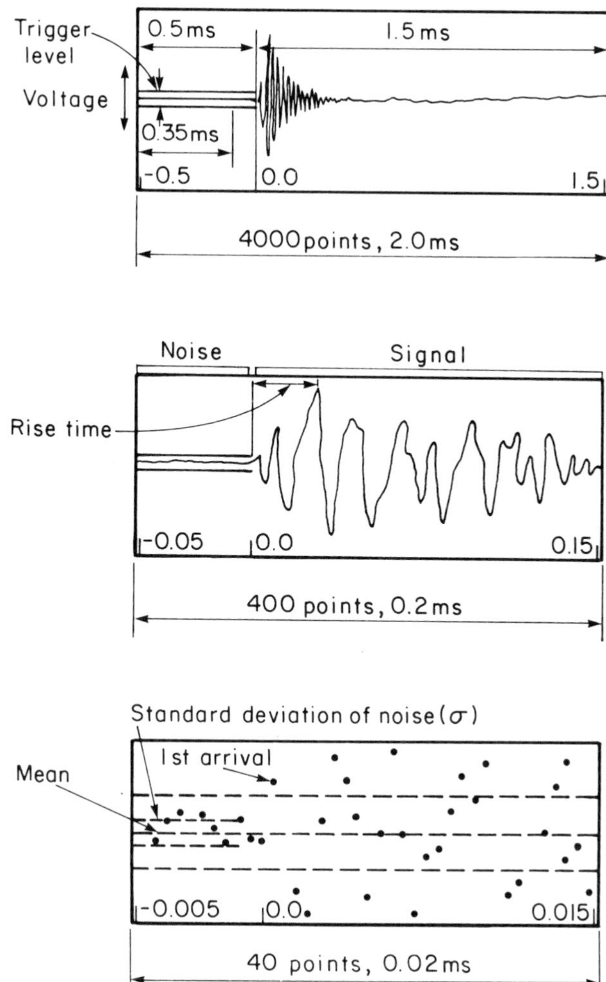

FIGURE 3. Typical view of an acoustic emission event as displayed on an oscilloscope screen. (From Maji, A. and Shah, S. P., *Exp. Mech.*, in press. With permission.)

time since the last count (i.e., since the most recent crossing of the threshold). In Figure 4, for instance, the number of events is three. This method records the number of events, which may be very important, but provides no information about the amplitudes involved.

Rise time — This is the interval between the time of first occurrence of signals above the level of the background noise and the time at which the maximum amplitude is reached. This may assist in determining the type of damage mechanism.

Signal duration — This is the duration of a single acoustic emission event; this too may be related to the type of damage mechanism.

Amplitude distribution — This provides the distribution of peak amplitudes. These may assist in identifying the sources of the emission events which are occurring.

Frequency analysis — This refers to the frequency spectrum of individual acoustic emission events. This technique, generally requiring a fast Fourier transformation analysis of the acoustic emission waves, may help to discriminate between different types of events. Unfortunately, a frequency analysis may sometimes simply be a function of the response of the transducer, and thus reveal little of the true nature of the pulse.

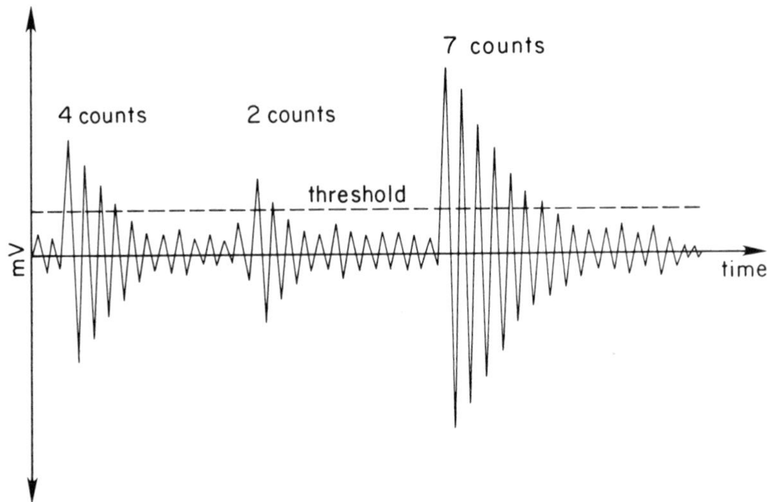

FIGURE 4. The principle of acoustic emission counting (ring-down counting).

Energy analysis — This is an indication of the energy released by an acoustic emission event; it may be measured in a number of ways, depending on the equipment, but is essentially the area under the amplitude vs. time curve (Figure 4) for each burst. Alternatively, the area under the envelope of the amplitude vs. time curve may be measured for each burst.
Defect location — By using a number of transducers to monitor acoustic emission events, and determining the time differences between the detection of each event at different transducer positions, the location of the acoustic emission event may be determined by using triangulation techniques. Work by Maji and Shah,[13] for instance, has indicated that this technique may be accurate to within about 5 mm.
Analysis of the wave-form — Most recently, it has been suggested[14,15] that an elaborate signals processing technique (deconvolution) applied to the wave-form of an acoustic emission event can provide information regarding the volume, orientation, and type of microcrack.

Ideally, since all of these methods of data analysis provide different information, one would wish to measure them all. However, this is neither necessary nor economically feasible. In the discussion which follows, it will become clear that the more elaborate methods of analysis are useful in fundamental laboratory investigations, but may be inappropriate for practical applications.

INSTRUMENTATION AND TEST PROCEDURES

Instrumentation (and, where necessary, the associated computer software) is available, from a number of different manufacturers, to carry out all of the methods of signal analysis described above. It might be added that advances in instrumentation have outpaced our understanding of the nature of the elastic waves resulting from microcracking in concrete. The main elements of a modern acoustic emission detection system are shown schematically in Figure 5. A brief description of the most important parts of this system is as follows:

1. *Transducers*: Piezoelectric transducers (generally made of lead zirconate titanate, PZT) are used to convert the surface displacements into electric signals. The voltage output from the transducers is directly proportional to the strain in the PZT, which depends in turn on the amplitude of the surface waves. Since these transducers are high impedance devices, they yield relatively low signals, typically less than 100 μV. There are

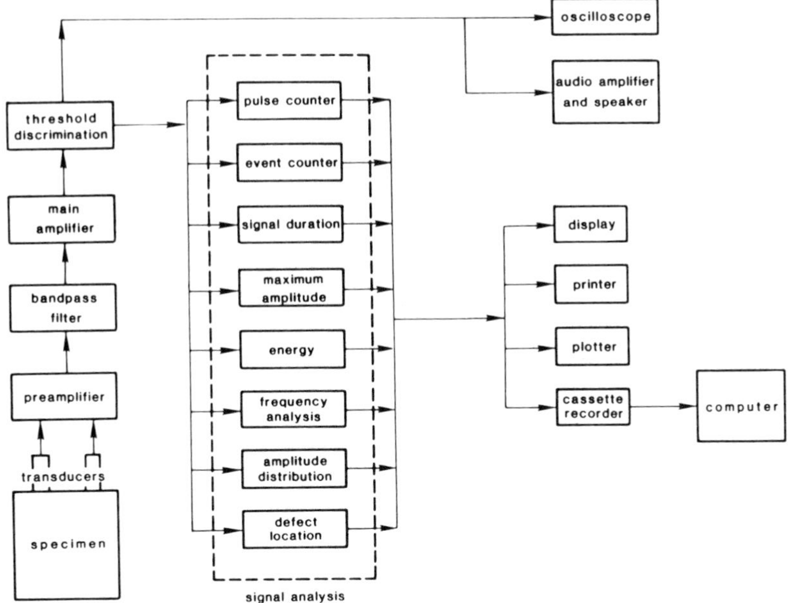

FIGURE 5. The main elements of a modern acoustic emission detection system.

basically two types of transducers. *Wide-band* transducers are sensitive to acoustic events with frequency responses covering a wide range, often several hundred kHz. *Narrow-band* transducers are restricted to a much narrower range of frequencies, using bandpass filters. Of course, the transducers must be properly coupled to the specimen, often using some form of silicone grease as the coupling medium.

2. *Preamplifier*: Because of the low voltage output, the leads from the transducer to the preamplifier must be as short as possible; often, the preamplifier is integrated within the transducer itself. Typically, the gain in the preamplifier is in the range 40 to 60 dB. (Note: The decibel scale measures only *relative* amplitudes. Using this scale.

$$\text{Gain (dB)} = 20 \log \frac{V}{V_i}$$

where V is the output amplitude and V_i is the input amplitude. That is, a gain of 40 dB will increase the input amplitude by a factor of 100; a gain of 60 dB will increase the input amplitude by a factor of 1000, and so on.)
3. *Passband filters* are used to suppress the acoustic emission signals which lie outside of the frequency range of interest.
4. The *main amplifier* further amplifies the signals, typically with a gain of an additional 20 to 60 dB.
5. The *discriminator* is used to set the threshold voltage above which signals are counted.

The remainder of the electronic equipment depends upon the way in which the acoustic emission data are to be recorded, analyzed and displayed.

Acoustic emission testing may be carried out in the laboratory or in the field. Basically, one or more acoustic emission transducers are attached to the specimen. The specimen is then loaded slowly, and the resulting acoustic emissions are recorded. There are generally two categories of tests:

1. To use the acoustic emission signals to learn something about the internal structure of the material, and how structural changes (i.e., damage) occur during the process of loading. In this case, the specimens are generally loaded to failure.
2. To establish whether the material or the structure meet certain design or fabrication criteria. In this case, the load is increased only to some predetermined level (''proof'' loading). The amount and nature of the acoustic emissions may be used to establish the integrity of the specimen or structure, and may also sometimes be used to predict the service life.

PARAMETERS AFFECTING ACOUSTIC EMISSIONS FROM CONCRETE

THE KAISER EFFECT

The earliest acoustic emission studies of concrete, such as the work of Rüsch,[3] indicated that a true Kaiser effect (see above) exists for concrete; that is, acoustic emissions were found not to occur in concrete that had been unloaded until the previously applied maximum stress had been exceeded on reloading. This was true, however, only for stress levels below about 75 to 85% of the ultimate strength of the material; for higher stresses, acoustic emissions began again at stresses somewhat lower than the previous maximum stress. Subsequently, a number of other investigators have also concluded that concrete exhibits a Kaiser effect, at least for stresses below the peak stress of the material.[10,13,16-18] Spooner and Dougill[16] confirmed that this effect did not occur beyond the peak of the stress-strain curve (i.e., in the descending portion of the stress-strain curve), where acoustic emissions occurred again before the previous maximum strain was reached. It has also been suggested that a form of the Kaiser effect occurs as well for cyclic thermal stresses in concrete[19], and for drying and wetting cycles.[20] On the other hand, Nielsen and Griffin[21] have reported that the Kaiser effect is only a very *temporary* effect in concrete; with only a few hours of rest between loading cycles, acoustic emissions are again recorded during reloading to the previous maximum stress. They therefore concluded ''that the Kaiser effect is not a reliable indicator of the loading history for plain concrete''. Thus, it is unlikely that the Kaiser effect could be used in practice to determine the previous maximum stress that a structural member has been subjected to.

EFFECT OF LOADING DEVICES

As is well known, the end restraint of a compression specimen of concrete due to the friction between the ends of the specimen and the loading platens can have a considerable effect on the apparent strength of the concrete. These differences are also reflected in the acoustic emissions measured when different types of loading devices are used.[22] For instance, in compression testing with stiff steel platens, most of the acoustic emission appears at stresses beyond about half of the ultimate stress; with more flexible platens, such as brush platens, significant acoustic emission appears at about 20% of the ultimate stress. This undoubtedly reflects the different crack patterns that develop with different types of platens, but it nonetheless makes interlaboratory comparisons, and indeed even studies on different specimen geometries within the same laboratory, very difficult.

SIGNAL ATTENUATION

The elastic stress waves which are generated by cracking attenuate as they propagate through the concrete. Thus, large acoustic emission events which take place in the concrete far from a pick-up transducer may not exceed the threshold excitation voltage due to this attenuation, while much smaller events may be recorded if they occur close to the transducer. Very little information is available on acoustic emission attenuation rates in concrete. It has been shown that more mature cements show an increasing capacity to transmit acoustic emissions.[20] Related to this, Mindess[23] has suggested that the total counts to failure for concrete specimens in compression are much higher for older specimens, which may also be explained by the better transmission through older concretes.

As a practical matter, the maximum distance between piezoelectric transducers, or between the transducers and the source of the acoustic emission event, should not be very large. Berthelot and Robert[24] required an array of transducers arranged in a 40-cm square mesh to locate acoustic emission events reasonably accurately. They found that for ordinary concrete, with a fifth transducer placed in the center of the 40 × 40-cm square mesh, only about 40% of the events detected by the central transducer were also detected by the four transducers at the corners; with high strength concrete, this proportion increased to 60 to 70%. Rossi[25] also found that a 40-cm square mesh was needed for a proper determination of acoustic emission events. Although more distant events can, of course, be recorded, there is no way of knowing how many events are ''lost'' due to attenuation. This is an area which requires much more study.

SPECIMEN GEOMETRY

It has been shown that smaller specimens appear to give rise to greater levels of acoustic emission than do larger ones.[17] The reasons for this are not clear, though the observation may be related to the attenuation effect described above. After an acoustic emission event occurs, the stress waves not only travel from the source to the sensor, but also undergo reflection, diffraction, and mode conversions within the material. The basic problem of wave propagation within a bounded solid certainly requires further study, but there have apparently been no comparative tests on different specimen geometries.

TYPE OF AGGREGATE

It is not certain whether the mineralogy of the aggregate has any effect on acoustic emission. It has been reported that concretes with a smaller maximum aggregate size produce a greater number of acoustic emission counts than those with a larger aggregate size;[10] however, the total energy released by the finer aggregate concrete is reduced. This is attributed to the observation that concretes made with smaller aggregates start to crack at lower stresses; in concretes with larger aggregate particles, on the other hand, individual acoustic events emit higher energies. For concretes made with lighweight aggregates, the total number of counts is also greater than for normal weight concrete, perhaps because of cracking occurring in the aggregates themselves.

CONCRETE STRENGTH

It has been shown that the total number of counts to the maximum load is greater for higher strength concretes.[23] However, as was mentioned earlier,[23] for similar strength levels the total counts to failure appears to be much higher for older concretes.

LABORATORY STUDIES OF ACOUSTIC EMISSION

By far the greatest number of acoustic emission studies of concrete have been carried out in the laboratory, and have been largely ''theoretical'' in nature:

1. To determine whether acoustic emission analysis could be applied to cementitious systems
2. To learn something about crack propagation in concrete

FRACTURE MECHANICS STUDIES

A number of studies have shown that acoustic emission can be related to crack growth or fracture mechanics parameters in cements, mortars, and concretes. Evans et al.[26] showed that acoustic emission could be correlated with crack velocity in mortars. Morita and Kato[27,28] and Nadeau, Bennett, and Mindess[20] were able to relate total acoustic emission counts to K_c (the fracture toughness). In addition, Lenain and Bunsell[29] found that the number of emissions could be related to the sixth power of the stress intensity factor, K. Izumi et al.[30] showed that acoustic emissions could also be related to the strain energy release rate, G. In all cases, however, these correlations are purely empirical; no one has yet developed a fundamental relationship between acoustic emission events and fracture parameters, and it is unlikely that such a relationship exists.

TYPE OF CRACKS

A number of attempts have been made to relate acoustic events of different frequencies, or of different energies, to different types of cracking in concrete. For instance, Saeki et al.,[31] by looking at the energy levels of the acoustic emissions at different levels of loading, concluded that the first stage of cracking, due to the development of bond cracks between the cement paste and the aggregate, emitted high energy signals; the second stage, which they termed "crack arrest", emitted low energy signals; the final stage, in which cracks extended through the mortar, was again associated with high energy acoustic events. Similarly, Tanigawa and Kobayashi[32] used acoustic energies to distinguish the onset of "the proportional limit, the initiation stress and the critical stress". On the other hand, Tanigawa et al.[11] tried to relate the fracture type (pore closure, tensile cracking, and shear slip) to the power spectra and frequency components of the acoustic events. The difficulty with these and similar approaches is that they tried to relate differences in the recorded acoustic events to preconceived notions of the nature of cracking in concrete; direct cause and effect relationships were never observed.

FRACTURE PROCESS ZONE (CRACK SOURCE) LOCATION

Perhaps the greatest current interest in acoustic emission analysis is its use in *locating* fracture processes, and in monitoring the damage that concrete undergoes as cracks progress. Okada et al.[33,34] showed that the location of crack sources obtained from differences in the arrival times of acoustic emissions was in good agreement with the observed fracture surface. At about the same time, Chhuy et al.[35] and Lenain and Bunsell[29] were able to determine the length of the damaged zone ahead of the tip of a propagating crack using one-dimensional acoustic emission location techniques. In subsequent work, Chhuy et al.,[36] using more elaborate equipment and analytical techniques, were able to determine damage both before the initiation of a visible crack and after subsequent crack extension.

Berthelot and Robert[24,37] and Rossi[25] used acoustic emission to monitor concrete damage as well. They found that, while the number of acoustic events showed the progression of damage both ahead and behind the crack front, this technique alone could not provide a *quantitative* description of the cracking. However, using more elaborate techniques, including amplitude analysis and measurements of signal duration, Berthelot and Robert[24] concluded that "acoustic emission testing is practically the only technique which can provide a quantitative description of the progression in real time of concrete damage within test specimens".

Recently, much more sophisticated signal processing techniques have been applied to acoustic emission analysis. In 1981, Michaels et al.[15] and Niwa et al.[38] developed deconvolution techniques to analyze acoustic waveforms, in order to provide a stress-time history of the source of an acoustic event. Similar deconvolution techniques were subsequently used by Maji and Shah[13,39] to determine the volume, orientation and type of microcrack, as well as the source of the acoustic events. Such sophisticated techniques have the potential eventually to be used to provide a detailed picture of the fracture processes occurring within concrete specimens.

STRENGTH VS. ACOUSTIC EMISSION RELATIONSHIPS

Since concrete quality is most frequently characterized by its strength, many studies have been directed towards determining a relationship between acoustic emission activity and strength. For instance, Tanigawa and Kobayashi[32] concluded that "the compressive strength of concrete can be approximately estimated by the accumulated AE counts at relatively low stress level." Indeed, they suggested that acoustic emission techniques might provide a useful nondestructive test method for concrete strength. Earlier, Fertis[40] had concluded that acoustic emissions could be used to determine not only strength, but also static and dynamic material behavior. Rebic,[41] too, found that there is a relationship between the "critical" load at which the concrete begins to be damaged, which can be determined from acoustic emission measurements, and the ultimate strength; thus, acoustic emission analysis might be used as a predictor of concrete strength. Sadowska-Boczar et al.[42] tried to quantify the strength vs. acoustic emission relationship using the equation

$$F_r = a F_p + b$$

where F_r is the rupture strength, F_p is the stress corresponding to the first acoustic emission signal, and a and b are constants for a given material and loading conditions. Using this linear relationship, which they found to fit their data reasonably well, they suggested that the observation of acoustic emissions at low stresses would permit an estimation of strength, as well as providing some characterization of porosity and critical flaw size.

Unfortunately, the routine use of acoustic emissions as an estimator of strength seems to be an unlikely prospect, in large part because of the scatter in the data, as has been noted by Fertis.[40] As an example of the scatter in data, Figure 6[23] indicates the variability in the strength vs. total acoustic emission counts relationship; the within-batch variability is even more severe, as shown in Figure 7.[23]

DRYING SHRINKAGE

Acoustic emission has been used to try to monitor shrinkage in cement pastes and mortars. Nadeau et al.[20] found that, in hardened pastes, the acoustic emission resulted from cracking due to the unequal shrinkage of the hydration products. Mortar gave less acoustic emission than hardened paste, suggesting that the fracture processes at the sand/cement paste interface are not an important source of acoustic emission. Jeong et al.[43] also suggested that, in autoclaved aerated concrete, the acoustic emissions during drying could be related to microcracking. Again, however, it is unlikely that acoustic emission measurements will be able to be used as a means of predicting the shrinkage as a function of time.

FIBER REINFORCED CEMENTS AND CONCRETES

A number of acoustic emission studies have been carried out on fiber reinforced cements and concretes. Lenain and Bunsell,[29] in a study of asbestos cement, found that acoustic emissions resulted both from cracking of the matrix and fiber pullout. They noted that the Kaiser effect was not found for this type of fiber reinforced composite, since on unloading of a specimen the partially pulled out fibers were damaged, and particles of cement attached to them were crushed, giving rise to acoustic emissions on unloading. Because these damaged fibers were then less able to resist crack growth, on subsequent reloading cracks started to propagate at lower stress levels than on the previous cycle, thus, giving off acoustic emissions below the previously achieved maximum load. Akers and Garrett[44] also studied asbestos cement; they found that acoustic emission monitoring could be used to detect the onset and development of prefailure cracking. However, they concluded that "there is no basis whatsoever for using amplitude discrimination in acoustic emission monitoring for distinguishing between the various failure modes which occur in this material". On the other hand, Faninger

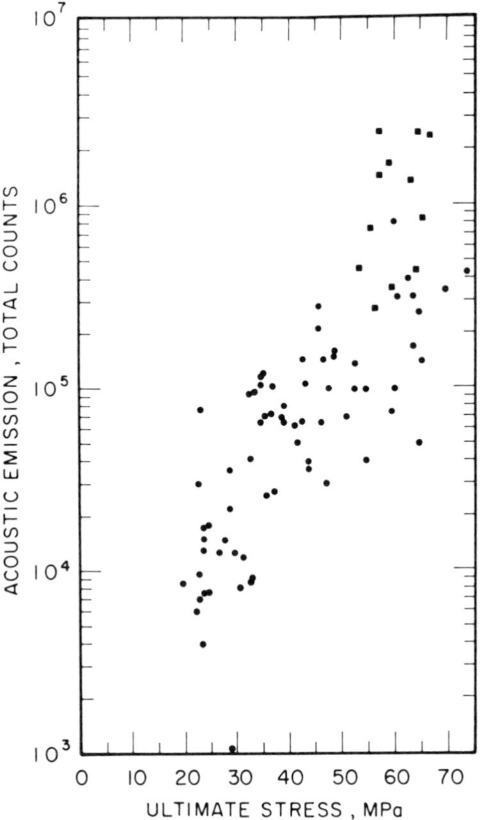

FIGURE 6. Logarithm of total acoustic emission counts vs. compressive strength of concrete cubes. (From Mindess, S., *Int. J. Cem. Comp. Lightweight Concr.*, 4, 173, 1982. With permission.)

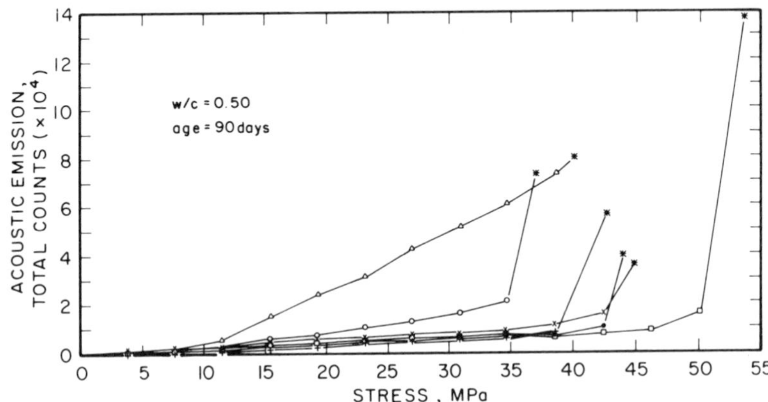

FIGURE 7. Within-batch variability of total acoustic emission counts vs. applied compressive stress on concrete cubes. (From Mindess, S., *Int. J. Cem. Comp. Lightweight Concr.*, 4,173, 1982. With permission.)

et al.[45] argued that in fiber reinforced concrete the amplitude pattern of the acoustic emission signals did make it possible to distinguish whether fracture had occurred in the fibers or between them. Similarly, Jeong et al.[43] stated that acoustic emission frequency analysis could distinguish between different micro-fracture mechanisms in fiber reinforced autoclaved aerated concrete.

HIGH ALUMINA CEMENT

In concretes made with high alumina (calcium aluminate) cement, the conversion from CAH_{10}* to C_3AH_6* on prolonged aging can lead to a large increase in porosity and therefore a large decrease in strength. There has thus been considerable interest in finding a non-destructive technique to monitor high alumina cement concrete (HAC) members. Parkinson and Peters[46] concluded that the conversion process itself is not a source of acoustic emission activity, since no acoustic emissions were generated during the accelerated conversion of pastes at the critical w/c ratio of 0.35. However, at the high w/c ratio of 0.65, conversion was accompanied by a high level of acoustic emission activity, due to the fracture processes taking place during conversion, associated perhaps with the liberation of excess water. Arrington and Evans[47] suggested that the structural integrity of HAC could be evaluated from the shape of the acoustic emission vs. load plot, the emissions recorded while the specimens were held under a constant load, and the decay of emission activity with time.

Perhaps the most extensive series of tests on HAC, carried out at the Fulmer Research Institute in the U.K., was reported by Williams.[48] Apart from observing that the Kaiser effect existed up to the point at which the beams cracked, some tentative suggestions were made for monitoring HAC beams with acoustic emissions:

1. If, on loading a beam, no acoustic emission is noted, then the applied load is still less than about 60% of the ultimate load; if acoustic emission occurs, then this percentage of the ultimate load has been exceeded. If, upon unloading such a beam, further acoustic emission activity is recorded, then the beam is cracked. The amount of acoustic emission during this unloading could indicate the degree to which the cracking load had been exceeded.
2. If a beam is under its service load, it would behave similarly on application of a superimposed load. The presence or absence of acoustic emissions during this further loading and unloading might indicate the condition of the beam.
3. If a beam under service load showed no acoustic emission activity during further loading, but did so at a later date when loaded to the same level, then the strength must have decreased during that time interval.

As well, Williams[48] noted similar behavior on testing of ordinary prestressed concrete beams, and suggested that these techniques could be used to evaluate any type of concrete structure, as long as acoustic emissions not connected with beam damage could be eliminated.

THERMAL CRACKING

Relatively little work has been carried out on acoustic activity when concrete is subjected to high temperatures, such as those that may be encountered in fires. However, Hinrichsmeyer et al.[19] carried out tests up to temperatures of 900°C. They claimed that acoustic emission analysis during heating enabled them to distinguish the different types of thermally induced cracking that occurred. They noted a thermal Kaiser effect in the temperature range 300 to 600°C, which might help in determining the maximum temperature reached in a previous

* Note that cement chemistry notation is being used: C = CaO; A = Al_2O_3; H = H_2O.

heating cycle. The technique was even sensitive enough to record the acoustic emissions from the quartz inversion at 573°C.

BOND IN REINFORCED CONCRETE

A number of acoustic emission studies of debonding of reinforcing bars in reinforced concrete have been carried out. Kobayashi et al.[49] tested simulated beam-column connections with a 90° hooked reinforcing bar subjected to various cyclic loading histories. They found that the penetration of a surface crack down to the level of the bar gave rise to only one or two acoustic events; most acoustic emission signals were generated by the internal cracking around the bar due to fracture at the lugs (ribs) of the bars. Acoustic emission signals were able to indicate, with reasonable accuracy, the degree of debonding. They suggested that acoustic emission techniques could be used to determine the amount of bond deterioration in concrete structures during proof testing, or due to overloads.

In addition, several studies of bond degradation at elevated temperatures have been carried out. Royles et al.[50] studied simple pullout specimens at temperatures up to 800°C. They found that acoustic emissions were associated with the adhesive failure at the steel-concrete interface, followed by local crushing under the ribs of the reinforcing bars. They suggested that acoustic emissions could be used to identify the point of critical slip. In further work, Royles and Morley[51] suggested that acoustic emission techniques might be useful in estimating the quality of the bond in reinforced concrete structures that had been subjected to fires.

CORROSION OF REINFORCING STEEL IN CONCRETE

The deterioration of concrete due to corrosion of the reinforcing steel is a major problem, which is usually detected only after extreme cracking has already taken place. Weng et al.[52] found that measurable levels of acoustic emission occurred even during the corrosion of unstressed reinforced concrete. They suggested that, at least in the laboratory, acoustic emission monitoring would assist in characterizing corrosion damage. In subsequent work, Dunn et al.[53] developed a relationship between the observed damage and the resulting acoustic emissions. Damage could be detected in its early stages, and by a combination of total counts and amplitude measurements, the nature of the corrosion damage could be determined.

FIELD STUDIES OF ACOUSTIC EMISSION

As has been seen in the previous section, acoustic emission analysis has been used in the laboratory to study a wide range of problems. Unfortunately, its use in the field has been severely limited; only a very few papers on field application have appeared, and these are largely speculation on future possibilities. The way in which acoustic emission data might be used to provide information about the condition of a specimen or a structure has been described by Cole;[54] his analysis may be summarized as follows:

1. Is there any acoustic emission at a certain load level? If *no*, then no damage is occurring under these conditions; if *yes*, then damage is occurring.
2. Is acoustic emission continuing while the load is held constant at the maximum load level? If *no*, no damage due to creep is occurring; if *yes*, creep damage is occurring. Further, if the count rate is increasing, then failure may occur fairly soon.
3. Have high amplitude acoustic emissions events occurred? If *no*, individual fracture events have been relatively minor; if *yes*, major fracture events have occurred.
4. Does acoustic emission occur if the structure has been unloaded and is then reloaded to the previous maximum load? If *no*, there is no damage or crack propagation under low cycle fatigue; if *yes*, internal damage exists and the damage sites continue to spread even under low loads.

5. Does the acoustic emission occur only from a particular area? If *no*, the entire structure is being damaged; if *yes*, the damage is localized.
6. Is the acoustic emission in a local area *very* localized? If *no*, damage is dispersed over a significant area; if *yes*, there is a highly localized stress concentration causing the damage.

Naus[55] used acoustic emissions to monitor intermediate pressure vessels and simple concrete structures and determine whether this technique could be applied to primary nuclear containment structures. He concluded that, for monitoring prestressed concrete members, acoustic emission results correlated with beam behavior and could be used to locate cracks in simple concrete structures. Robert and Brachet-Rolland[56] also suggested that acoustic emission surveys could be used to detect the presence of active cracks in prestressed concrete structures and to help provide an estimate of the load-carrying capacity of such structures.

Perhaps the first real field application was carried out by Woodward,[57] who used acoustic emission to monitor cracks in the anchor block of a prestressed concrete bridge. The signals were characterized by bursts of activity, followed by periods of inactivity. Since the bridge was subjected to dead-weight loading, longitudinal and vertical prestress, and environmentally induced loads, it was not clear which of these effects gave rise to the acoustic emissions. At least three different sources of acoustic emission were identified: direct crack propagation, cracking due to the redistribution of strains within the concrete, and cracking due to thermal expansions and contractions, but no quantitative results could be obtained.

CONCLUSIONS

From the discussion above, it appears that acoustic emission techniques may be very useful in the laboratory to supplement other measurements of concrete properties. However, their use in the field remains problematic. Many of the earlier studies held out high hopes for acoustic emission monitoring of structures. For instance, McCabe et al.[17] suggested that, if a structure was loaded, the absence of acoustic emissions would indicat that it was safe under the existing load conditions; a low level of acoustic emissions would indicate that the structure should be monitored carefully, while a high level of acoustic emission could indicate that the structure was unsafe. But this is hardly a satisfactory approach, since it does not provide any help with quantitative analysis. In any event, even the sophisticated (and expensive) equipment now available still provides uncertain results when applied to structures, because of our lack of knowledge about the characteristics of acoustic emissions due to different causes, and because of the possibility of extraneous noise (vibration, loading devices, and so on). Another serious drawback is that acoustic emissions are only generated when the loads on a structure are increased, and this poses considerable practical problems. Thus, one must still conclude, with regret, that ''acoustic emission analysis has not yet been well developed as a technique for the evaluation of phenomena taking place in concrete in structures''.[18]

REFERENCES

1. ASTM E610-82, Standard definitions of terms relating to acoustic emission, American Society for Testing and Materials, Philadelphia, Vol. 2.03, 1987, 373.
2. **Kaiser, J.,** Untersuchungen uber das auftreten Geraushen beim Zugversuch, Ph.D. thesis, Technische Hochschule, Munich, 1950.
3. **Rüsch, H.,** Physical problems in the testing of concrete, *Zement-Kalk-Gips,* 12, 1, 1959.

4. **L'Hermite, R. G.,** What we know about the plastic deformation and creep of concrete?, *RILEM Bull.,* 1, 21, 1959.
5. **L'Hermite, R. G.,** Volume changes of concrete, in *Proc. 4th Int. Symp. on Chemistry of Cement,* Vol. II, Washington, National Bureau of Standards, Washington, D.C., NBS, Monograph, No. 43, 1960, 659.
6. **Robinson, G. S.,** Methods of detecting the formation and propagation of microcracks in concrete, in *Proc. Int. Symp. on the Structure of Concrete,* Brooks, A. E. and Newman, K., Eds., London, 1965, Cement and Concrete Association, London, 1968, 131.
7. **Wells, D.,** An acoustic apparatus to record emissions from concrete under strain, *Nuclear Engineering and Design,* 12, 80, 1970.
8. **Green, A. T.,** Stress wave emission and fracture of prestressed concrete reactor vessel materials, in *Proc. 2nd Interamerican Conf. on Materials Technology,* American Society of Mechanical Engineers, Vol. I, 1970, 635.
9. **Diederichs, U., Schneider, U., and Terrien, M.,** Formation and propagation of cracks and acoustic emission, in *Fracture Mechanics of Concrete,* Wittmann, F. H., Ed., Elsevier, Amsterdam, 1983, chap. 3.5.
10. **Malhotra, V. M.,** Testing hardened concrete: non-destructive methods, Monogr. No. 9, American Concrete Institute, Detroit, MI, 1976, chap. 12.
11. **Tanigawa, Y., Yamada, K., and Kiriyama, S.-I.,** Power spectra analysis of acoustic emission wave of concrete, in *Proc. 2nd Austr. Conf. on Engineering Materials,* Sydney, 1981, 97.
12. **Berthelot, J. M., Robert, J. L., Bruhat, D., and Gervais, J. P.,** Damage evaluation of concrete test specimens related to failure analysis, private communications, 1987.
13. **Maji, A. and Shah, S. P.,** Process zone and acoustic emission measurements in concrete, *Exp. Mech.,* in press.
14. **Seruby, C. B., Baldwin, G. R., and Stacey, K. A.,** Characterization of fatigue crack extension by quantitative acoustic emission, *Int. J. Fracture,* 28, 201, 1985.
15. **Michaels, J. E., Michaels, T. E., and Sachse, W.,** Applications of deconvolution to acoustic emission signal analysis, *Mater. Eval.,* 39, 1032, 1981.
16. **Spooner, D. C. and Dougill, J. W.,** A quantitative assessment of damage sustained in concrete during compressive loading, *Mag. Conc. Res.,* 27, 151, 1975.
17. **McCabe, W. M., Koerner, R. M., and Lord, A. E., Jr.,** Acoustic emission behaviour of concrete laboratory specimens, *Am. Concr. Inst. J.,* 73, 367, 1976.
18. **Mlaker, P. F., Walker, R. E., Sullivan, B. R., and Chiarito, V. P.,** Acoustic emission behaviour of concrete, in *In Situ/Nondestructive Testing of Concrete,* Malhotra, V. M., Ed., ACI SP-82, American Concrete Institute, Detroit, 1984, 619.
19. **Hinrichsmeyer, K., Diederichs, U., and Schneider, U.,** Thermal induced cracks and acoustic emission in cement paste, mortar and concrete, in *Sci. Ceram.,* 12, 1984, 667.
20. **Nadeau, J. S., Bennett, R., and Mindess, S.,** Acoustic emission in the drying of hardened cement paste and mortar, *J. Am. Ceram. Soc.,* 64, 410, 1981.
21. **Nielsen, J. and Griffin, D. F.,** Acoustic emission of plain concrete, *J. Test. Eval.,* 5, 476, 1977.
22. **Schickert, G.,** Critical reflections on non-destructive testing of concrete, *Mater. Construct.,* 17, 217, 1984.
23. **Mindess, S.,** Acoustic emission and ultrasonic pulse velocity of concrete, *Int. J. Cement Composites Lighweight Concr.,* 4, 173, 1982.
24. **Berthelot, J. M. and Robert, J. L.,** Modeling concrete damage by acoustic emission, *J. Acoust. Emission,* 6, 43, 1987.
25. **Rossi, P.,** Fissuration du Beton: Du Materiaux a la Structure, Application de la Mecanique Lineaire de la Rupture, Ph.D. thesis, Ecole National des Ponts et Chaussees, Paris, 1986.
26. **Evans, A. G., Clifton, J. R., and Anderson, E.,** The fracture mechanics of mortars, *Cement Conc. Res.,* 6, 535, 1976.
27. **Morita, K. and Kato, K.,** Fundamental study on fracture toughness and evaluation by acoustic emission technique of concrete, in *Rev. 32nd General Meet.,* Cement Association of Japan, Tokyo, 1978, 138.
28. **Morita, K. and Kato, K.,** Fundamental study on evaluation of fracture toughness of artificial lightweight aggregate concrete, in *Rev. 33rd General Meet.,* Cement Association of Japan, Tokyo, 1979, 175.
29. **Lenain, J. C. and Bunsell, A. R.,** The resistance to crack growth of asbestos cement, *J. Mater. Sci.,* 14, 321, 1979.
30. **Izumi, M., Mihashi, H., and Nomura, N.,** Acoustic emission technique to evaluate fracture mechanics parameters of concrete, in *Fracture Toughness and Fracture Energy of Concrete,* Wittmann, F. H., Ed., Elsevier, Amsterdam, 1986, 259.
31. **Saeki, N., Takada, N., and Hataya, S.,** On studies for cracking and failure of concrete by acoustic emission techniques, in *Rev. 33rd General Meet.,* Cement Association of Japan, Tokyo, 1979, 234.
32. **Tanigawa, Y. and Kobayashi, H.,** A study on acoustic emission of concrete, in *Rev. 33rd General Meet.,* Cement Association of Japan, Tokyo, 1979, 159.

33. **Okada, K., Koyanagi, W., and Rokugo, K.,** Energy transformation in the fracture process of concrete, in *Memoirs of the Faculty of Engineering,* Kyoto University, Kyoto, Japan, 34(3), 1977, 389.
34. **Okada, K., Koyanagi, W., and Rokugo, K.,** Energy approach to flexural fracture process of concrete, *Trans. Jpn. Soc. Civil Eng.,* 11, 301, 1979.
35. **Chhuy, S., Baron, J., and Francois, D.,** Mecanique de la rupture appliquee au beton hydraulique, *Cement Conc. Res.,* 9, 641, 1979.
36. **Chhuy, S., Cannard, G., Robert, J. L., and Acker, P.,** Experimental investigations on the damage of cement concrete with natural aggregates, personal communication, 1987.
37. **Berthelot, J.-M. and Robert, J.-L.,** Application de l'emission acoustique aux mecanismes d'endommagement du beton, *Bull. Liasion Ponts Chaussees,* 140, 101, 1985.
38. **Niwa, Y., Ohtsu, M., and Shiomi, H.,** Waveform analysis of acoustic emission in concrete, in *Memoirs of the Faculty of Engineering,* Kyoto University, 43(4), 319, 1981.
39. **Maji, A. and Shah, S. P.,** Initiation and propagation of bond cracks as detected by laser holography and acoustic emission, in *Bonding in Cementitious Materials,* Mindess, S. and Shah, S. P., Eds., Materials Research Society Symposium Proceedings, Vol. 114, Materials Research Society, Pittsburgh, 1988, 55.
40. **Fertis, D. G.,** Concrete material response by acoustic spectra analysis, *J. Struct. Div.,* ASCE, 102, 387, 1976.
41. **Rebic, M. P.,** The distribution of critical and rupture loads and determination of the factor of crackability, in *In Situ/Nondestructive Testing of Concrete,* Malhotra, V. M., Ed., ACI SP-82, American Concrete Institute, Detroit, 1984, 721.
42. **Sadowska-Boczar, E., Librant, Z., Ranachowski, J., and Ciesla, Z.,** Application of acoustic emission to investigate the mechanical strength of ceramic materials, in *Sci. Ceram.,* 12, 1984, 639.
43. **Jeong, H. D., Takahashi, H., and Teramura, S.,** Low temperature fracture behaviour and AE characteristics of autoclaved aerated concrete (AAC), *Cement Concr. Res.,* 17, 743, 1987.
44. **Akers, S. A. S. and Garrett, G. G.,** Acoustic emission monitoring of flexural failure in asbestos cement composites, *Int. J. Cement Composites Lightweight Concrete,* 5, 97, 1983.
45. **Faninger, Von G., Grünthaler, K. H., and Schwalbe, J. H.,** Faserverstarkter beton und schallemission [Fibre-reinforced concrete and acoustic emission], *Betonwerk + Fertigteil-Technik,* 2, 82, 1977.
46. **Parkinson, R. W. and Peters, C. T.,** Acoustic emission activity during accelerated conversion of high alumina cement pastes, *J. Mater. Sci.* (letters), 12, 848, 1977.
47. **Arrington, M. and Evans, B. M.,** Acoustic emission testing of high alumina cement concrete, *Non-Destructive Testing Int.,* 10, 81, 1977.
48. **Williams, R. V.,** *Acoustic Emission,* Adam Hilger, Bristol, 1980, 110.
49. **Kobayashi, A. S., Hawkins, N. M., Chan, Y.-L. A., and Jin, I.-J.,** A feasibility study of detecting reinforcing-bar debonding by acoustic-emission technique, *Exp. Mech.,* 20, 301, 1980.
50. **Royles, R., Morley, P. D., and Khan, M. R.,** The behavoiur of reinforced concrete at elevated temperatures with particular reference to bond strength, in *Bond in Concrete,* Bartos, P., Ed., Applied Science Publishers, London, 1982, 217.
51. **Royles, R. and Morley, P. D.,** Acoustic emission and bond degradation in reinforced concrete due to elevated temperatures, *Matr. Construct.,* 17, 185, 1984.
52. **Weng, M. S., Dunn, S. E., Hartt, W. H., and Brown, R. P.,** Application of acoustic emission to detection of reinforcing steel corrosion in concrete, *Corrosion,* 38, 9, 1982.
53. **Dunn, S. E., Young, J. D., Hartt, W. H., and Brown, R. P.,** Acoustic emission characterization of corrosion induced damage in reinforced concrete, *Corrosion,* 40, 339, 1984.
54. **Cole, P. T.,** Using acoustic emission (AE) to locate and identify defects in composite structures, *Compos. Struct.,* 3, 259, 1985.
55. **Naus, D. J.,** Acoustic emission monitoring of steel and concrete structural elements with particular reference to primary nuclear containment structures, in *Advances in Acoustic Emission,* Dunhart, 1981, 249.
56. **Robert, J. L. and Brachet-Rolland, M.,** Survey of structures by using acoustic emission monitoring, in *IABSE Symposium: Maintenance, Repair and Rehabilitation of Bridges, Final Report,* International Association for Bridge and Structural Engineering, Washington, D.C., 1982, 33.
57. **Woodward, R. J.,** Cracks in a concrete bridge, *Concrete,* 17, 40, 1983.

INDEX

A

B

C

D

I

K

L

M

N

O

P

R

S

T

U

V

W